T5-CQQ-721

Prentice-Hall, Inc.
Englewood Cliffs, New Jersey 07632

Warren L. Flock

University of Colorado
Boulder, Colorado

ELECTROMAGNETICS AND THE ENVIRONMENT: Remote Sensing and Telecommunications

Library of Congress Cataloging in Publication Data

Flock, Warren L
Electromagnetics and the environment.

Bibliography: p.
Includes index.
1. Electromagnetism. 2. Telecommunication.
3. Remote sensing. I. Title.
QC760.F56 621.36′7 78-14356
ISBN 0-13-248997-X

Editorial/production supervision and interior design
by Barbara A. Cassel
Jacket Design by Jayne Conte
Manufacturing buyer: Gordon Osbourne

Printed in the United States of America

10 9 8 7 6 5 4 3 2 1

PRENTICE-HALL INTERNATIONAL, INC., *London*
PRENTICE-HALL OF AUSTRALIA PTY. LIMITED, *Sydney*
PRENTICE-HALL OF CANADA, LTD., *Toronto*
PRENTICE-HALL OF INDIA PRIVATE LIMITED, *New Delhi*
PRENTICE-HALL OF JAPAN, INC., *Tokyo*
PRENTICE-HALL OF SOUTHEAST ASIA PTE. LTD., *Singapore*
WHITEHALL BOOKS LIMITED, *Wellington, New Zealand*

CONTENTS

This volume differs from most environmental literature in the attention it gives to the environmental role of electromagnetic radiation, and it is rather unusual among literature treating technical topics in the attention it gives to the environment. The material is aimed at persons with interests in remote sensing of the environment, radar, telecommunications, or solar energy; those interested in applications of electromagnetic theory; and those with potential or established interests in the environment itself.

The title, *Electromagnetics and the Environment*, was chosen in recognition of the role of electromagnetic radiation in the environment and in our lives. The subtitle, *Remote Sensing and Telecommunications*, indicates specifically two of the principal topics that are considered to fall under the heading of the title. A third important aspect considered is solar energy, with emphasis on the effects of the propagation medium (the atmosphere) and on the availability of the resource at ground level. Other aspects of solar energy, such as flat-plate and concentrating collectors, are considered briefly as well. Important features of climate depend on the balance between incoming solar radiation and outgoing infrared radiation, which balance may be affected by man's activities. Also there is concern that the climate may be changing adversely at present because of poorly understood factors of natural origin. Such topics are complex and difficult to analyze but are discussed qualitatively in sections of Chapters 1, 3, and 5. The atmosphere is important to all the subjects considered and affects radiation of a given frequency in the same way regardless of whether the radiation is a feature of natural processes or is used for remote sensing or telecommunications or for providing energy for man's needs.

PREFACE

The electromagnetic aspects of remote sensing and telecommunications receive emphasis. It has been necessary, except for the inherent, inevitable overlap that occurs, to omit or minimize the interesting and important subjects of information theory and data processing. Furthermore, not all applications of remote sensing could be covered; subsurface sensing and extraterrestrial sensing have been omitted, and a number of individuals may find that some topic of interest to them is missing or does not receive proper attention. Certain major topics are treated in sufficient detail, however, to be useful to engineers and scientists in industry and to be suitable for a one-semester or longer upper-division engineering or science course. By providing both overall breadth and depth in certain areas, it is believed that this volume is suitable for different types of readers, including those who wish to concentrate on breadth and those who wish to devote effort to quantitative topics in some depth. Some persons interested in the environment, remote sensing, or telecommunications but lacking a strong technical background, for example, may wish to concentrate on the less technical sections and accept as fact the conclusions of the more highly technical sections. The electrical engineer or physicist will have had, or will want, the kind of background provided in a beginning, conventional course on electromagnetic fields and will give considerable attention to the technical material. This volume is suitable also for the important case of the person who is not an engineer or physical scientist, and does not plan to become one, but wishes to have an exposure to technical topics and wants an understanding of the language and techniques of remote sensing and telecommunications.

The material is an outgrowth of an undergraduate course entitled Environmental Electromagnetics that was developed for electrical engineering students. The course has the purpose of serving as an elective course in the electromagnetic fields area. As such it includes basic subject matter on atmospheric effects on electromagnetic waves, thermal radiation, and applications and extensions of material in introductory courses on electromagnetic fields. Any topic involving electromagnetic radiation, including radiation at optical frequencies, is considered to fall in the area of electromagnetic fields. The course is intended to provide experience with interdisciplinary applications of electromagnetic theory and with environmental and resource problems, rather than to serve as another pure conventional electromagnetics course emphasizing boundary value problems. The material of portions of Chapters 2, 3, and 4 has also been used in a course in an interdisciplinary program in telecommunications. Problems are included in the text for class use and for those who wish to obtain experience in applying the concepts presented. Cited references are listed at the end of each chapter, and an additional short bibliography of selected publications on remote sensing is placed at the end of the volume. Whereas the body of the text tends to emphasize technical principles and gives considerable attention to remote sensing of the atmosphere, a number of treatments of remote sensing omit the atmosphere and emphasize applications to the land, of interest to geographers, geologists, biologists, and many other users of remote sensing. The bibliography at the end of this volume is oriented in the latter direction.

One way for an instructor to conduct a technical course based on the material of this volume but to give also an exposure to environmental and resource problems is to emphasize certain of the technical sections and the problems in class generally but to spend a few periods at least on Chapter 1, assign the chapter for reading, and call for as an assignment a critical review of a nontechnical environmental reference, or references, such as those listed for Chapter 1. Alternatively, Chapter 1 and other nontechnical sections and references could receive more emphasis in a less technical course. The engineering instructor may need to make a choice between emphasizing Chapters 3 and 4 or Chapters 5 and 6, with the technical portion of Chapter 2 being essential in either case.

Knowledge is commonly compartmentalized in our educational institutions, with practical justification to a certain extent. The compartmentalization is often artificial, however, and the solution of real-world problems often needs and is facilitated by an interdisciplinary approach. This volume draws strongly upon the area of electromagnetics but is devoted also to an interdisciplinary viewpoint.

I would like to express my appreciation to the many individuals whose work is reported here or who have provided information of value. These persons are too numerous to mention individually, but I do wish to give special thanks to colleagues associated with the International Union of Radio Science (URSI), the Institute of Electrical and Electronics Engineers (IEEE), the American Geophysical Union (AGU), the Geophysical Institute of the University of Alaska, the EROS Data Center of the U.S. Geological Survey, and Goodyear Aerospace Corp. Also I am particularly grateful to colleagues in Boulder, Colorado, who are with the Environmental Research Laboratories of the National Oceanic and Atmospheric Administration (NOAA), the Institute of Telecommunications Sciences (ITS), the National Bureau of Standards (NBS), the National Center for Atmospheric Research (NCAR), and the University of Colorado. In addition I appreciate the attentiveness and helpful questions of the students who have taken the course on which the material is based and the support of the Department of Electrical Engineering of the University of Colorado.

WARREN L. FLOCK

1-1 ESSENTIAL NATURAL RESOURCES

Belatedly, in the late 1960's, the environment became a popular cause. The widespread concern that developed about the environment was overdue and most welcome to persons who had been concerned about conservation in earlier years when it was not so popular. By 1973 the energy crisis posed severe problems for the environmental movement but it still had momentum.

In the following pages we shall consider such well-known environmental topics as air pollution and energy and also the less familiar concepts of the electromagnetic spectrum as a natural resource and the environmental role of electromagnetic radiation. Remote sensing of the environment receives considerable emphasis; indeed all of the chapters contribute to that subject. The topics of the various chapters are also important, however, in their own right. The technical bases for remote sensing and those aspects of telecommunications considered here are similar, and for some purposes remote sensing might be considered to fall under the heading of telecommunications. In any case Chaps. 2, 3, and 4 are clearly applicable to both topics, and Chaps. 5 and 6 clearly apply to remote sensing. Before treating these topics, however, natural resources and the environment are discussed in general terms.

Concern over natural resources and the environment has been of at least two types—concern over the adequacy of natural resources for the survival of increasing numbers of persons and concern over the quality of the environment. In the first case one might question whether food, energy, and mineral resources will be adequate or not; in the second case one might ask

NATURAL RESOURCES AND THE ENVIRONMENT

1

whether the environment will be so crowded and ravaged that life will be no more than mediocre, if bearable, even though food and other supplies are adequate to support life. The two aspects cannot be completely separated, but one aspect or the other can be emphasized at a time. There are serious questions about the adequacy of food and energy resources, but the potential role of technology in alleviating shortages of essential resources seemed to many, until recently at least, to provide a basis for being less concerned about these than about the quality of the environment. The environmental movement has focused attention on the quality of environment, or on those aspects of environment which contribute to or detract from the quality of life. To many persons, the availability of natural, wild, scenic, and uncrowded areas is a highly desirable, if not essential, component of a high quality of life.

Some of the resources essential to life in a modern society are air, water, food, energy, minerals, and forest products. Of these essential resources much attention has been focused on food for obvious reasons. The Reverend Thomas R. Malthus in 1798 raised the discouraging prospect that the human race would run out of food, and ever since then writers have argued either that Malthus was wrong or that he was right. Since his time much has been accomplished in increasing food production by the application of machinery, fertilizers, pesticides, and plant breeding. The Rockefeller and Ford Foundation programs for increasing wheat and rice production were so highly successful that the term Green Revolution was coined for the leap forward in food production that was under way by about 1966. Highly optimistic statements about the ability to feed the peoples of the world were made at that time. By about 1974, however, the great optimism engendered in some quarters about food resources by the Green Revolution lessened, and the actual and potential role of adverse climatic conditions on food production began to attract more attention. Lester Brown [16, 17] has been an informative and prolific writer about food and related matters, and a thought-provoking treatment of food and climate has been presented by Stephen Schneider [77]. *Science* magazine has been a good source of information on food [2], and a study of research on food and nutrition has been made by the National Academy of Science [67]. Some developments and considerations concerning food and climate are now outlined briefly.

In 1972 the U.S.S.R. experienced a severe drought, the monsoon was delayed by several weeks in India, the Sahelian drought continued in Africa, and the Midwest of the United States experienced floods. As a result world food production fell in 1972 for the first time in two decades. Two years later, 1974 was a year of productivity decline in North America. Corn production in the United States was especially hard hit by drought and early frost and was down 20% from the trend in 1974 [74]. It was the 1972 drought in the U.S.S.R. that led to the large Soviet grain purchases from the United States and Canada that contributed to dramatic increases in the cost of wheat and other foods. The agricultural problems of 1972 and 1974 drew attention to the importance of climate and to uncertainty about future climatic trends.

Another unfavorable event in 1972 was the failure of the anchovy fishery off the coast of Peru, due to the invasion of the cold upwelling waters off Peru by the warm current, El Niño, from the north. Recovery of the anchovy fishery has been slow. More intensive development of the world's fisheries has been often mentioned as one of the means of feeding the growing world population, but the production of food from the sea, after increasing by roughly 5% per year from 1950 to 1970, dropped in 1970 and continued to fall for three consecutive years [16], presumably due to overfishing and pollution as well.

The formerly large reserves of grains in the United States became small by about 1974, due to the demands upon them and because of a Department of Agriculture policy that government should get out of the food business. Memory of the depressing effect on prices of food surpluses and stores in the past was a factor in forming attitudes. The world reserves in terms of days of grain consumption decreased from 105 in 1961 to 35 in 1975 [77]. The 1961 figure included about 31 days as the grain equivalent of idled U.S. cropland, but the idled cropland had dropped to zero by 1974. Replenishment of the reserves has been advocated by a number of writers [16, 60, 74, 77]. Since the low point of 1974, inventories of grain increased, however, from the lowest point in 25 years in 1974 to the highest values in 14 years in 1977, in spite of the drought that occurred in large portions of the United States. Thus the immediate crisis in grain supplies eased, but the low market prices for wheat and corn received by farmers in 1977, problems of food distribution, and uncertainties of climate were such that hunger and the threat of future food shortages remained. Careful planning and strenuous efforts are needed if food supplies are to be adequate for all hungry peoples in the future.

North America has come to dominate exportable grain production to essentially the same extent that the Middle East dominates with respect to oil exports. In the 1930's North America exported only a moderate amount of grain, and Eastern Europe and the U.S.S.R., Africa, Asia, and Latin America also exported. By the mid-1970's, exports from North America had risen sharply, while all the other areas mentioned had increasingly become deficit areas, though to a lesser extent for Latin America where Argentina is a wheat producing area [17] (Table 1-1). Figure 1-1 is included to suggest the agricultural technology and fertile productive land of the United States and to illustrate also land as a potential object for remote sensing.

Part of the Great Plains of North America is subject to drought that tends to recur about every 22 years. The Dust Bowl of the 1930's was a well-known result of the occurrence of severe drought in the area. A factor that is pertinent to considerations of the food resources of the world is that nearly

Table 1-1 World Grain Trade (Million Metric Tons).

Region	*1934–1938*	*1948–1952*	*1960*	*1970*	*1976**
North America	+ 5	+23	+39	+56	+94
Latin America	+ 9	+ 1	0	+ 4	− 3
Western Europe	−24	−22	−25	−30	−17
Eastern Europe, U.S.S.R.	+ 5	—	0	+ 1	−25
Africa	+ 1	0	− 2	− 5	−10
Asia	+ 2	− 6	−17	−34	−47
Australia, New Zealand	+ 5	+ 3	+ 6	+12	+ 8

*Preliminary. Plus sign indicates exports; minus sign, imports.
After L. Brown [17].

FIGURE 1-1. Farm land near Garnavillo, Iowa. (U.S. Department of Agriculture photograph by W. H. Lathrop.)

half of the present century has been unusually warm, in terms of climatic history. Peak average temperatures were reached around 1940 or 1945, and average temperatures declined following that time. The period from 1930 to 1960, for example, should not be regarded as a normal period but rather as the most abnormally warm period in 1000 years [19]. The variations of average temperature are not impressive in magnitude but are significant. From 1830 to 1930 the average temperature increased by about 1.7°C (3°F) near Philadelphia, while the drop from 1940 to 1975 was about 0.6°C (1°F) [84]. Superimposed on the general trend are decade-to-decade fluctuations of about 2.2°C (4°F). Cooling affects the higher latitudes most severely and results in shortened growing seasons and reduced productivity there. Because of its high-latitude location and the long period of records kept, Iceland has an especially useful record of climatic history. Based on reports of drift ice in earlier years, Bergthorsson constructed a 1000-year history of temperature in Iceland [20]. His results show alternating periods of relative warmth and cold—warmth from about 900 to 1200, cold from 1200 to 1400, warmth around 1500, cold from 1600 to 1900, and warmth in the first half of the present century. With respect to corn, wheat, and soybean production at lower latitudes, cooling itself is not so very harmful as somewhat lower temperatures and somewhat higher rainfall than normal tend to be optimum for these crops. Such conditions were generally present in North America from about 1955 to 1972. Further cooling could adversely affect food production in Canada and the U.S.S.R. significantly, but at lower latitudes weather variability (drought, floods, and changes in wind patterns) is a greater hazard [84]. It has been proposed, however, that cooling tends to be accompanied by "unusual" weather, greater weather variability, and failure of monsoons. Reid Bryson [20] of the University of Wisconsin is a leading proponent of this viewpoint. Whatever the explanation, the winter of 1976–1977 was characterized by unusual cold and severity in the eastern United States, mildness in Alaska, and drought in the western United States. Also the summer of 1977 was unusually warm in the eastern United States, and drought was widespread throughout the United States.

While the possibility that the earth may be returning to a more nearly normal, cool climatic period appears to pose a threat to food supplies, the steady buildup in CO_2 content of the atmosphere may cause the climate of the earth to become warmer. Some climatologists place considerable emphasis on this possibility and anticipate that this man-made change in climate will be greater in magnitude than natural fluctuations in the foreseeable future [55]. In any case it appears that change of climate is something mankind will have to deal with and that careful monitoring and analysis of climate are essential. The atmospheric CO_2 content is considered further in Sec. 3-4.

Energy resources are crucial resources because they can be used to supply most other resources. With unlimited energy, fresh water can be obtained from seawater, minerals can be extracted from low-grade ores and from the sea, and food can be produced. In 1972–1973 serious energy shortages were experienced in the United States for the first time, and much discussion has been devoted since to the energy crisis, especially since the Arab oil embargo of 1973–1974. With the end of the embargo and with some easing of the economic downturn that seemed to be triggered by the embargo and high prices for oil, the public's concern about energy apparently decreased. The trend toward buying smaller cars, for example, lessened, and General Motors found in early 1976 that a new compact model was not selling so well as expected. Energy is a major problem, however, and failure to take effective measures to solve the problem following the embargo made the problem worse. The dependence of the United States on imported oil increased, and it was reported that the United States imported more oil than it produced for the first time in the second week of March 1976. In 1977, President Carter presented a national policy for energy and provided a basis for hope that much-needed actions might finally be taken. It has become apparent that much of the prosperity of the United States and much of its productivity have been based in good part on extensive supplies of low-priced fuel, but such low-priced supplies cannot be expected to be available in the future. Some of the useful references on energy are those by Holdren and Herrera [52], the *Scientific American* [79], the AAAS [1, 3], Clark [25], the Ford Foundation [43], Krenz [59], and Lovins [62]. Among the approaches to the energy problem are attempting to increase supply and increasing efficiency of use. Krenz has emphasized increased efficiency of use and has pointed out that increasing efficiency has the same effect as increasing production but tends to require less capital. The subject of energy supply and demand has been discussed recently in terms of the use of "soft" and "hard" technology. Lovins is among those who have addressed the subject in these terms, and he has espoused soft technology (energy conservation, solar energy, and related measures and expanded use of coal, with individual facilities tending to be relatively small and widely dispersed) as contrasted to hard technology (such as breeder reactors and/or other large, centralized, complex facilities). Lovins has met criticism, and even some advocates of solar energy, for example, believe he has overstated the case for solar energy, but he has been effective in promoting discussion.

Water and land are closely related to food and energy. Water is needed for food production, and production of oil from shale in Colorado, for example, would require large amounts of water and contribute to a conflict between the use of water for producing food and for producing energy. Especially in the western United States, water has been transported for large distances, as from northern to southern California, to sustain populations without adequate local supplies. Such transport of water is justifiable to a point, but the concept of massive diversions from one area to another has been questioned and opposed increasingly in recent years [14,80]. A related matter is the controversy engendered in 1977 when President Carter sought to eliminate funds for a number of water projects on the basis that they were not cost effective and represented traditional pork-barrel politics.

One of the major problems concerning land use is that, as the need for food increases because of the growth of population, good agricultural land is being made unavailable because of residential and commercial construction and because of waterlogging and the accumulation of salts resulting from inadequate drainage [36]. Deforestation and overgrazing contribute to erosion, floods, silting of reservoirs and irrigation canals, and the conversion of productive land to desert (desertification). Such desertification has taken place extensively in northern and central Africa, and widespread application of slash-and-burn agriculture continues to take a toll of tropical forests in Latin America and elsewhere. Strip mining of coal and production of oil from shale threaten to seriously damage large areas of land, and much debate has taken place as to what requirements should be formulated for restoration of such land. Urban sprawl proliferates near cities, and wild and natural areas, including the U.S. National Parks, are under increasing pressure and abuse. While the use of fertilizers and pesticides has increased food production, it has contributed to eutrophication of lakes [12,75] and other adverse ecological effects [18]. The annual reports of the Council on Environmental Quality are valuable sources of information on environmental matters, including land use [31].

Warnings have been given from time to time of impending shortages of materials (nonenergy minerals, forest products, etc.), but the warnings have been countered by pointing out that the predictions of the past have often proved to be unduly pessimistic. Also the ability to work with lower-grade ores, to substitute new or more plentiful materials for previously employed materials, and to recycle materials can be pointed to. All of these measures have merit. The problem of incompatibility between mineral production and environmental quality, however, must be taken into account [23], and while we shall not run out of geologic resources, they will become more expensive [30]. A fundamental consideration is that processing lower-grade ores, substituting plentiful aluminum for steel, etc., require energy, and energy costs are increasing. Energy is a critical factor in considering the adequacy of mineral resources [30,47]. The AAAS has assembled a compendium on material resources [4].

Mankind lives in an environment permeated by electromagnetic radiation of natural and man-made origin. The electromagnetic radiation from the sun is the prime essential for life, and electromagnetic waves of man-made origin provide the basis for telecommunications. The concept of the electromagnetic spectrum as a natural resource is considered in Chap. 2.

General references on natural resources not already mentioned include Ehrlich and Ehrlich [39,40], *Resources and Man* (National Academy of Sciences–National Research Council) [66]; Brown et al. [15], and Huberty and Flock [53].

1-2 POPULATION AND ECONOMIC GROWTH

The present rapid increase in the world's population puts obviously corresponding pressures on the environment, and population is considered by many to be the basic overriding environmental problem. Of course the environment can be badly managed in sparsely settled areas. Also harmonious living patterns and good management have resulted in reasonably pleasant conditions in some rather heavily populated areas such as portions of Europe. Thus population is by no means the only factor involved in determining the state of the environment. But it is certainly a basic factor, and it seems clear that any and all efforts to improve the condition of man can be frustrated by excessive population.

For a number of years it was widely recognized that other parts of the world—portions of Asia, South America, and Africa—were, or were becoming, overpopulated, but it was considered that the United States did not have a population problem. After experiencing conditions in the Los Angeles, San Francisco Bay, New York, and Chicago areas, etc., however, many have felt that these and other areas were excessively congested and in deteriorating condition with respect to the quality of life, in good measure because of too many people. Persons living in other yet less congested areas have sensed a trend in the same direction in their own areas and considerable antigrowth sentiment has developed throughout the country. In Colorado, for example, people are concerned about excessive population in the Front Range area, along the base of the mountains from Fort Collins to Pueblo, and the Colorado Environmental Commission, under the chairmanship of Dean Max Peters, College of Engineering, University of Colorado, suggested in 1971 that the population of the Denver metropolitan area be limited to 1.5 million. A factor that needs to be taken into account is that the inhabitants of the United States consume much more and pollute much more per capita than persons in other portions of the world. With only 6% of the world's population, the United States has been said to use something like 35% of the world's resources that are consumed each year. One resident of the United States may be equivalent to, say, 30 Indians insofar as the use of resources and pressure on the environment is concerned. Some might say that Americans consume too much and that they should change their living habits and patterns in order to consume less. Some such changes may be both necessary and desirable; many wasteful practices could well be eliminated with little or no loss to the quality of life. On the other hand, many features of life in the United States are quality features, and abandoning these only to allow increasing numbers of people to lead increasingly mediocre lives should be resisted. Also it is pertinent to recall that with only 6% of the world's population, the United States was responsible in 1975 for 40.4% of the world's wheat exports, 60.8% of the corn exports, 76.6% of the soybean exports, etc. [83]. Thus it appears that at least some of the energy used in the United States is put to good purpose.

The birth rate in the United States has dropped recently to very close to the replacement level. This development has surprised demographers and is encouraging to persons who have been concerned about the growth of the U.S. population. Because of the distribution of population with age, however, the number of people in the United States will continue to increase for some time even if births do occur only at a replacement level. The western and southern sections of the country are presently attracting people, and population growth is taking place there more rapidly than elsewhere. The world population in 1970 was about 3.5 billion and estimates for the year 2000 have ranged between about 5.5 and 8.0 billion. It seems clear that excessive and rapidly growing population is a serious problem in parts of Latin America, Africa, and Asia. Brazil, Mexico, Peru, and Venezuela have population growth rates of about 3%. That population is a serious problem is still questioned by some who point out that population tends to stabilize as nations become industrialized and relatively affluent. It is true that populations in the highly developed nations have stabilized to a degree in this way. Populations in many of the less developed countries, however, have already grown to such an extent and are continuing to increase so very rapidly, with little in the way of capital available, that there seems to be little basis for optimism that industrial growth and affluence can develop soon enough to accomplish what is needed. Counting on improvements in standard of living alone, without attempting other means as well, to solve the population problem can only be regarded as an impractical policy. Echols [35] feels that, without drastic population control measures, high fertility in the less developed countries will overtake food production by the twenty-first century. India has come to feel the necessity for strong action, and a program involving involuntary sterilization was one of the reasons for the downfall of Indira Ghandi's government. The new government apparently decided to continue efforts to limit population but without the involuntary sterilization provision. Another compendium by the AAAS [5] deals with population. *The Population Bomb* by Paul Ehrlich [38] is well known for pointing out the dangers of excessive population.

The less developed countries will suffer the most from uncontrolled growth of their populations. In addition to the physical deprivation of hunger and want, severe restrictions on personal freedom are and will be experienced. The more highly developed nations will experience serious harmful effects as well; it would be naive to suppose that they can exist comfortably in peace while the rest of the world is in turmoil and hunger is rampant. The highly developed countries can and should provide technical assistance and some short-term aid to the less developed countries, but the ability of the developed nations to provide aid is limited, contrary to opinions in some quarters, and should not be counted on unduly. Also little incentive for embarking on aid programs exists unless some reasonable probability of their contributing to a long-term solution exists. Among the less repressive measures which could contribute to a solution of the population problem [16] are widespread provision of family planning services, a strong effort to achieve literacy for entire populations, improved health services, equality for women, establishment of social security systems, and suitable tax and benefit policies, such as supplying benefits for no more than a certain limited number of children. Kahn et al. [54], in an optimistic assessment of world prospects, pointed out a number of factors favoring economic growth of presently underdeveloped countries and emphasized the view that economic growth will provide the solution to population problems.

In discussions of population problems it is often pointed out that a large part of the difficulty is that impoverished, poorly educated individuals generally are not aware of and cannot be expected to react on their own initiative to the perils of population. If ignoring and minimizing the extent of the population problem are a shortcoming, however, it is a fault that educated persons in positions of leadership have also indulged in all too often. William and Paul Paddock [72] have asserted that excessive optimism about the world capability for food production has contributed to the failure to take the population problem as seriously as it should be taken.

A matter closely related to population is growth of the economy or of the gross national product (GNP), or, for short, simply growth. In the decade from 1960 to 1970, before the energy and other problems of the 1970's, the per capita GNP in the United States increased from $4078 to $5248 in 1972 dollars. At the same time the population increased from about 180.6 million to about 204.9 million. The population increase was about 6.3%, but the increase in GNP per capita was about 28.7%. In 1974 and 1975 the GNP actually decreased by 1.7 and 1.8%, respectively [83]. Growth was long regarded as obviously desirable, but some dissenting viewpoints have been heard. Fuller [46], in discussing this subject, noted that the President's Materials Policy Commission of 1951–1952 reached sobering conclusions about the use of resources in the United States but nevertheless made the following statement, which Fuller quoted.

"We share the belief of the American people in the principle of Growth. Granting that we cannot find any absolute reason for this belief we admit that to our Western minds it seems preferable to any opposite, which to us implies stagnation and decay."

Fuller asked, however, "What really is the principle of growth to which the American people are said to be dedicated? By whom and by what means are its pace and its acceleration determined? Are stagnation and decay the only alternatives to growth? Is it prudent to push further into the depths of resource uncertainty in order to maximize the certainty of contemporary prosperity?" This questioning of growth came at a rather early date and may not be widely known. Boulding [13] has suggested that it may be necessary to move from what he referred to as a cowboy economy to a spaceman economy. In the cowboy economy it is desirable to maximize the throughput (basically the gross national product), whereas in the spaceman economy success is not measured by the amount of production and consumption but by the nature and quality of the total capital stock.

The desirability of growth has been questioned increasingly in recent years, but growth has its ardent supporters as well as a much larger number of passive supporters. Indeed the predominant attitude of government and industry seems to favor growth. One of the reasons for this attitude is the opinion that growth is necessary for prosperity. The argument in favor of growth seems especially impelling in a time of recession. It has been asserted by some, however, that enhanced development of services should take the place of increasing consumption of resources as a mainstay of employment and economic well-being. A collection of essays on the theme that economic growth for growth's sake is destructive and unsustainable has been provided by Daly [33]. A strong defense of growth has been presented by Beckerman [10]. Economic growth very likely will continue to some degree whether we like it or not, of course, as long as population continues to increase. This being the case, attention needs to be given to guiding and controlling growth in ways that are most acceptable environmentally and socially. And while blind allegiance to economic growth will hopefully become a characteristic of the past, blind opposition to all development is not desirable either.

The situation is different for the developed and underdeveloped countires. In the latter an increase in gross national product can be more obviously beneficial than in the former, *if* population is stabilized at the same time. The energy-intensive industry of the developed nations may not be suitable, however, for the underdeveloped nations. The intermediate technology promoted by Schumacher [78] has attracted considerable attention. This technology requires a high degree of ingenuity but utilizes more human labor and less energy than the large-scale industry of the developed nations. The term *soft technology* has also been applied to what is referred to above as intermediate technology.

In discussions of population and economic growth, frequent reference is made to exponential growth. Exponential growth or decay of population, for example, takes place in the case where there is a constant fertility rate r, where $r = b - d$ and b is birth rate and d is death rate. The following differential equation then applies to the population N:

$$\frac{dN}{dt} = (b - d)N = rN \qquad (1\text{-}1)$$

The solution for the equation, when r is constant, is

$$N = N_0 e^{r(t-t_0)} \qquad (1\text{-}2)$$

where N_0 is the population at t_0. It is of interest to solve this equation for the doubling time. If $N/N_0 = 2$, $t - t_0 = 0.693/r$. If the fertility rate is 1.9%, for example, the doubling time is 36.5 years. It is impossible for exponential growth to continue indefinitely, of course, and although exponential growth may occur over a period of time, it should not be projected indefinitely into the future. The possibly dangerous aspects of exponential growth should not be ignored either. In particular it should be kept in mind that a population can grow exponentially through several doubling periods without apparent difficulty but encounter serious problems during one or two additional doubling periods. Al Bartlett [9] has treated the exponential function and its ramifications in detail and has pointed out that, although it is actually an extremely simple mathematical function, the function and its implications are very poorly understood by many people. The exponential function will be encountered on several occasions later in this volume (see Sec. 3-2-1, for example).

Because of the obvious limitations of the expression describing exponential growth, efforts have been made to develop other relations that describe population growth more adequately. A short review of various approaches has been given by Austin and Brewer [8]. One modification involves assuming an upper limit M for the population. In this case

$$\frac{dN}{dt} = rN\left(1 - \frac{N}{M}\right) \tag{1-3}$$

which is known as the logistics relationship and was proposed by Pearl and Reed in 1920. Another approach has been to assume a variable r, having an upper limit, such that

$$r = A(1 - e^{[-(a/A)N^{1/k}]}) \tag{1-4}$$

where a, A, and k are constants, with k less than or equal to unity. The combination of the two modifications to the exponential relationship tends to give better results in some situations than either alone. The prediction and even the description of population, however, are very difficult, and no expression can be expected to be entirely adequate.

Interesting but controversial attempts to quantify relationships involving population, quality of life, natural resources, pollution, and capital investment have been made by Forrester [44] and Meadows et al. [64] of M.I.T. with support from the Club of Rome. The authors used computer analysis based on specified models and asserted that the technique gives valid, often counterintuitive results which are more reliable than those achieved by mental analysis using fuzzy, poorly defined models. The *Limits to Growth* by Meadows et al. points out potential dangers of continued growth and indicates that important social and technological changes are needed to avoid serious difficulty in the future. Their work, however, has met severe criticism on the basis that it is said to be an overly pessimistic, doomsday prophecy that fails to allow for man's ingenuity, and that lumps all of the regions of the world together in one [10, 11, 26, 63]. A further study sponsored by the Club of Rome [65] takes into account regional differences and points out the difference between undifferentiated growth and organic growth. These terms are defined by analogy with cell division, resulting in merely the multiplication of identical cells (undifferentiated growth), and processes resulting in cells that differ in structure and function (organic growth). The authors assert that unqualified arguments for and against growth are naive until the "location, sense, and subject of growing and the growth process itself are defined." In this study and in a more recent meeting sponsored by the Club of Rome in Philadelphia, a more hopeful view of the future is said to be presented than in previous Club-of-Rome-sponsored studies and meetings.

For a highly optimistic view of the future that minimizes the dangers of growth and population, one can turn to Kahn et al. [54]. They feel that 200 years from now people almost everywhere will be numerous, rich, and in control of the forces of nature, and they contrast this condition with the small numbers and poverty of 200 years ago. The discrepancy between such optimistic views and contrasting pessimistic views is confusing to the layman. The important question, however, it appears to the writer, is not whether optimism or pessimism should prevail but whether problems will be faced and acted on or ignored. Furthermore, the classification of persons into optimists and pessimists is often arbitrary and meaningless.

1-3 THE CONSERVATION AND ENVIRONMENTAL MOVEMENTS

The inclusion of some historical material about the conservation and environmental movements in the United States seems desirable at this point, as an aid in understanding our present condition.

In the nineteenth century population pressure as we know it was not such a problem, but the public domain suffered from governmental laxness toward raids on resources and from what Stewart Udall called the Myth of Superabundance, namely the idea that the forests and other resources were without limit and inexhaustible. It was George Perkins Marsh who, in his book *Man and Nature*, published in 1864, first articulated well a concern over man's treatment of nature and a program of action to care properly for natural resources. In some respects the story of conservation in the United States begins with him, although the term was popularized later in the Theodore Roosevelt administration.

Before Marsh there were stirrings of conscience and some dissenters to man's treatment of the forests and soils. Henry David Thoreau, the famed naturalist and nonconformist of Massachusetts who lived until 1861, was prominent among these. Marsh, however, was a more versatile man—a naturalist, geographer, historian, and practical politician (a senator from Vermont and U.S. Ambassador to Italy, etc.) in one person. Giveaways and mismanagement of the nation's forests were a serious problem in his time, and in 1873 some members of the American Association for the Advancement of Science opened a campaign to save the forests. The American Forestry Association came into being 2 years later and pursued this same goal. Carl Schurz, appointed Secretary of the Interior by President Hayes in 1877, also fought abuses of the forests, proposed unsuccessfully the establishment of federal forest reserves, and espoused conservation causes generally. John Wesley Powell explored the length of the Colorado River in 1869, studied the arid west in following years, and later served under Carl Schurz. His paper of 1878, "A Report on the Lands of the Arid Region of the United States," is a classic paper and presented a broad conservation plan. In 1891 Congress passed an act that allowed the President to set apart or reserve timber lands and Presidents Harrison and Cleveland made effective use of the act. It was in the administration of Theodore Roosevelt, beginning in 1901, however, that vigorous widespread action was mounted to correct past abuses. Gifford Pinchot, Roosevelt's right-hand man, organized the U.S. Forest Service and arranged for the timber reserves to be administered by the Forest Service. Roosevelt also made large additions to the reserves that the previous presidents set aside. Another important feature of the Roosevelt administration was the convening of a White House Conference on Conservation in 1908.

The next major surge of conservation activity was during the presidency of Franklin Roosevelt. Among the developments of that period, starting in 1933, were the Soil Conservation Service, the Civilian Conservation Corp., the Tennessee Valley Authority, the Taylor Grazing Act, and additions to the national forests. More recently, former Secretary of the Interior Stewart Udall has been much concerned about the environment. His book *The Quiet Crisis* [87] and *Man's Dominion* [48] are good histories of conservation in the United States. Among the important legislation of the 1960's was the Wilderness Act of 1964, which provided for establishing wilderness areas in national parks, national forests, and wildlife refuges. The Secretary of the Interior following Udall, Walter Hickel, also took some firm actions in the conservation cause (such as opposing a Florida jetport which threatened Everglades National Park) and largely allayed the earlier doubts of conservationists. *Who Owns America?* by Hickel [51] is a fascinating account of his actions and dismissal. With respect to wilderness areas, mention should be made of Aldo Leopold and his classic and beautiful presentation of an ethic for wild lands in *A Sand County Almanac* [61]. The concept of wilderness is also discussed by Kilgore [56]. A worldwide account of conservation is given in *The Environmental Revolution* by Nicholson [68].

Our national parks and wildlife refuges preserve major features of our national heritage [22, 85]. The first national park, Yellowstone, was established in 1872, and the Yosemite, Sequoia, and General Grant areas of California became federal reserves in 1890. Parts of the Yosemite area, including its famous valley, had been held previously in reserve by the state of California since 1865. It was not until 1916 that the National Park Service was formed. Stephen Mather, the first director, served until 1929 and put the service on a firm foundation. The Fish and Wildlife Service, which administers the national wildlife refuges, had its beginnings in the Bureau of Fisheries, established in the Department of Commerce in 1871, and in the Division of Economic Ornithology and Mammalogy, set up in the Department of Agriculture in 1886. A decade later the division became the Division of Biological Survey, and in 1905 it became the Bureau of Biological Survey. The first federal wildlife refuge, Pelican Island in Florida, was established in 1903. In 1940 the Bureaus of Fisheries and Biological Survey were combined into the Fish and Wildlife Service, which was put into the Department of the Interior. Two other federal agencies, the Forest Service and Bureau of Land Management, manage important parts of the public domain on a multiple-use basis. Controversy exists between commercial and conservationist interests as to the relative priority of the multiple uses of public lands and as to what rates and policies of exploitation are appropriate. Conservationists feel that the Forest Service has given undue emphasis to timber production in southeastern Alaska and other areas of the western states and has dragged its feet on studies of potential wilderness areas. Disagreement has also arisen over the advisability of clear cutting on forest lands. Sixty percent of the trees harvested in the United States are provided by clear cutting. The argument has not been strictly one of esthetics or interest in recreational use versus commercial use and practicality but involves dangers of erosion, silting, and loss of nutrients due to clear cutting as well.

The National Environmental Policy Act (NEPA) of 1970 established the Council on Environmental Quality (1974) in the Executive Office of the President and set forth requirements for environmental impact statements by federal agencies concerning most of their actions that affect the environment in a major way. Two important decisions that drew attention to NEPA and its influence required the Department of the Interior to formulate an environmental impact statement concerning the Alaska pipeline and required the Atomic Energy Commission to establish more effective procedures for forming environmental impact statements than it had planned to use in licensing the Calvert Cliffs nuclear plant in Maryland [31]. The Environmental Protection Agency, formed by combining some 15 programs scattered throughout several agencies, was established as an independent agency in 1970. It has duties of establishing and enforcing environmental safeguards. The Air Quality Act of 1967, amended in 1970, and The Federal Water Quality Act of 1972 are important acts of environmental legislation.

A number of nongovernmental organizations and foundations have made important contributions to conservation. The first Audubon Society was established in New York in 1886, and from this developed a National Audubon Society for the Protection of Birds in 1905. Now known simply as the Audubon Society, this organization was instrumental in promoting the 1918 treaty for the protection of migratory birds. It maintains wildlife sanctuaries and sponsors education programs. The Sierra Club, founded in California in 1892 by John Muir, is one of the largest and most influential conservation groups. The Wilderness Society specializes in the preservation of wilderness, as the name suggests. It was founded in 1934. The Nature Conservancy, which acquires or finances the acquisition of natural areas having unique characteristics, was started in 1951. Among other rather long-established conservation organizations of national stature are the National Parks & Conservation Association, National Wildlife Federation, and Save-the-Redwoods League. The Wildlife Management Institute, Conservation Foundation, and Resources for the Future are among the better known foundations and institutes that have carried out effective programs of environmental education and research.

The federal government has supplied leadership in conservation at times and in various ways, but on other occasions it has not supplied the needed leadership or has taken actions that were detrimental to the environment. The conservation groups have often supplied leadership and at times have forced the government into positive action or to abandon unwise programs. The Sierra Club, for example, played a major role in preventing projects that would have flooded parts of the Grand Canyon and Dinosaur National Monument in Colorado. The publications of the conservation organizations (*Audubon, Sierra Club Bulletin, Living Wilderness, Nature Conservancy News*, etc.) are in many cases very attractive and interesting, and they are excellent sources of information on legislative action affecting conservation. Of course they present a certain point of view, and not everyone will agree with their position on all questions. To receive the publication of one of the groups, simply become a member. In addition to the national organizations mentioned there are many local groups and local chapters of the national groups. Large numbers of Americans appreciate the magnificent wilderness and wild areas which have been preserved as a result of strenuous efforts of conservation organizations and representatives of government. Opposition to the establishment of these preserves has commonly been based on arguments against "locking up" the areas for the use of a privileged few, but overuse is now a major problem.

Developers and preservationists have frequently come into conflict about environmental matters. Some development schemes have been brazen and unwise, and certainly all such developments cannot be looked on as progress. Neither are conservationists or preservationists on solid ground at all times, although an error in favor of preservation is usually more readily reversible than an error in favor of unwise development. In many conflicts between preservation and development, sound arguments are presented by both sides, and mutual respect and toleration are called for. John Muir, the founder of the Sierra Club, mountaineer, and naturalist, and Gifford Pinchot, the efficient forester, conservationist, and organizer, were on opposite sides in their outlooks on development and preservation and came to a falling out for that reason. That occurrence was regrettable because both made splendid contributions to conservation in their own way.

The interests of the conservation groups and much of the activity of the federal government in conservation also have tended in the past to be directed toward the rural and wild lands, but it is now recognized that inner cities and suburbs sorely need attention as well. Our metropolitan areas have been growing rapidly, and the nation has been becoming more extensively urbanized, although recently there has been a tendency for a reversal in the trend. Many of the nation's most pressing problems are in the cities, and many of these problems are interrelated. Surely the deterioration of the physical environment contributes to crime and to other social and racial problems. State and local governments generally in the past have exerted little leadership in dealing with the environment, although progress has been made in recent years. The State of Washington, for example, established a Department of Ecology, Hawaii has set up strict environmental zoning, etc. Additional local and regional parks are badly needed in many areas. In Colorado the cities of Boulder and Aspen have set good examples by starting greenbelt programs. Urban sprawl, lack of strong zoning, and inappropriate taxing policies are still major problems in the areas surrounding the cities. Many of the conservation groups have broadened their outlooks and consider the cities, population, and environmental problems of all kinds rather than those of the wild lands alone. In addition new environmental organizations have been formed, including the Scientists Institute for Public Information, which publishes *Environment*, and the Foundation for Environmental Conservation, headquartered in Geneva, Switzerland, which publishes *Environmental Conservation*. The Environmental Defense Fund specializes in legal actions for environmental causes. The Friends of the Earth was founded in 1969 by David Brower, former Executive Director of the Sierra Club. Student groups are active on many campuses, and grass roots interest in the environment and in solar energy has developed. Among the well-known writers, not yet mentioned, of the environmental movement are Rachel Carson [24], author of *Silent Spring*, which pointed out the pesticide problem; Barry Commoner [28, 29]; and Garret Hardin [49, 50].

1-4 ECOSYSTEMS AND BIOTIC COMMUNITIES

Ecology is concerned with ecosystems, such a system being a basic functional unit including living organisms and their abiotic (nonliving) environment, the two interacting with each other and achieving some sort of at least temporary stability. The living portion of an ecosystem is known as a biotic community. Ecosystems may be of various sizes. A pond is an example of a small aquatic ecosystem. A tract of grassland or forest is an example of a terrestrial ecosystem. Ecosystems are composed of four constituents: the abiotic materials; the producers or autotrophic organisms which are plants that carry out photosynthesis; the large consumers or macroconsumers which are heterotrophic organisms, chiefly animals which utilize the food produced by the autotrophs; and the decomposers or microconsumers, bacteria and fungi which decompose material produced by the autotrophs. Ecological research may be devoted to particular processes or features of ecosystems such as energy flow, biogeochemical cycles, population ecology, and limiting factors. Other studies are directed to the overall organization or structure and dynamics of particular freshwater, marine, and terrestrial ecosystems or communities. Laboratory ecosystems can be useful for certain purposes. A spacecraft to be used to carry men to one of the planets would be safer and more stable for a longer time if it were a self-sustaining ecosystem.

Some of the material on energy and biogeochemical cycles in later portions of this volume is pertinent to ecology. The lemmings of the arctic tundra provide an interesting example of the subject matter of population ecology. The population of lemmings, small rodents, is subject to cyclical oscillations having periods of 3 or 4 years. Snowy owls, which feed on lemmings, tend to be abundant at times when lemmings are abundant. A basic principle concerning limiting factors is Leibig's law, enunciated in 1840, which states that growth will be limited by the concentration of any factor or factors present in minimum amount, regardless of the concentration of other necessary factors. Ecosystem dynamics may involve, for example, the changes in plant and animal life following fire, volcanic activity, or the abandonment of cultivated fields. The gradual filling in of a pond provides another example. The occurrence of successive stages of vegetation in such cases is known as ecological succession. The community which will result eventually, at least in theory, from a process of succession is known as the climax community. In Colorado aspen groves are seldom a climax community and aspen tends to be replaced gradually by spruce. In addition to the one climatic climax community which characterizes an area, however, edaphic climaxes, determined by the substrate, topography, etc., may occur. For example, the climatic climax community in Florida may be the broad-leaved evergreen subtropical forest, but swamps are found instead in low wet areas. The pine forests of the Atlantic and Gulf coastal plains are apparently maintained by fire, and in the absence of fire it is believed that they would eventually develop into temperate deciduous forests [71]. An interesting development is the application of information theory to the subject of diversity [58]. It is well known that tropical areas have a higher degree of diversity (more species of trees, birds, etc.) than locations closer to the poles. Diversity is a function not only of the number of species, however, but of the probability of occurrence of each species if the information theory measure H is used for diversity, where $H = -\sum_i p_i \log_2 p_i$ and p_i is the probability of occurrence of the ith species.

The major terrestrial ecosystems of the world include the seas, estuaries and seashores, streams and rivers, lakes and ponds, freshwater marshes, deserts, tundras, grasslands, and forests of various types. Ecological studies commonly emphasize the living portions of the ecosystems or the biotic communities; some emphasize plant ecology or plant communities, and some emphasize animal ecology. The major terrestrial biotic communities are known as biomes. Listings of biomes by different writers differ in terminology, amount of detail, and sometimes as to how certain areas are classified as well. One listing of major biomes is shown in Table 1-2.

The major biomes tend to occur in corresponding climatic regions, but with some exceptions. The temperate rain forest has a marine west coast climate, but western Europe, which has a marine west coast climate, has been

Table 1-2 The Major Biomes and Their Locations.

Biome	*Locations*
1. Tundra	Circumpolar (Arctic)
2. Northern coniferous forest (taiga, boreal forest)	Circumpolar, south of the tundra
3. Temperate rain forest (marine west coast climate)	N. Pacific coast of N. America (Alaska to California), w. coast New Zealand, Tasmania and s.e. Australia, s. Chile
4. Temperate deciduous forest	Eastern N. America, Europe, China, Japan, e. coast of Australia
5. Broad-leafed evergreen subtropical forest (humid subtropical climate)	S.e. U.S., central and s. Japan
6. Temperate grasslands and steppes	Interior regions of N. America, e. Europe, central and w. Asia, Argentina, Australia
7. Tropical savanna	Central Africa, C. and S. America (n. and s. of Amazon basin), n. Australia, s.e. Asia
8. Desert	S.w. and intermountain N. America; n., s.w., and e. Africa; Mid East; central Asia; Peru and n. Chile
9. Broad-sclerophyll vegetation, chapparal (Mediterranean climate)	Mediterranean area, California, central Chile, Cape of Good Hope, s.w. and s. Australia
10. Tropical forest	
a. Rainforest	Amazon basin, Brazilian coast of S. America, C. America; Congo, Niger, and Zambezi basins of Africa; Indonesia, Malaysia, Borneo, New Guinea, and Pacific islands
b. Scrub and deciduous forest	C. and S. America, Africa, Australia, s. Asia

greatly modified and deforested by man and does not support a temperate rain forest. Western Europe is included in Table 1-2 along with much of the rest of Europe in the category of temperate deciduous forest, which elsewhere (e. Europe and e. U.S., etc.) occurs in a humid continental climate.

Figure 1-2 shows a generalized pattern which indicates roughly the distribution of the biomes on a theoretical flat continent, which might correspond to the combination of North and South America or Eurasia and Africa. The numbers in Fig. 1-2 correspond generally to the entries in Table 1-2. Note, for example, that there is a region of chapparal vegetation in both the northern and southern hemispheres (no. 5, California and Chile in North and South America) with desert equatorward and temperate rain forest poleward from the chapparal area, which has a Mediterranean climate. The American reader might be puzzled by the region of temperate deciduous forest (no. 4), shown in Fig. 1-2 to the west of the grassland area (no. 6). In

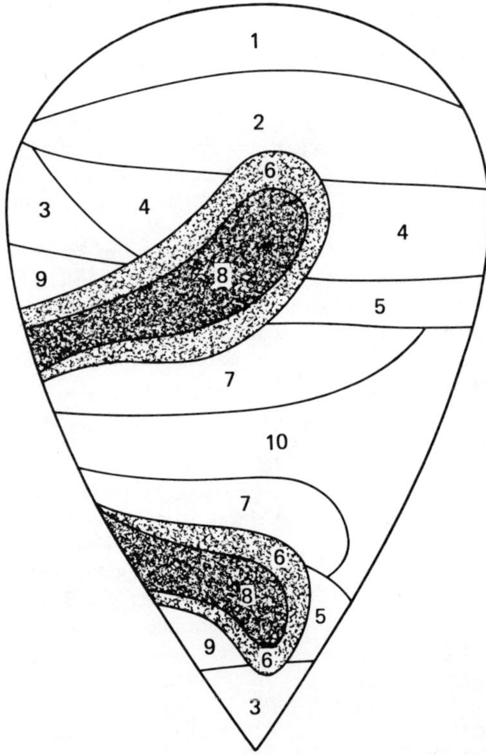

FIGURE 1-2. Generalized distribution of major biomes. (Adapted from *An Introduction to Climate*, 4th ed., by Trewartha [86]. Copyright © 1968 by McGraw-Hill, Inc. Used with permission of McGraw-Hill Book Company.)

North America the area is mountainous and a temperate deciduous forest has not developed, but conditions in Europe do correspond to Fig. 1-2 in this respect and central Europe does have a temperate deciduous forest cover. A large part of India falls in the tropical deciduous forest region, listed in Table 1-2 as a subdivision of tropical forests. Figure 1-2 follows generally the classification of climates by Köppen [86] but, having been introduced to illustrate the distribution of biomes, does not use his letter designations or terminology. The *Aldine University Atlas* [45] includes maps of the biomes or natural regions of all the continents.

In areas where mountains occur an increase in altitude tends to be accompanied by a change in vegetation similar to that encountered in traveling toward the poles, and local altitudinal variations are superimposed on and found closely adjacent to the characteristic vegetation of the lowlands. Coniferous forest similar to the northern coniferous forest, for example, extends south along the Appalachian Mountains to Great Smoky Mountains National Park in North Carolina and Tennessee. Northern types of coniferous forest and eventually alpine tundra may be reached as one travels to higher altitudes in the mountains of the western United States.

The biomes are mentioned here because of their importance to ecology and to mankind but also because they represent potential applications for the remote sensing techniques discussed in Chap. 6. The units are of interest not only in their natural state, which is sometimes difficult to find, but with respect to their use and abuse by man. As indicated, the biomes tend to have closely corresponding climates, and it is primarily the climate that affects telecommunications. The nature of the biome, however, has some effect as well. The problem of mobile communications, for example, is different in a tropical rain forest than in a desert.

The temperate prairies and grasslands of North America have been highly productive for corn, wheat, soybeans, and other foods. In its original state, the grasslands consisted, proceeding from Illinois to the west in the United States, of tall-grass prairie, a mixed-grass transitional region, and short-grass steppe grassland. The tall-grass prairie, especially, has been almost entirely converted to agriculture, and only tiny replicas of the original vegetation can be found. It extends eastward as a wedge into the temperate deciduous forest and is commonly classified as a feature of a humid continental climate, rather than a phenomenon of dry climates. The short-grass steppe region of the western Great Plains does have a dry climate and is subject to recurring drought. The United States is blessed with a number of other areas of highly productive land, including the grasslands of eastern Washington and Oregon (a wheat-producing area), irrigated grasslands and desert areas of California and other western states, the Willamette and Puget Sound valleys of the Pacific Northwest, and lowlands of the Gulf and Atlantic coasts and Florida. California has a dry-summer Mediterranean climate and supports chapparal, oak woodlands, grassland, and desert at lower elevations. Deserts in the United States are of two general types, the cool desert, typified by sagebrush (*Artemisia*), and the hot desert, characterized by creosote bush (*Larrea*) and also by a variety of cacti and other interesting desert plants. Both types of desert may be productive if water is available for irrigation. The southeastern United States has a subtropical humid climate, with precipitation distributed throughout the year and with temperate deciduous forests and broad-leaved evergreen subtropical forests (magnolia and live-oak forests, etc.) as the climatic climax communities. (In some classifications the entire region except for the southern tip of Florida is considered to be part of the temperate deciduous forest biome.) The temperate deciduous forest lands of Europe, China, and eastern North America have been highly utilized and modified by man and have agricultural land of varying productivity. The biomes and their subdivisions in North America are described in detail by Shelford [81]. The tropical deciduous forest areas of India are highly modified by man. Delay of the moisture-bearing monsoon in the warm season can be disastrous as the area depends on the monsoon for its moisture.

The large areas of tropical rain forest of the Amazon and Orinocco basins of South America are facing rapid development and modification, and rain-forest areas elsewhere in South America and Central America are in the process of rapid dimunition due to the effects of the burgeoning populations of these areas. Slash-and-burn subsistence farming methods are having serious detrimental consequences. Changes in albedo (reflectivity) for solar radiation and in soil moisture are likely consequences of deforestation, and these changes may cause climatic change. There is concern that once the tropical rain forest is disrupted it may have the capability to return to a natural or beneficial state only very slowly if at all. The possible effects of deforestation are not well understood, but caution is called for [42]. At one time, one heard much talk of the potential of the Amazon and similar areas for food production, but the high productivity that characterizes the rain forest in its natural state usually does not continue for more than a few seasons when conventional agriculture is attempted. A result in some locations is the development of lateritic soil (iron-rich, humus-poor soil which hardens on repeated wetting and drying). Sanchez and Buol [73], however, in a discussion of the potential of tropical soils for food production, point out that only a fraction of tropical soils becomes lateritic. Some high-base tropical soils (basic as opposed to acidic) that developed from alluvium, sediments, or volcanic ash are fertile. The more widespread low-base soils are inherently less productive and require careful choice of crop varieties and efficient utilization of fertilizer. Highland tropical areas having fertile volcanic soil, coconut-lined beaches, climates requiring no use of energy for space heating, and tropical fruits, flowers, and birds are all attractive features of the tropics, but population pressures and fragile rain-forest ecosystems present problems.

Wetlands, swamps, and estuaries play important ecological roles but have often been regarded merely as wasteland and have been subject to abuse and neglect. The mangrove swamps of Florida, for example, are important to the shrimp industry, but ill-advised dredging and filling operations have destroyed areas of mangroves to an alarming degree.

Some general remarks about ecology now conclude this section. Problems may arise from the tendency of pesticides and pollutants to be concentrated in certain organisms rather than to be dispersed and degraded as in the case of energy flow in a food chain. In a study of food chains in a Long Island estuary, for example, Woodwell [89] found concentrations of DDT to increase from as low as 0.04 parts per million in plankton to as high as 26.4 parts per million in cormorants (fish-eating birds). Another major problem, eutrophication of lakes, of which Lake Erie is a notable example (although its condition has improved lately), has resulted from the discharge of phosphates, nitrates, and pollutants of various kinds into lakes. Eutrophication results in excessive growth of algae, depletion of oxygen, and death of animals that have high oxygen requirements.

Ecology is an interdisciplinary subject involving meteorology, geophysics, etc., as well as subjects that are more readily recognized as biological. A very important characteristic of ecology is that it places emphasis on the systems approach and on interactions among the various components of systems. Like ecology, engineering is also basically interdisciplinary in nature and uses the systems approach, although the education and training of engineers have not generally prepared them to deal with natural systems. Ecologists and engineers could contribute much to each other, and in many cases could well work together in teams. An academic program for bridging the gap between ecology and engineering has been described by Werner and Goodman [88].

Study of ecology under a trained ecologist is, of course, highly desirable. However, there are good references which can be read profitably by most persons, and ecological material can be introduced into engineering and other courses. Among the references on general ecology are those by Buchsbaum and Buchsbaum [21], Kormondy [57], Odum [69, 70], Krebs [58], and Colinvaux [27]. Daubenmire [34] and Oosting [71] treat plant ecology, and Allee et al. [6] and Elton [41] are concerned with animal ecology. Examples of publications treating applied ecology and related subjects are those by Cox [32] and Ehrenfeld [37]. Among the professional organizations concerned with ecology are the Ecological Society of America, which publishes *Ecology* and *Ecological Monographs*, and the American Institute of Biological Sciences (AIBS), which publishes *BioScience.*

1-5 ENGINEERING AND PHYSICAL SCIENCE

Engineering and science have contributed to man's well-being and increased his standard of living and quality of life in obvious ways. Today, however, there is realization that all of the results of the application of technology have not been beneficial. Lack of confidence in science and technology and outright suspicion have not been uncommon in recent years. Some of this feeling has been part of a general mood of anti-intellectualism, but some has been based on evidence of association between environmental deterioration and technological developments. A careful examination of the present situation certainly reveals many shortcomings in the application of science and technology. Thoughtful consideration of environmental problems reveals also, however, that the efforts of engineers and scientists are needed very much to help solve the problems. Furthermore, the environment is still poorly understood in many ways, and engineers and scientists are needed to carry out the necessary research on the environment. Of course the environment needs the attention and efforts of people in all walks of life. Engineers and scientists are included in this general observation, and they play a crucial role as well because of their special know-how and training.

If engineering and science are to fulfill their proper and essential role in helping to solve the problems of society, they will need the understanding and support of society. In this respect, engineers and scientists probably need to devote more effort to communication with other members of society. Also lawmakers and society need to be better educated in the area of science and engineering. Just as an engineering student with little or no interest in the humanities or in social problems is apt to be a narrow individual, so a liberal arts, social science, or business major or a law school student with little or no education or interest in the sciences or engineering is also apt to be a narrow individual who poorly comprehends important aspects of the world around him. Persons in the fields mentioned do not need detailed knowledge, design ability, or facility with higher mathematics or complex theories, but they should be familiar with basic concepts and capabilities.

The basic function of science is to search for truth and understanding without regard for immediate application, although many scientists are interested in applying their knowledge as well. Engineering is by definition devoted to applications of science for the benefit of mankind, although some individual engineers may basically be scientists and mathematicians. It is pertinent to quote a definition of engineering that has been given by the Engineers Council for Professional Development. The definition reads, "Engineering is the profession in which a knowledge of the mathematical and natural sciences, gained by study, experience, and practice, is applied with judgment to develop ways to utilize, economically, the materials and forces of nature for the benefit of mankind." It is suggested that this definition could well be modified slightly to read as follows. "Engineering is the profession in which a knowledge of the physical and biological sciences and of society and its problems is applied with judgment to develop ways to utilize, economically and with regard for the quality of the environment, the materials and forces of nature for the benefit of mankind." The changes are the addition of the phrases "and society and its problems" and "with regard for the quality of the environment" and the specific reference to the biological sciences. The specific mention of these additional factors makes it clearer that they should be considered.

One feature that should be expected of engineers and scientists is that they should tell the public what engineering and science can and cannot accomplish. The engineering profession needs enthusiasts who will tell the public of the valuable services it does and can provide in dealing with the environment and in meeting other needs. The public should also be told not to put reliance on remarkable technological or scientific developments which will eliminate the need for rational and realistic population attitudes and policies. And just as a good surgeon can be expected to recommend either for or against surgery, a good engineer should be expected to recommend either for or against a particular technological undertaking. Engineering organizations should not support any and all technical projects and should not be perennially at odds with conservationists. While engineers should criticize specific errors in the statements of environmentalists, they should be careful not to give the impression that engineering is in an opposing camp.

Much can be said for emphasizing engineering as a profession and for de-emphasizing the importance of the divisions into which engineering is commonly divided. The late Dean L. M. K. Boelter, College of Engineering, U.C.L.A., was an articulate spokesman for this view. It is necessary, however, for practical reasons to have effective, active proponents for the various areas of interest within engineering, and, whether one considers the traditional or other branches of engineering, most are concerned with the environment. Civil engineering deals with water supply and pollution, waste disposal, and transportation. Chemical engineering is concerned with the processing of fossil fuels, air pollution, etc. Mechanical engineering is intimately involved with energy. Electrical engineering is obviously concerned with energy in the electrical form. Solar energy has received attention from all of the above branches of engineering; in one university one engineering department may be most prominent in this respect, and in another university it may be a different department. Electromagnetic radiation (both natural and man-made) and its utilization for communications and remote sensing of the environment and as the basic energy source are fundamental but often neglected aspects of the environment. Electrical engineers have a suitable background for dealing with these topics. The journals of the various professional engineering societies frequently have papers on environmental and natural resource topics.

The interests of the physical sciences, especially physics, geophysics, and chemistry, and engineering overlap. The sciences tend to emphasize basic understanding, and engineering tends to emphasize applications, but the distinction is not always clear. The AAAS (American Association for the Advancement of Science, which includes an engineering section) has given considerable attention to environmental and resource problems in *Science.* Recently it has prepared four compendiums on population, food, energy, and materials [2–5]. Similarly the American Geophysical Union has shown an active interest in energy and the environment in its conferences and publications, including EOS. The American Chemical Society has given attention to pollution problems and energy in the periodical *Environmental Science and Technology* and in the book *Cleaning our Environment* [7]. Questions concerning man's effects on climate have attracted the attention of meterologists and other scientists and have resulted in publications of general interest [76, 77, 82].

1-6 CONCLUDING REMARKS

Debate over the world predicament has often been shrill. Experts and scholars are found on each side of most major questions, and for every argument there seems to be a counterargument. An unfortunate consequence may well be failure to take the necessary actions to ameliorate future and present problems. The future is uncertain, and heated arguments for one point of view or the other do not make it less so. Should we not prepare for an uncertain future, with actual or potential hazards of unknown degree, by careful study of the hazards and by trying to allow some margin of safety, some reserve of resources, and some degree of flexibility in our plans? Such policies are proposed by Stephen Schneider as the Genesis strategy, based on the warning to Pharaoh that during seven prosperous years preparations should be made for seven lean years.

Environmentalists are sometimes accused of being prophets of doom and gloom. Their critics apparently feel greater optimism about the benefits of conventional growth and development and what technology can accomplish. Actually it appears that some people like to be optimistic and are irritated by pessimism and that some people enjoy being pessimistic and are irritated by optimism. But both qualities have their virtues if they are not carried to extreme. One needs a certain amount of optimism in undertaking any program. Otherwise there is little incentive for action, and a person may simply "drop out" and do nothing constructive. It has been suggested also that pessimism tends to be self-fulfilling. On the other hand, optimism needs to be tempered with realism, as it may otherwise result in a lack of a needed sense of urgency and in failure to take necessary actions.

Are the environmentalists guilty of undue pessimism, or are the environmentalists and their critics comparable to two political parties, each of which is optimistic about its own program and pessimistic about the program of the other party? If the environmentalists are pessimistic about the benefits of conventional growth and development, are the critics of the environmentalists any less pessimistic about the value of the environmental movement and what it might accomplish? It seems to be overlooked by many critics that a number of persons who are concerned about the environment and about growth advocate constructive programs, are reasonably optimistic about what the results of these programs might be, and are pessimistic primarily only about what will transpire if society continues to blindly follow outmoded concepts of the past. Working for the preservation of natural, wild areas for enjoyment and inspiration, for example, is a positive action and not necessarily an indication of pessimism. The argument between advocates of soft technology (frequently environmentalists) and hard technology (frequently ardent advocates of continued growth), as another example, is not properly described as an argument between pessimists and optimists.

Broad sweeping statements are seldom completely accurate. It is not correct to accuse all environmentalists of being prophets of doom and gloom or foes of technology or even of being pessimistic. Nor is it correct to accuse all representatives of business and industry or the establishment of not caring about the environment. Neither should environmentalists be looked on as liberals or radicals and their critics be regarded as conservatives. Environmentalists tend to embrace the old-fashioned virtue of thrift, and many persons who are conservative politically are strong environmentalists. Note that the words conservative and conservationist have common roots. The environment is one area that many conservatives and liberals can agree on, although not all of either group, by any means, supports the environmental cause.

In attempting to downgrade somewhat the optimist-versus-pessimist argument and in pointing out that the terms are often used carelessly, it is not intended to deny that many persons are discouraged by what they regard as serious deterioration of the environment and of the quality of life. Let people enjoy either optimism or pessimism as suits their temperament, it might be said, however, as long as they avoid unreasonable extremism, temper their feelings with realism, and promote taking the necessary constructive actions that are needed to solve our problems. Persons considering themselves to be environmentalists should not merely oppose everything but should seek to further constructive programs and emphasize the positive nature of the environmental movement. Persons considering themselves to be optimists should not place their heads in the sand and avoid dealing with the problems we face but should likewise support needed realistic programs. The important argument is not between optimists and pessimists but between constructive optimists and constructive pessimists on the one hand and, on the other hand, do-nothing optimists, oppose-everything pessimists, persons basing their attitudes on ideological dogmas, and those who think that we can continue indefinitely into the future doing things the same way as in the past.

The environment, food, and energy are subjects of concern to everyone. All three subjects are considered here but only briefly in many respects, and most attention is given to techniques for utilizing the electromagnetic spectrum for remote sensing and telecommunications and as an energy source. The techniques involve engineering and science, and training in these fields can prepare one to design and analyze radar systems and radiometers, for example, and to use them to obtain environmental data. This text can serve as a reference for persons with professional interests in such fields. Many of the specific topics included here, such as telecommunications, the earth's atmosphere and air pollution, solar energy, and remote sensing of the sea and land, are of actual or potential interest, however, to a wide group of people in all walks of life. Some may wish to concentrate on ideas and capabilities rather than mathematical details.

REFERENCES

[1] AAAS (Hammond, A. L., W. D. Metz, and T. H. Maugh II), *Energy and the Future*. Washington, D.C.: AAAS, 1973.

[2] AAAS (Abelson, P. H., ed.), *Food*. Washington, D.C.: AAAS, 1976.

[3] AAAS (Abelson, P. H., ed.), *Energy: Use, Conservation, and Supply*. Washington, D.C.: AAAS, 1976.

[4] AAAS (Abelson, P. H., and A. L. Hammond, eds.), *Materials: Renewable* and *Nonrenewable Resources*. Washington, D.C.: AAAS, 1976.

[5] AAAS (Reining, P., and I. Tinker, eds.), *Population: Dynamics, Ethics, and Policy*. Washington, D.C.: AAAS, 1976.

[6] ALLEE, W. C., A. E. EMERSON, O. PARK, and K. P. SCHMIDT, *Principles of Animal Ecology*. Philadelphia: Saunders, 1949.

[7] AMERICAN CHEMICAL SOCIETY, *Cleaning Our Environment. The Chemical Basis for Action*. Washington, D.C.: American Chemical Society, 1969.

[8] AUSTIN, A. L., and J. W. BREWER, "World population growth and related technical problems," *Spectrum*, vol. 7, pp. 43–54, Dec. 1970.

[9] BARTLETT, A. A., "The exponential function, Parts I–V," *The Physics Teacher*, vols. 14 and 15, 1976, 1977.

[10] BECKERMAN, W., *In Defence of Economic Growth*. London: Jonathan Cape Ltd., 1974.

[11] BECKMANN, P., *Eco-hysterics and the Technophobes*. Boulder, CO: Golem Press, 1973.

[12] BEETON, A. M., "Eutrophication of the St. Lawrence Great Lakes," *Limnology and Oceanography*, vol. 10, pp. 240–254, 1965. (Also in Cox, G. W., *Readings in Conservation Ecology* [32] and T. R. DETWYLER, *Man's Impact on Environment*. New York: McGraw-Hill, 1971.)

[13] BOULDING, K. E., "The economics of the coming spaceship earth," in *Environmental Quality in a Growing Economy* (H. JARRETT, ed.), pp. 3-20. Baltimore: The Johns Hopkins Press, 1966.

[14] BOYLE, R. H., J. GRAVES, and T. H. WATKINS, *The Water Hustlers*. San Francisco: Sierra Club, 1971.

[15] BROWN, H., J. BONNER, and J. WEIR, *The Next Hundred Years*. New York: Viking, 1963.

[16] BROWN, L. R., *By Bread Alone*. New York: Praeger, 1974.

[17] BROWN, L. R., *The Politics and Responsibility of the North American Breadbasket, Worldwatch Paper 2*. Washington, D.C.: Worldwatch Institute, Oct. 1975.

[18] BRUBAKER, S., *To Live on Earth*. Baltimore: The Johns Hopkins Press, 1972.

[19] BRYSON, R. A., "A perspective on climatic change," *Science*, vol. 184, pp. 753–760, May 17, 1974.

[20] BRYSON, R. A., and T. J. MURRAY, *Climates of Hunger*. Madison, WI: University of Wisconsin Press, 1977.

[21] BUCHSBAUM, R., and M. BUCHSBAUM, *Basic Ecology*. Pittsburgh: Boxwood Press, 1958.

[22] BUTCHER, D., *Seeing America's Wildlife*. New York: Devin Adair Co., 1955.

[23] CARPENTER, R. A., "Tensions between materials and environmental quality," *Science*, vol. 191, pp. 665–668, Feb. 20, 1976.

[24] CARSON, R., *Silent Spring*. Boston: Houghton Mifflin, 1962.

[25] CLARK, W., *Energy for Survival*. Garden City, NY: Anchor Press/Doubleday, 1974.

[26] COLE, H. S. D., C. FREEMAN, M. JAHODA, K. L. R. PAVITT (eds.), *Models of Doom*. New York: Universe Books, 1973.

[27] COLINVAUX, P. A., *Introduction to Ecology*. New York: Wiley, 1973.

[28] COMMONER, B., *Science and Survival*. New York: Viking, 1967.

[29] COMMONER, B., *The Closing Circle*. New York: Knopf, 1971.

[30] COOK, E., "Limits to exploitation of nonrenewable resources," *Science*, vol. 191, pp. 677–682, Feb. 20, 1976.

[31] Council on Environmental Quality, *Environmental Quality, 1974*. Washington, D.C.: U.S. Government Printing Office, 1974. (Issued annually; land use is the subject of Chapter 1 of the 1974 report.)

[32] COX, G. W. (ed.), *Readings in Conservation Ecology*. New York: Appleton, 1969.

[33] DALY, H. E. (ed.), *Toward a Steady-State Economy*. San Francisco: W. H. FREEMAN, 1973.

[34] DAUBENMIRE, R. F., *Plants and Environment*. New York: Wiley, 1947.

[35] ECHOLS, J. R., "Population vs. the environment: a crisis of too many people," *American Scientist*, vol. 64, pp. 165–173, March–April 1976.

[36] ECKHOLM, E. P., *Losing Ground*. New York: Norton, 1976.

[37] EHRENFELD, D. W., *Biological Conservation*. New York: Holt, Rinehart and Winston, 1970.

[38] EHRLICH, P. R., *The Population Bomb*. New York: Ballantine Books, 1968.

[39] EHRLICH, P. R., and A. H. EHRLICH, *Population Resources Environment*. San Francisco: W. H. FREEMAN, 1970.

[40] EHRLICH, P. R., and A. H. EHRLICH, *The End of Affluence: A Blueprint for Your Future*. New York: Ballantine Books, 1974.

[41] ELTON, C., *The Ecology of Animals*. London: Methuen, 1933.

[42] FARNWORTH, G. A., and F. B. GOLLEY (eds.), *Fragile Ecosystems*. New York: Springer-Verlag, 1974.

[43] FORD FOUNDATION, *A Time To Choose*. Cambridge, MA: Ballinger Press, 1974.

[44] FORRESTER, J. W., *World Dynamics*. Cambridge, MA: Wright-Allen Press, 1971.

[45] FULLARD, H., and H. C. DARBY, *Aldine University Atlas*. Chicago: Aldine, 1969.

[46] FULLER, V., "Natural and human resources," in *Natural Resources* (M. R. HUBERTY and W. L. FLOCK, eds.), pp. 1-28. New York: McGraw-Hill, 1959.

[47] GOELLER, H. E., and A. M. WEINBERG, "The age of substitutability," *Science*, vol, 191, pp. 683–689, Feb. 20, 1976.

[48] GRAHAM, F., Jr., *Man's Dominion*. New York: M. EVANS & Co. (distributed in association with Lippincott, Philadelphia and New York), 1971.

[49] HARDIN, G., "The tragedy of the commons," *Science*, vol. 162, pp. 1243–1248, 1968. (Also in Hardin [50].)

[50] HARDIN, G. (ed.), *Population, Evolution, and Birth Control*, 2nd ed. San Francisco: W. H. FREEMAN, 1969.

[51] HICKEL, W. J., *Who Owns America?* Englewood Cliffs, NJ: Prentice-Hall, 1971.

[52] HOLDREN, J., and P. HERRERA, *Energy*. San Francisco: Sierra Club, 1971.

[53] HUBERTY, M. R., and W. L. FLOCK (eds.), *Natural Resources*. New York: McGraw-Hill, 1959.

[54] KAHN, H., W. BROWN, and L. MARTEL, *The Next 200 Years*. New York: Morrow, 1976.

[55] KELLOGG, W. W., "Global influences of mankind on the climate," in *Climate Change* (J. GRIBBIN, ed.). Cambridge, England: Cambridge University Press.

[56] KILGORE, B. M. (ed.), *Wilderness in a Changing World*. San Francisco: Sierra Club, 1966.

[57] KORMONDY, E., *Concepts of Ecology*. Englewood Cliffs, NJ: Prentice-Hall, 1969.

[58] KREBS, C. J., *Ecology: The Experimental Analysis of Distribution and Abundance*. New York: Harper & Row, 1972.

[59] KRENZ, J. H., *Energy: Conversion and Utilization*. Boston: Allyn and Bacon, 1976.

[60] LAUR, T. M., "The world food problem and the role of climate," *EOS*, vol. 57, pp. 189–195, April 1976.

[61] LEOPOLD, A., *A Sand County Almanac*. New York: Oxford University Press, 1949.

[62] LOVINS, A., "Energy strategy: the road not taken?," *Foreign Affairs*, vol. 55, pp. 65–96, Oct. 1976.

[63] MADDOX, J., *The Dooms-day Syndrome*. New York: McGraw-Hill, 1972.

[64] MEADOWS, D. H., D. L. MEADOWS, J. RANDERS, and W. W. BEHRENS, *The Limits to Growth*. New York: Universe Books, 1972.

[65] MESAROVIC, M., and E. PESTEL, *Mankind at the Turning Point*. New York: Dutton, 1974.

[66] National Academy of Sciences–National Research Council, *Resources and Man*. San Francisco: W. H. FREEMAN, 1969.

[67] National Academy of Sciences–National Research Council, *World Food and Nutrition Study: The Potential Contributions of Research*. Washington, D.C.: National Academy of Sciences, 1977.

[68] NICHOLSON, M., *The Environmental Revolution*. New York: McGraw-Hill, 1970.

[69] ODUM, E. P., *Ecology*. New York: Holt, Rinehart and Winston, 1963.

[70] ODUM, E. P., *Fundamentals of Ecology*, 3rd ed. Philadelphia: Saunders, 1971.

[71] OOSTING, H. J., *The Study of Plant Communities*. San Francisco: W. H. FREEMAN, 1956.

[72] PADDOCK, W., and P. PADDOCK, *Time of Famines*. Boston: Little, Brown, 1976.

[73] SANCHEZ, P. A., and S. W. BUOL, "Soils of the tropics and the world food crisis," *Science*, vol. 188, pp. 598–603, May 9, 1975.

[74] SANDERSON, F. H., "The great food fumble," *Science*, vol. 188, pp. 503–509, May 9, 1975.

[75] SAWYER, C. N., "Basic concepts of eutrophication," *Journal Water Pollution Control Federation*, vol. 38, pp. 737–744, May 1966. (Also in Cox [32].)

[76] SCEP (Study of Critical Environmental Problems), *Man's Impact on the Global Environment*. Cambridge, MA: M.I.T. Press, 1970.

[77] SCHNEIDER, S. H., *The Genesis Strategy*. New York: Plenum, 1976.

[78] SCHUMACHER, E. F., *Small Is Beautiful*. New York: Harper & Row, 1973.

[79] *Scientific American, Special Issue on Energy and Power*, vol. 224, Sept. 1971.

[80] SECKLER, D., *California Water—A Study in Resource Management*. Berkeley: University of California Press, 1971.

[81] SHELFORD, V. E., *The Ecology of North America*. Urbana: University of Illinois Press, 1963.

[82] SMIC (Study of Man's Impact on Climate), *Inadvertent Climate Modification*. Cambridge, MA: M.I.T. Press, 1971.

[83] *Statistical Abstract of the United States, 1976*. Washington, D.C.: Supt. of Documents, U.S. Government Printing Office, 1976.

[84] THOMPSON, L. M., "Weather variability, climatic change, and grain production," *Science*, vol. 188, pp. 535–541, May 9, 1975.

[85] TILDEN, F., *The National Parks*. New York: Knopf, 1973.

[86] TREWARTHA, G. T., *An Introduction to Climate*. New York: McGraw-Hill, 1968.

[87] UDALL, S., *The Quiet Crisis*. New York: Holt, Rinehart and Winston, 1965.

[88] WERNER, P. A., and E. D. GOODMAN, "New academic training for ecologists and engineers," *AIBS Education Review*, vol. 3, pp. 1–5, Dec. 1974.

[89] WOODWELL, G. M., "Toxic substances and ecological cycles," *Scientific American*, vol. 216, pp. 24–31, March 1967.

THE ELECTROMAGNETIC SPECTRUM AND ITS UTILIZATION

2

2-1 THE ELECTROMAGNETIC SPECTRUM AND ELECTROSPACE AS NATURAL RESOURCES

Electromagnetic waves can exist over a tremendously wide range of frequencies extending from nearly zero as a lower limit to as high as 10^{23} or more. Included in this range are radio waves and light. Electromagnetic radiation is a term that refers to electromagnetic waves; it is common to speak of electromagnetic waves as being radiated from a source, such as the sun or an antenna. One may also speak of electromagnetic radiation as being emitted from or incident upon a structure or surface or as existing in free space. The term spectrum refers to a class or group of similar entities arranged in the order in which they possess a certain characteristic. Thus one can refer to a group of people whose political opinions range from extremely conservative to extremely radical as representing a wide political spectrum. The term electromagnetic spectrum is used here to refer to electromagnetic radiation or waves of all possible frequencies arranged or displayed as a function of frequency.

The availability of electromagnetic radiation is beneficial to man in various ways, and thus radiation or the electromagnetic spectrum can be looked on as a major natural resource. Indeed, the earth receives its energy from the sun in the form of electromagnetic radiation, and electromagnetic radiation is essential to all life on the earth. In addition the electromagnetic spectrum is used for communications and various other purposes, either in free space or in transmission lines or systems. Some previous references to the spectrum as a natural resource have referred primarily to the radio-frequency portion of the spectrum and to transmission in the earth's atmosphere. In this restricted sense, transmission of signals in a cable does not make use of the radio spectrum as it does not involve transmission in the earth's atmosphere and does not have the potential for interfering with other uses. The broader view of the electromagnetic spectrum is taken in this volume, but considerable attention is also given to the radio spectrum in particular, especially in this chapter.

The electromagnetic spectrum and the environment are closely related in various ways, in addition to the fact that the spectrum is essential to life on the earth. Mankind lives in an environment permeated by electromagnetic radiation of both natural and man-made origin, and it is important to understand as well as possible just what effects the radiation has on people and biological matter. This topic is discussed in Sec. 2-5. The designer and user of telecommunication and other electronic equipment, on the other hand, must consider the electromagnetic environment in which the equipment must operate. He must, for example, consider both the natural and man-made radiations in the frequency range of the equipment. Also he must take care to minimize interference caused by his equipment. The earth's atmosphere is part of our environment and affects our ability to use the electromagnetic spectrum. Electromagnetic waves that are incident upon the earth's atmosphere from the outside, as from the sun, tend to be sufficiently attenuated or reflected that they have negligible intensity at the earth's surface, except in two wavelength bands that are referred to as the optical and radio windows (about 0.4 to 0.8 μm and 1 cm to 10 m, respectively). Wavelengths falling within the windows are affected also in some degree by the atmosphere but usually not so drastically as those without. Other ways in which the environment and the electromagnetic spectrum are related involve the poles or towers and wires that may be employed to utilize the electromagnetic spectrum for telecommunications. For example, the early years of telecommunications were characterized by very large numbers of unsightly telephone and telegraph wires, but these were reduced in number by development of underground cables and carrier systems. Further developments in communications involving more extensive utilization of broadband telecommunication services–cable television, closed-circuit television, video teleconferencing, digital data transmission, etc.—may also have effects on the environment. For example, travel for some business and educational purposes might be reduced if more reliance were placed on the use of versatile broadband telecommunication facilities.

Some of the characteristics of the radio spectrum as a natural resource, as specified by the Joint Technical Advisory Committee (JTAC) of the Institute of Electrical and Electronic Engineers (IEEE) and the Electronic Industries Association [30,31], are

1. The radio spectrum is utilized but not consumed. Present usage does not interfere with future usage or cause any degree of deterioration of the resource. A portion of the spectrum that is assigned or otherwise occupied for one use at a given place and time, however, may not be available for other uses in the same area and at the same time.
2. The resource has dimensions of space, time, and frequency, and all three dimensions are interrelated.
3. The spectrum is an international resource.
4. The resource is wasted when assigned to tasks that can be done more easily in other ways or when it is not correctly applied to a task.
5. The spectrum is subject to pollution by man-made radio noise.

It has been argued in certain quarters that the radio resource should be placed on the open market to be bought and sold as any other commodity. A counterargument is that telecommunications has the potential for providing social benefits which are not proportional to market value.

The above listing of characteristics mentions space and time as well as spectrum itself. Hinchman [24] has pursued this view further and has pointed out that efficient use of the spectrum involves time, wave polarization, radiated power, antenna directivities, and terminal locations as well as frequency. This reasoning led him to propose the term *electrospace* for the radio resource, electrospace being a quantity that can be represented by an eight-dimensional matrix involving frequency, time, polarization, power, direction of propagation, and three spatial dimensions. The term has since been used by other authors in a general sense rather than as a term referring specifically to an eight-dimensional matrix [17]. Another variation is the treatment presented in the JTAC report *Radio Spectrum Utilization in Space* [32]. In this case, frequency, orbit, and number of earthward beams per satellite are regarded as the orthogonal axes pertinent to satellite communications. It is suggested here that efficient use of the spectrum requires, among other factors, suitable atmospheric characteristics (or a suitable medium or waveguide structure of some kind). Thus the concept of electrospace can be enlarged to encompass the subject of atmospheric effects on electromagnetic waves. If this version of electrospace is accepted, it can be said that Sec. 2-3 of this volume emphasizes one aspect of electrospace (spectrum) and Chap. 4 emphasizes another aspect of electrospace (atmospheric characteristics).

Utilization of the spectrum and electrospace requires governmental regulation and licensing and international cooperation so that users can operate without interfering with other parties and without being interfered with. Obviously chaos would result without regulation and licensing procedures. The first international conference on the use of the radio spectrum was held in 1903. Following several other conferences, the International Telecommunications Union (ITU) was formed in 1932 [10]. This organization has its headquarters and a permanent staff in Geneva, Switzerland. A technical body which makes recommendations for standards concerning the use of the radio spectrum, the International Radio Consultative Committee (CCIR) is also located in Geneva. A technical arm of the ITU, the CCIR has an extensive list of publications on the use of the radio spectrum. The Consultative Committee on International Telegraphy and Telephony (CCITT) plays a similar role to that of the CCIR. A third agency of the ITU is the International Frequency Registration Board (IFRB), and the ITU publishes a monthly periodical entitled *Telecommunications*. Major decisions concerning the use of the spectrum are made in World Administrative Radio Conferences (WARCs) arranged by the ITU. An especially important WARC, considering most aspects of telecommunications, is scheduled for 1979. The International Scientific Radio Union (URSI) and the national committees of URSI promote research and hold regular meetings concerning radio-wave propagation and related topics. Jointly with professional groups of the Institute of Electrical and Electronics Engineers (IEEE), the U.S. National Committee of URSI sponsors meetings every spring and fall. Within the U.S. government are an Office of Telecommunications in the Department of Commerce and the Federal Communications Commission, which is responsible for licensing, enforcement of regulations, etc. An Office of Telecommunications Policy was formerly in the White House. JTAC, mentioned earlier, has made a number of studies of the use of radio spectrum for the government. A President's Task Force on Communications Policy completed a significant report in 1968 [47].

Management of telecommunications involves economic, political, sociological, and technical factors and is becoming increasingly complex. The President's Task Force was impressed by the lack of data relating to the economic characteristics and performance of the telecommunications industry. It concluded that the importance of telecommunications justified a much larger amount of serious policy research than it had received. Formulation of policy has been handicapped by a shortage of qualified personnel, among other factors, and universities have not in the past provided the type of interdisciplinary traning that is needed for dealing effectively with telecommunications policy. The President's Task Force therefore urged "governmental, foundation, and business support for increased interdisciplinary research and training in telecommunications policy." It further stated that "to ensure that the government is exposed to a steady flow of independent, critical, and creative ideas, we believe that an institute and preferably more than one institute, for communication training and research, should be developed outside the government." In furtherance of this goal the University of Colorado in 1972–1973 commenced an interdisciplinary M.S. in Telecommunications program.

From the practical viewpoint, utilization of the electromagnetic spectrum involves communication systems, and these must have adequate signal-to-noise ratios for successful operations. Signals are generally man-made, but there are practical limits to the signal intensities which can be provided. Consequently noise may be the factor which limits the use of the electromagnetic spectrum in a particular application. Noise is generated in the receivers of communication systems and externally to the receivers. The external noise may be of natural origin or man-made. In this respect what is noise to one party may be the desired signal to another party and vice versa. Man-made signals that are not in accordance with regulations and good practice or are otherwise unwanted may be considered to be a form of pollution. Because of its importance to the utilization of the radio spectrum, radio noise is considered further in Sec. 2-3-2.

When the public thinks of the environment it perhaps thinks of air pollution or water pollution but very likely not of the electromagnetic spectrum and its pollution. The electromagnetic spectrum is an essential natural resource, nevertheless, and the electromagnetic environment is exceedingly important to electrical engineering, which has the major technical responsibility for communications. It is largely the public which uses the communication facilities developed by engineers, and the electromagnetic spectrum and its utilization are thus important to society as well as to engineers. Whether or not society thinks of telecommunications as involving environmental or pollution problems, it does appear to appreciate the importance and impact of telecommunications.

The electromagnetic spectrum and its utilization is in many ways the main theme of this entire volume, the present chapter emphasizing use by man for communication. The world's population is increasing rapidly, and its use of communication services is increasing more rapidly than population, as has been the case for energy and other resources. Arthur D. Little in May 1977 predicted the telecommunications equipment market to double (corresponding to an 8% growth rate) in the decade ending in 1985. The communications industry appears to be in a better position to meet demands than in some other cases, but communication facilities use spectrum, energy, materials, and land. The demands placed upon natural resources and the environment by the expansion of communications facilities and possible beneficial environmental effects resulting from such expansion need careful analysis.

To consider the utilization of the electromagnetic spectrum further it seems desirable to present some material concerning the uses and allocations of the spectrum and the history and trends of the telecommunications industry. The next section is devoted to meeting that need. Technical principles involved in utilizing the radio spectrum are then discussed, and the types of radio noise are described. Microwave systems are then treated, partly because of their intrinsic interest and importance and partly to provide illustrations of considerations that are pertinent to all systems. Some knowledge of the earth's atmosphere and its effects is required to better understand some aspects of the utilization of the spectrum, and these topics are considered in Chaps. 3 and 4. Electromagnetic radiation from the sun is treated in Chap. 5. Most techniques for remotely sensing the environment involve the use of the electromagnetic spectrum, and some of these are considered in Chap. 6.

2-2 TELECOMMUNICATIONS—HISTORY AND PRESENT TRENDS

The story of telecommunications (the technology of utilizing the electromagnetic spectrum for communicating at a distance) is a fascinating one. Valuable descriptions of telecommunications systems, their use for computer data transmission, and possible future developments in telecommunications have been presented by Martin [39,40]. Certain historical highlights are mentioned in this section. The telegraph system was the first telecommunications system to be developed, and the name of Samuel Morse is the one most widely associated with the development of telegraphy. In 1845 he opened a telegraph line between Washington and Baltimore and sent the famous words "What hath God wrought?" Other developments in telegraphy had taken place at about the same time and earlier, and Wheatstone and Cooke's needle telegraph system was utilized in railroad operations out of Paddington Station, London in 1839. Telegraphy was first restricted to land, but submarine cables were laid under the English channel in the early 1850's and between Ireland and Newfoundland in 1858. The telephone, which transmits human speech rather than coded signals, was developed by Alexander Graham Bell in 1876. Elisha Gray applied for a telephone patent a few hours later than Bell and lost out in litigation over patent rights in the ensuing years. The development of radio came near the end of the nineteenth century, and the name of Guglielmo Marconi, who took out his first radio patent in 1896, is the one generally associated with its invention. In the U.S.S.R. credit is commonly given to Alexander Popoff.

The first commercial telephone exchange was opened in New Haven, Connecticut in 1878. An automatic electromechanical exchange was put in service in 1912. The first experimental electronic exchange was tried successfully by AT&T in 1960, and electronic exchanges were becoming a significant factor by the late 1960's. In the early days of the telephone a pair of wires was used for each circuit, and cities were festooned with huge numbers of wires. Underground cables, however, soon began to replace the open wires. Loading, namely the insertion of series inductances into wire lines, was used by about 1899 and resulted in reduced signal attenuation and distortion and in the ability to transmit signals over longer distances.

In 1907 Lee DeForest added a control grid to the vacuum-tube diode and formed the triode, which has the capability of amplifying signals. This development had a major impact on telecommunications. By 1911 voice communication as far as from New York to Denver had been carried out without the use of amplifiers, and coast-to-coast service utilizing vacuum-tube amplifiers was established in 1915. By 1918 electronics had been developed and applied to the point that a commercial multiplex system was put into operation. Multiplexing has allowed the development of high-capacity communication systems, and the development of such systems is outlined in the following section.

In 1947, the transistor was invented by John Bardeen, Walter Brattain, and William Shockley, and by the 1970's integrated circuits and microelectronics were playing major roles in telecommunications.

MULTIPLEXING AND HIGH-CAPACITY TELECOMMUNICATION SYSTEMS. Multiplexing, the transmission of two or more signals over the same channel, involves separating the signals in either frequency or time. The process utilizing separation in frequency is known as frequency-division multiplexing (FDM), and that using separation in time is called time-division multiplexing. In frequency-division multiplexing a bandwidth of 4 kHz is allotted for a voice channel of which close to 3 kHz may actually be used for voice transmission, with the remainder used as a guard band which provides separation between the voice channels. At the time of World War II the Western Electric 12-channel J open-wire and 12-channel K cable systems represented the state of the art. The J system used the frequency band from 36 to 84 kHz for transmission in one direction and that from 92 to 140 kHz for transmission in the other direction. Single-sideband suppressed carrier operation was used so that a bandwidth of 4 kHz was needed per voice channel. The K system used wire-pair cables and the frequency band from 12 to 60 kHz, with a different wire pair being used for each direction. Many pairs of wires can be placed in one lead or aluminum sheathed cable. Telegraph signals occupy less bandwidth than voice signals, and a voice channel can be divided into 24 telegraph channels.

Following World War II, attention was directed to the use of coaxial and microwave systems. Coaxial cables can transmit signals covering a much wider frequency range than open-wire and wire-pair cables, and coaxial cables provide shielding from noise and cross talk between adjacent cables. A Bell System transcontinental L-1 coaxial system having a capacity of 600 voice channels was put into operation in 1948. The later L-3 system covers the frequency band from 312 to 8284 kHz (8.284 MHz) and has a capacity of 1860 voice channels, and the L-4 covers roughly twice as great a frequency range, with an upper frequency of about 17.5 MHz and a capacity of 3600 voice channels. Numbers of coaxial cables can be grouped together within one large cable. One in use includes 12 coaxial cables and has a total capacity of 21,600 voice channels or 10,800 two-way channels. Coaxial lines can carry television signals as well as voice channels. A channel width of 6 MHz is required for television in the United States, and the L-3 system is used for either 1860 voice channels or one color television channel.

The development of surface microwave communication systems, which involve line-of-sight transmission in the earth's atmosphere between repeater stations that are spaced about 32 to 64 km apart, was based on experience with microwave radar during World War II. The first commercial microwave service began in May 1945 between New York and Philadelphia. This system, operated by Western Union, provided both telephone and telegraph service. AT&T first operated their original experimental system between New York and Boston in November 1947. By September 1950, they supplied network television service between New York and Chicago, and in September 1951, the first transcontinental television service was provided. By 1965, 80,000 route miles and 650,000 radio-frequency one-way channel miles of microwave radio relay were in service in the United States. The Bell System operates its TD-2 microwave relay system in the 3.7- to 4.2-GHz band with capacities of 600 to 1200 voice channels. The TD-2 is a long-haul system for transcontinental and other long-distance use. The TM and TH surface microwave relay systems operate in the 5.925- to 6.425-GHz range, with the TM being used for short-haul applications and the TH system, which has the capability of handling about 1800 channels, being used for long and medium hauls. The TJ and TL systems operate in the 10.7- to 11.7-GHz range and are used for short hauls. Further developments in high-capacity telecommunication systems are discussed under the headings of satellite systems, millimeter waves, and optical fiber communications. It should also be mentioned that submarine cables still have their attractive features and are not dying out. On the contrary, submarine bandwidths have increased, and the submarine cable industry has flourished [21]. The AT&T fourth-generation transatlantic cable system has a capacity of 4000 two-wire voice channels [39].

SATELLITE SYSTEMS. Operational communication satellites came on the scene in 1962 when the Telstar satellite transmitted signals across the Atlantic from Andover, Maine to Pluemeur, France and Goonhilly Downs, England. The first commercial satellite, however, was Early Bird (INTELSAT I), launched in June 1965, and this was a synchronous satellite which stayed in an essentially fixed position 35,900 km or 22,300 miles over the Atlantic, in contrast to Telstar and other experimental satellites which rotated around the earth. Early Bird had the capacity to carry 240 voice signals or one TV signal. It was launched by COMSAT (the Communications Satellite Corporation) for INTELSAT (the International Telecommunications Satellite consortium). Following Early Bird, INTELSAT II was placed in service in 1967, and this was succeeded by INTELSAT III and IV. INTELSAT IV, which was introduced into service in 1971, has a capacity for about 5000 telephone circuits. Four operational satellites each were in positions over the Atlantic, Pacific, and Indian Oceans to provide global service in 1976. In 1977 there were 174 ground antennas at 140 earth stations in 82 countries operating with the INTELSAT satellite system. The INTELSAT satellite system uses the 3.7- to 4.2-GHz band for transmission to satellites and the 5.9- to 6.4-GHz band for transmissions from satellites. Frequencies in the 7.2- to 8.4-GHz range are utilized in the United States for military satellite-to-earth and earth-to-satellite paths. The U.S.S.R. has used an elliptical 12-h synchronized orbit for its Molyna satellite system operating in the 0.8- to 1.0-GHz band and has also more recently employed a stationary synchronous satellite.

The INTELSAT system is an international civil system generally utilizing expensive earth stations with large (29.5-m) antennas, which are justified by the large volume of traffic. In addition, 11-m antennas are used with the INTELSAT system when the full 500-MHz bandwidth is not needed. A number of domestic or regional satellite systems tending to have lower-cost earth stations than INTELSAT were also operating or planned for the future by 1977. The Canadian Telesat system put its first Anik satellite into operation in 1973, and the first U.S. domestic satellite, the Western Union Westar, went into operation in 1974. By 1976 RCA Satcom and Comstar satellites (the latter used by AT&T and GTE) were in operation. Also in 1976 the Indonesian Palapa satellite system (same technology as Anik and Westar systems) went into operation, linking a 4800-km chain of islands.

The increasing use of small earth terminals and the increasing use of frequencies higher than the 4- to 6-GHz frequencies used in the INTELSAT and initial domestic systems appear to be important trends in satellite communications. The use of small earth terminals requires reduced bandwidths and/or increased satellite transmitter power or directivity. Operational or pending applications for rather small earth terminals include the transmission of news to local publishers and television signals to cable and pay TV operators. The U.S. 2.5-GHz ATS-F health-education experimental satellite employed a 9.1-m-diameter satellite antenna, with 15 W of power per channel, and receiving antennas only 3 m in diameter. In the 12- to 14-GHz bands, a Communications Technology Satellite (CTS), a joint NASA and Canadian project, involves use of some 4.6-m earth-station antennas. An all-digital Satellite Business Systems (SBS) satellite sponsored by COMSAT, IBM, and Aetna Casualty and Surety Co. is scheduled to operate in the 12- to 14-GHz bands with 5- and 7-m antennas located on customer premises. The Marisat satellite system for ships at sea employs 1.22-m shipboard antennas for communication with satellites, using 1.5- to 1.6-GHz frequencies. In Japan and Germany plans were proceeding in 1977 for broadcasting TV signals to individual households having antennas about 1 m in diameter. A futuristic view of the possibilities of satellite communications, including the use of wrist radiotelephones, has been presented by Bekey and Mayer [2]. With the large number of synchronous satellites placed in orbit and planned for the future, consideration must be given to the number which can be accommodated in the synchronous orbit and to the angular separation which must be maintained between satellites (or to the angular slot which can be assigned to a satellite). An angular slot of 4° has been considered suitable, but values from 3 to 5° have been mentioned at times. Although small earth-terminal antennas can theoretically be used if high satellite powers are employed, possible interference problems, associated with the relatively wide beamwidths and side lobes of the small antennas, may indicate otherwise. Satellite and terrestrial systems share the same frequencies, and potential interference between these two types of systems must be taken into account also. This consideration has dictated that major earth stations not be located too close to metropolitan areas where extensive usage of terrestrial links occurs. Possible interference problems associated with physically small antennas could be minimized to some extent by employing millimeter waves. Technical considerations concerning the performance of satellite links is considered further in Sec. 2-3.

Digital Systems. Time-division multiplexing is becoming increasingly important. The most effective form of time-division multiplexing is that used in pulse-code modulation (PCM) in which signals are sampled at a certain rate and in which a pulse code is then used to represent each signal sample. The Bell System T1 PCM system utilizes "frames" of 125 μs. Each frame contains one sample from each of 24 voice channels which are thus sampled once each 125 μs, or 8000 times a second. This sampling rate is sufficient for transmitting a voice signal having a bandwidth of 4000 Hz, the general rule being that the sampling rate must be at least $2F$ to reproduce a signal of frequency F. Each sample is represented by a seven-bit code (corresponding to 128 levels), where a bit refers to a binary representation of 0 or 1 and is a basic measure of information content. An eighth bit is used for supervisory or signaling information, and one additional bit per frame is used for synchronizing purposes. Thus each frame contains 193 bits $[(24)(8) + 1 = 193]$, and the system operates at a rate of $(193)(8000) = 1.544 \times 10^6$ bits per second (1.544 Mb/s). A bandwidth of about 1.5 MHz is needed for this service. A PCM system is thus rather extravagant of bandwidth, but the terminal costs for PCM are lower than for other modulation systems and PCM systems are less vulnerable to noise and interference.

A T2 system can be formed by combining four T1 systems. A T2 system can transmit 96 voice channels and has a capacity of 6.3 Mb/s. The Bell System Picturephone system uses PCM and can be accommodated by a T2 system. A T3 system, operating at a rate of 90 Mb/s and accommodating 1344 channels, has a capacity 14 times that of a T2 system and can be used for TV signals. Several similar systems having a data capacity of 90 Mb/s, operating at frequencies above 10 GHz, and utilizing 8 PSK modulation were expected to be in operation by 1978. The designation 8 PSK refers to phase shift modulation involving eight different phases that are 45° apart.

PCM systems are ideally suited for data transmission and for mixtures of different kinds of signals. Some interesting comments on PCM and the reasons it was not more widely adopted sooner were given by Bennett [4].

TELEGRAPHY, DATA TRANSMISSION, AND FACSIMILE. Although telegraphy was the first telecommunications system developed, it was rather overshadowed by telephony once the latter was well developed, and domestic telegraph service declined over a period of years. The development of the teletypewriter, however, has assured telegraphy of an important role in the telecommunications field. A worldwide switched teletypewriter system, designated as TELEX, is operated in the United States by Western Union. The Teletypewriter Exchange Service (TWX) is a competitive system operated by the Bell System in the United States and Canada. Unfortunately, telegraph codes are not standardized. The TELEX system, which operates at a rate of 50 bits/s uses the Baudot 5-bit code. The TWX system can operate at speeds up to 150 bits/s and uses either the Baudot code or a Data Interchange code, while the Friden Flexowriter code uses 6 bits. The ASCIE (American Standard Code for Information Exchange) uses 7 bits.

Developments in telecommunications and computing have been taking place at a rapid rate, and the two technologies are becoming closely interwoven. This theme has been expressed well by Martin [39], who pointed out that previously persons were familiar with one or the other field but that in the future there will be a need for persons familiar with both. Telegraph systems can be utilized for data transmission, but a number of data transmission machines, having advantages over telegraph machines, have been developed. The Bell System, for example, advertises its dataphone service for talking with computers. Visual display units are common in certain types of applications, as in stockbrokers' offices for displaying prices of stocks, etc.

Facsimile transmission is used for transmitting replicas of documents and photographs. A common procedure is to place a photograph to be transmitted on a drum. The photograph is then scanned rapidly by a narrow light beam, and the reflected light is focused onto a phototube. At the receiving end a photographic film or an electrothermal or electrolytic recording paper is mounted on a drum which is illuminated by a lamp which is moved in synchronism with the scanning beam at the transmitting end, while the intensity of the light from the lamp varies in accordance with the phototube signal at the transmitter end. In 1977 a digital higher-speed facsimile system capable of sending or receiving the data on an $8\frac{1}{2}$- to $11\frac{1}{2}$-in. sheet in 15 s, using a 16-kb/s rate of transmission, was being developed for the U.S. military services. The system could operate over voice-grade circuits and HF radio, and the service could be considered to be an early form of electronic mail.

BROADBAND CABLE SERVICES AND PICTURE TELEPHONES. The visual display and facsimile systems mentioned in the previous section generally utilize voice bandwidths. In this section some of the potential applications of broadband transmissions are considered.

In areas of poor television reception, users are frequently supplied with television signals via cable. New regulations announced in February 1972 for the cable television industry by the Federal Communications Commission were expected to have a profound effect on the industry. Among the new FCC regulations was a requirement for 20-channel capacity for stations in the top-100 markets, including a dedicated channel for educational use, a dedicated channel for local government use, and a dedicated channel for noncommercial public access. A requirement was included for two-way capability, namely the capacity for return communication on at least a nonvoice basis. The regulations appeared to presage a vastly increased role for cable television, but it takes times for such a role to develop. A leading cable television firm canceled a franchise that was rather uniquely favorable to the community in Boulder, Colorado in 1973, and the city was forced to look for new applicants for the franchise. Cancellations have taken place elsewhere. Perhaps the arrangement in which cities negotiate with cable firms for franchises is not an effective way to achieve the condition of a wired nation, as described by Smith [53].

Cable operations do have some technical advantages; the over-the-air radio spectrum has become crowded and congested, and cable systems can provide increased communication capacity without adding to the congestion of the spectrum. In this respect it should be pointed out, however, that some cable systems have not been sufficiently well engineered to be free from interference, or RF pollution, and thus have not provided the anticipated advantage to the degree that they should. The way in which the cables are terminated or connected to television receiver inputs seems to be an especially weak link, and this feature is a function of the type of television input terminal as well as of the cable system itself. Assuming, however, that such problems are corrected by using suitable coaxial cable connectors at the television receiver, etc., considerable advantage should accrue from cable systems. Among other features, a wider range of programs and information should be made available.

An additional level of cable application that has been discussed involves the installation of multipurpose broadband two-way cable terminals in the office and home. These could be used for entertainment, education, communication (including electronic mail), computation, etc. It has been pointed out, however, that some of the applications that are proposed such as billing, computation, etc., can be carried out by using the narrow-band telephone lines which are almost universally available at present.

Picture telephones are a related topic; it was not clear, several years ago, to what extent they would be used extensively and what the relative success of picture telephones and broadband cable would be. As it has developed, picture telephones have not achieved widespread acceptance, and progress toward broadband cable has not been rapid either. The size of the original U.S. picture telephone screen may have been too small, and the concentration on the display of the head and shoulders of users (rather than on information and data displays) may have been a negative factor [14].

BROADCASTING. Broadcasting is a major area of telecommunications, but we are unable to do justice to the subject here and must be content to make some brief historical remarks and to express some of the common complaints. The first regularly scheduled radio brodcast programs were those of station KDKA in East Pittsburgh, Pennsylvania in 1920. This station operated at a frequency of 833 kHz and used amplitude modulation (AM). Radio broadcasting quickly became widespread, and the AM broadcast band from 535 to 1605 kHz became very crowded and subject to interference from distant stations. The development of frequency modulation (FM) facilities which took place in the 1940's was a major step forward. FM systems operating near 100 MHz are inherently less susceptible to noise and provide higher fidelity due to their broader bandwidth. As they are essentially restricted to line-of-sight operation, they do not have the potential for long-range interference of the 535- to 1605-kHz band. Frequencies from about 6 to 20 MHz, in the HF range, have also been used for broadcasting, especially for long-distance use.

Radio broadcasting reached its peak in the late 1940's, and television came into widespread use and popularity after that time. Radio has continued to be important, but television has been in the limelight since then. Black and white television broadcasts began in the 1940's, and the first coast-to-coast color television broadcast was that of the Rose Festival in Pasadena, California on Jan. 1, 1954.

The commercial TV networks and stations in the United States have provided excellent service in some ways and have furnished some exceptionally fine programs, but the quality of programming in general has left much to be desired in the opinion of many. Among the features which have been questioned are the widespread portrayal of violence, the emphasis on the sensational, and the low intellectual level of many of the programs. An additional feature of commercial TV and radio which many people find objectionable is the nature of the advertising. It seems that most do not object to the advertising as such but rather to the tasteless nature of many advertisements, the successive presentation of advertisements for several different products during station breaks, etc. It is to be hoped, as the President's adviser on consumer affairs has stated, that advertising can become more truly informative and factual. The banning of cigarette advertising was one favorable move.

A major encouraging development of recent years is the increase in quality and stature of the noncommercial educational TV stations and networks. On the technical side, the large bandwidth allotted exclusively for television has been questioned.

LAND MOBILE RADIO. The term land mobile radio may be used in the restricted sense employed by the FCC or in a more general sense to include some amateur and CB (citizen's band) radio operations as well. The different types of standard mobile radio services are public safety service, land transportation service, maritime mobile and aeronautical mobile service, industrial service, domestic public radio service, and governmental service. The demand for land mobile services and CB radio has increased rapidly in recent years. The frequencies allocated for land mobile radio in the general sense include the 25- to 50-, 132- to 174-, 400- to 512-, and 806- to 947-MHz bands (with some gaps within these limits). Frequencies above and near 150 MHz have been especially heavily used, and reference is commonly made to the 150-MHz band.

The interest in CB radio began a spectacular increase during the oil-embargo crisis of 1973–1974. As of Jan. 1, 1977, 40 CB channels, from 26.96 to 27.405 MHz, were available. Before that time 23 channels had been available from 26.96 to 27.30 MHz. The CB band has suffered from congestion and misuse, and it appears that expansion from 23 to 40 channels may not comprise a permanent solution. By 1977, however, the growth in CB usage had slowed, and sales of CB radios were in a slump. The explosive growth of CB radio created a shortage of quartz crystals that are used to establish precise transmitter or local oscillator frequencies. The development of digital frequency synthesizers, however, allows the use of only one quartz crystal, as compared to a total of 14 for 23 channel units. Such synthesizers are also highly useful for other mobile operations, as they can allow one receiver to take the place of several. The land mobile bands below 470 MHz have become highly congested in major metropolitan areas, sometimes to the point of being unusable, especially in New York City but also in Chicago and Los Angeles to a lesser degree [11]. The addition of 115 MHz of spectrum in the 806- to 947-MHz band will provide some relief when this band becomes widely used.

Other approaches to minimize problems of congestion are to use tone-coded squelch (so each listener hears only messages intended for him), trunked systems, and cellular systems. Trunking involves automatically scanning a group of channels in order to find an unused channel. Cellular systems involve dividing the area to be served into small cells, possibly as small as half a mile in diameter. The same frequencies can then be reused at a distance of a few cell diameters. Paging systems constitute a high-growth, low-cost segment of land mobile radio, and they reduce the need for two-way mobile channels.

2-3 ELECTROMAGNETIC SPECTRUM UTILIZATION

2-3-1 General Considerations

Two types of considerations in utilizing the electromagnetic spectrum are the following:

1. How should the spectrum be allocated among the various possible uses and users, and how can the congestion of the spectrum be alleviated?
2. Given that a particular frequency band is available for a given application, what steps must be taken to ensure operation with a satisfactory signal-to-noise ratio?

The first question is partially an engineering one and partially an economic and political question. The second question is primarily an engineering matter. The two aspects are closely related in that allocations should take into account what is feasible and practical technically, and the necessary design features are influenced strongly by what frequency allocations have been made.

Some allocations made previously need to be updated and revised, as some frequency bands are underused and others are very crowded. Mobile service and CB bands, for example, tend to be very crowded, while bands used for forestry are said to be lightly used in urban areas. Frequencies need to be shared more commonly, rather than assigned exclusively on a nationwide scale. Sharing has been adopted as the policy for satellite and terrestrial microwave links. Narrower channels could be used in many cases. The policy of a user being able to select from several channels instead of having to rely on one could be expanded. Low-power transmitters can be used in localized areas, as mentioned previously for the case of land mobile service. Antenna characteristics can be improved, maximum use can be made of cables, and higher frequencies can be employed.

Every communication system requires some minimum signal-to-noise ratio at its receiver input terminals. For successful operation the available signal-to-noise ratio must be greater than the required (minimum) signal-to-noise ratio. A communications engineer may need to determine what these ratios are or will be, given or assuming system characteristics such as transmitter power, antenna gain, terminal spacings, etc. In system design or modification, an engineer may need also to take measures to provide an adequate available signal-to-noise ratio at the receiver terminals or to improve upon an existing ratio. Likewise he may need to provide a receiving system having as low a required signal-to-noise ratio as practical. In considering this type of problem it may be desirable to distinguish between broadband noise, interfering or unwanted narrow-band signals, and intermodulation or fading effects. Thus Gierhart et al. [17] assume that satisfactory service is possible only when

$$\left(\frac{S}{N}\right)_a > \left(\frac{S}{N}\right)_r$$

$$\left(\frac{D}{U}\right)_a > \left(\frac{D}{U}\right)_r$$

$$(R_{IM})_a > (R_{IM})_r$$

where S/N = signal power-to-broadband noise power ratio,
D/U = desired signal power-to-undesired signal power ratio,
R_{IM} = signal power-to-intermodulation power ratio,
and the subscripts a and r denote available and required ratios, respectively. Elsewhere in these notes, S/N is used in a more general sense, with N taking into account undesired signals as well as random noise. Also we shall use C for the average power of a wanted *radio frequency* signal and X for the corresponding average power due to all other sources of signal and noise. Intermodulation may be due to equipmental limitations or may be caused by fading in the propagation path. Discussion of such propagation effects is deferred until Chap. 4.

Determination of C involves the values of transmitter power, antenna gain, and path loss and will be illustrated for the case of microwave systems in Sec. 2-3-4. X is a measure of noise, which is obviously very important to telecommunications, especially from the environmental viewpoint, and is considered in Sec. 2-3-3. The required or minimum C/X ratio is a function of the type of modulation employed and the type of detection circuitry employed and is also considered briefly in Sec. 2-3-3.

The comments made above refer to the requirements for successful operation of a telecommunications system. A related important topic is that of the measures which can be taken to minimize interference between systems. Some of the obvious, basic considerations are that a sufficient degree of frequency stability be maintained, that both the desired signals and spurious and harmonic emissions be kept to acceptable levels, and that antennas be suitably located, directed, and designed. Attention may need to be given also to grounding, bonding, shielding, and filtering techniques. The problem of interference between the electronic facilities on ships, on aircraft, and at military bases has received considerable attention, and the Department of Defense has established an Electromagnetic Compatibility Analysis Center. A reference which discusses the practical measures mentioned in this paragraph, man-made sources of interference, etc., is that by Ficchi [16].

2-3-2 Survey of the Spectrum as a Function of Frequency

In this section the various portions of the electromagnetic spectrum and their uses are described briefly. The discussion of the spectrum and its uses begins with the lowest frequencies and proceeds toward the highest. Further information about frequency uses and allocations in the radio-wave portion of the spectrum can be found in *Radio Spectrum Utilization* [30] and in *Reference Data for Radio Engineers* [26].

VLF, ELF, AND LOWER FREQUENCIES. The VF (voice-frequency) band extends from 300 to 3000 Hz, and the ELF (extra-low-frequency) band extends from 30 to 300 Hz, but some apply the term ELF to the entire 3- to 3000-Hz band and use the designation of ULF (ultra low frequency) for still lower frequencies. Geomagnetic pulsations, a subject of interest to geomagneticians and atomospheric geophysicists [8,27], fall in the ULF range. Geomagnetic pulsations are naturally occurring variations in magnetic flux density and are measured by the use of induction coils and magnetometers of various kinds. Corresponding electric field variations at the earth's surface are recorded by measuring the potential difference between electrodes placed in the ground. The term *telluric current* or *earth current* is applied to such electric field measurements. Telluric currents can interfere with telephone and, possibly, electric power services at high geomagnetic latitudes where they are most intense.

ELF and VF signals are generated largely by lightning discharges. Resonances of the earth-ionosphere waveguide occur at frequencies of about 7.8, 14.1, 20.3, 26.4, and 32.5 Hz. ELF and lower frequencies are utilized in some geophysical exploration methods [34]. The ELF band also has a limited potential for communication, and the U.S. Navy's controversial Project Seafarer (formerly Sanguine) has been devoted to this possible application [5,55]. The frequency of commercial electric power in the United States is 60 Hz, and consequently significant 60-Hz electromagnetic fields are encountered almost everywhere in this country. The prospect of high-voltage direct-current transmission lines has caused concern to geomagneticians who measure the earth's magnetic field, as direct currents can cause errors in their measurements.

Whistlers, mentioned further in Sec. 2-3-3, because of their contribution to atmospheric noise, typically occur in the VF and VLF ranges, and particular varieties of whistlers are also encountered at lower frequencies. An interesting use of the VF and lower VLF portions of the spectrum has been in a mine rescue communication system [36]. VF electromagnetic waves propagate reasonably well through the earth, and voice signals can be transmitted in this frequency range without the necessity of modulation and demodulation.

3–200 kHz (VLF, Lower LF). Frequencies in the 10- to 50-kHz range were used for transatlantic communication in the early days of radio, but the VLF and LF bands, especially at frequencies below 200 kHz, have a number of disadvantages which have limited their application for communication. The most basic limitation is that they can accommodate only a very limited bandwidth of transmissions. The high level of atmospheric noise and the necessity for large expensive antenna structures are other drawbacks. However, these frequencies are relatively free from the effects of ionospheric disturbances, and signals in these frequency ranges propagate reliably for long distances. Also frequencies at the lower end of the VLF band are low enough to penetrate usefully below the surface of the ocean. In general, frequencies below 200 kHz are useful for long-range, highly reliable service where the ability to transmit information at a high rate is not required.

Present uses include maritime mobile long-range service, standard time and frequency broadcasts (as from WWVL and WWVB in Fort Collins, Colorado at 20 and 60 kHz), and the LORAN-C (90- to 110-kHz) and Decca (70- to 90-kHz and 110- to 130-kHz) radionavigation systems.

A useful reference on the VLF (3- to 30-kHz) band is that by Watt [56].

200–3000 kHz (MF, Upper LF). This crowded frequency band includes the AM broadcast band (535–1605 kHz). It would be advantageous from the viewpoint of spectrum utilization if local broadcasting could be transferred to the FM band, leaving the present AM band primarily for regional and clear channel broadcasting. Interference tends to be severe at night because the *D* region that attenuates the sky wave (the wave reflected from the ionosphere) disappears at night, and signals propagate for long distances by reflection from higher ionospheric layers. The problem is partially solved, but at the expense of reduced service, by restricting hours of some local stations to daylight hours.

Other uses of this band include radionavigation (285–325 and 405–415 kHz for maritime radionavigation and direction finding and the rest of the 205- to 415-kHz band for aeronautical radionavigation), the maritime mobile service (415–510 kHz including the 500-kHz distress frequency), the LORAN-A navigation system (1800–2000 kHz), and miscellaneous marine and land mobile services, etc., in the 1605- to 2850-kHz band. Most of the police mobile services formerly in the 1605- to 2850-kHz band have moved to VHF frequencies (30–300 MHz).

3–30 MHz (HF). Known as the HF band, the 3- to 30-MHz range of frequencies is crowded and subject to mutual interference between users. Radio amateurs were the first to make use of the band, employing it to communicate between the United States and Europe in 1921. Equipment for HF communications over long distances is relatively simple and low in cost, but propagation is subject to outage during ionospheric disturbances. Even during undisturbed periods different frequencies have to be used during different times of the day and different portions of the sunspot cycle as propagation depends on reflection from the *E* and *F* layers of the ionosphere, especially the *F* layer, and electron densities in the layers vary diurnally and with the sunspot cycle.

One measure that is helpful in alleviating the crowding of the spectrum is to use single-sideband (SSB) transmissions. In simple amplitude modulation, the carrier frequency and frequencies above and below the carrier are transmitted. Thus a 4-kHz voice band occupies 4 kHz above the carrier frequency and 4 kHz below for a total bandwidth of 8 kHz. All of the information is carried, however, in one sideband alone, and circuits have been devised to suppress one of the sidebands and the carrier itself.

Before the advent of satellites HF transmissions supplied the bulk of long-range communication needs, but the need for HF for such service has diminished considerably. Nevertheless, HF is still useful for certain purposes, especially for mobile operations and communication with isolated locations which do not require high-capacity communication links, and the HF band is still crowded. This crowding can be alleviated by shifting to cable or higher frequencies wherever possible for communication between fixed points, by using higher frequencies for mobile communication where possible, by using SSB transmission for essential mobile service, and by using other portions of the spectrum for short-range broadcasting.

The higher frequencies within the HF band are too high for reliable ionospheric communication. The 26.96- to 27.405-MHz band is used for CB, but the choice was unfortunate in the respect that such frequencies can propagate sporadically for long distances by ionospheric reflection. The resulting interference can be expected to be especially severe during periods of sunspot maxima.

30–1000 MHz (VHF, Lower UHF). This frequency range includes the popular FM and television bands, FM operation covering the range from about 88 to 108 MHz and television having a number of assigned bands between 41 and 960 MHz. Land, aeronautical, and marine mobile services also have assigned bands in this frequency range (including 118–136 MHz for air-ground communications). The 136- to 137-MHz band is used for telemetry and tracking functions associated with space research, and radio astronomy has several small assigned bands. Aeronautical and radionavigation aids have assigned bands, and some radars operate in this band (the 400-MHz Ballistic Missile Early Warning System or BMEWS radars, for example). Tropospheric scatter communication systems, operating over longer distances than line-of-sight relay systems, have been used in Alaska and the Arctic, etc. (The White Alice system in Alaska operates at frequencies near 950 MHz.) Tropospheric scatter systems depend on weak signals scattered from irregularities or layers of the lower atmosphere.

Broadcasters, amateurs, land mobile users, and CB users tend to have competing interests within the 30- to 1000-MHz band. The VHF television bands (54–88 MHz for channels 2 through 6 and 174–216 MHz for channels 7 through 13) have been quite extensively utilized, but other potential users have questioned the wide bandwidth formerly available exclusively for television in the UHF range (470–890 MHz). As a result, frequencies in the 470- to 512-MHz band and above 800 MHz are now shared with the mobile services in some areas, and certain land mobile groups would like to see sharing extended to the entire UHF television band. The frequencies allocated for forestry and conservation have not always been used extensively in certain urban areas, and pressure has developed for the forestry-conservation users to share some of their frequencies with others in urban areas. The expansion of CB service from 23 to 40 channels involved the loss of certain 27-MHz channels for land mobile use. CB users complain that a large number of them are still crowded into a narrow frequency range, however, whereas a smaller number of radio amateurs have a wider bandwidth available. It has been proposed that part of the 220- to 225-MHz amateur band be used for CB operations, but amateurs feel that their contributions to radio since its inception, their technical expertise, licensing requirements, and responsible operating procedures entitle them to keep the band. Harmonics of some of the CB transmissions have interfered with television reception, expecially in the case of channel 2, and CB operators tend not to be popular with television broadcasters and users.

1000–12,000 MHz (Upper UHF, Lower SHF). Most of the present surface and satellite microwave communication systems operate in this frequency range. These systems are considered further in Secs. 2-2 and 2-3-4. Several bands are assigned for radar. The 1215- to 1400-MHz range is used for the long-range FAA and Air Force radars, and the 2700- to 2900-MHz band is used for air traffic control at airports. The FAA long-range Air Route

Surveillance Radars (ARSRs) transmit their signals to Air Route Traffic Control Centers scattered around the nation. The airport radars are designated as Airport Surveillance Radars (ASRs). Aircraft radars operate near 5400 and 9400 MHz, and shipboard radar operates in the 2900- to 3100-, 5460- to 5650-, and 9300- to 9500-MHz bands. The 1400- to 1427- and 1664.4- to 1668.4-MHz bands are reserved for radio astronomy, for observations of hydrogen atom and OH radical emissions in space. The 2690- to 2700- and 4990- to 5000-MHz bands are also assigned to radio astronomy for continuous emission spectra.

This 1000- to 12,000-MHz band has been rather well managed, but congestion is increasing. Possible interference between surface and satellite microwave links is considered in Sec. 2-3-4. Applications of radars and microwave radiometers, which also operate in the band, are considered in Chap. 6. New developments in microwave solid-state devices (silicon and GaAs diode oscillators, etc.) can be expected to have a major impact on this frequency band in the future.

MILLIMETER WAVES. Most microwave communication and radar systems have operated at frequencies below about 14.5 GHz, in fact below 10 GHz, up to now, but the wide bandwidths available in the millimeter band are attracting attention, as the lower frequency bands become increasingly crowded and congested. It would seem reasonable to classify as millimeter waves only those having lengths less than a centimeter (or frequencies higher than 30 GHz). However, the frequency band between 10 and 300 GHz is sometimes considered to be the millimeter band. In a treatment of barriers to telecommunications growth [11], Crombie took 14.5 GHz as a dividing line between lower, rather well-utilized frequencies and higher frequencies not yet exploited extensively.

Attenuation due to rain, water vapor, and oxygen need to be taken into account above about 12–16 GHz, as discussed further in Sec. 4-4, but it should be mentioned here that interruptions in service on millimeter-wave earth-satellite paths due to rain need not necessarily inhibit unduly the use of millimeter waves. It has been generally assumed that extreme reliability (99.99% or greater) must be maintained on all commercial telecommunication links, and the margins required to provide such reliability on millimeter links would appear prohibitive. However, Feldman and Dudzinsky [15] have pointed out that studies of telephone usage show that delays in reaching the desired party are so common and so widely accepted, partly in order to take advantage of cheaper grades of service, that the quite infrequent delays due to rainfall on millimeter links may be of little consequence, as long as one has a choice in grade of service. Thus it appears logical to exploit the wide bandwidths available in the millimeter wavelength range and the potentially large volume of traffic that can be accommodated. For many years there has been interest in the TE_{01} mode of propagation in circular guide, which mode has the useful and unique characteristic that attenuation in the guide decreases with frequency up to about 100 GHz or more. The potential of this mode has prompted the Bell System to develop a circular waveguide about 5 cm in diameter that operates in the frequency range from 40 to 110 GHz, allowing about 240,000 two-way voice channels. A problem is that the TE_{01} mode does not have the lowest cutoff frequency of circular waveguide modes. The other modes, however, are inhibited by using a helical winding to form the inner metallic wall of the guide [50]. Although the circular waveguide operating in the TE_{01} mode has shown much potential, it is possible that it may be overshadowed by more recent developments in optical fibers.

Optical Fiber Communication. The invention of the laser spurred interest in the use of light for the transmission of information, and the production of optical fibers having losses under 20 dB/km in 1970 at the Corning Glass Works caused heightened interest in the use of such fibers. By 1974 a loss as low as 2 dB/km had been achieved in the laboratory, and the application of fibers appeared imminent [42,54]. Laboratory attenuations of a fraction of a decibel per kilometer had been achieved by 1978, and fibers were available commercially from several sources, including Corning Glass Works, which sold fibers having an attenuation $\leq$ 10 dB/km.

Optical fibers for communications consist of a center glass core having an index of refraction of about 1.5, surrounded by cladding (concentric and cylindrical) having a slightly lower index of refraction. The need for cladding arises from the fact that electromagnetic fields are not confined entirely to the interior of a single fiber but occur as evanescent fields outside the fiber [50]. Thus anything in contact with a single fiber would disturb its transmission characteristics. When cladding is employed, however, the fields are negligible at the outer surface of the cladding.

Although circularly symmetric TE and TM modes, like those in hollow metallic waveguides, can occur in fibers, the modes of most importance have three E and three H field components and are designated as EH_{lm} and HE_{lm} modes. All exhibit a cutoff frequency except the HE_{11}, which is called the dominant mode and is the mode used in single-mode operation. Analysis shows [42] that the number N of modes that can be transmitted in a fiber is given by

$$N \approx \frac{v^2}{2} \approx \left(\frac{ka}{\lambda}\right) n_1^2 \, \Delta$$

where $v = (2\pi a/\lambda)(n_1^2 - n_2^2)^{1/2}$. a is the radius of the core, λ is the wavelength, n_1 is the index of refraction of the core, n_2 is the index of refraction of the cladding, and Δ is defined by $n_2 = n_1(1 - \Delta)$. To achieve single-mode operation a must be small, a value of 5 μm being suitable. For multimode operation a is typically near 60 μm. For both single and multimode operation the outer diameters of fibers are usually in the 75- to 100-μm range. If the values of n_1 and n_2 are constant as a function of radius in the multimode case, significant dispersion can occur in fibers, with the result that when a single pulse is transmitted, for example, a smeared out version is received because the different modes have different velocities. If n_1 is made to vary appropriately with radius, however, the multimode dispersion is considerably reduced. The term *graded index* is applied to the variation of n_1 with radius. Waveguide dispersion (variation of velocity within a mode) and material dispersion (variation of index of refraction of glass with frequency) also contribute to total dispersion in both single-mode and multimode fibers. Whether it is practical to reduce multimode dispersion sufficiently to achieve total dispersion as low for multimode fibers as for single-mode fibers is not clear. The relative merits of single-mode and graded-index multimode operation were a subject of discussion in 1978.

Light-emitting diodes and injection lasers are the most prominent optical sources used with fibers; the former is an incoherent source, and the latter is a coherent source. Photodiodes are used as detectors. Multimode fibers must be used when incoherent sources are employed, and lasers must be used for single-mode operation. Single-mode fibers pose more severe requirements upon connectors than do multimode fibers. Light-emitting diodes and injection lasers may be modulated internally, but external modulation using electro-optic or acousto-optic techniques allows higher bandwidths and bit rates, up to GHz and Gb/s values. By 1977 a number of optical fiber installations were being put into service in North America, Europe, and Japan, including a Bell Telephone System in downtown Chicago that carries voice, video, and data signals, GTE installations in California and Hawaii, and 140-Mb/s links in England and the Netherlands.

2-3-3 Radio Noise

All communication systems must function with signals that are contaminated to some degree by noise. Since some minimum signal-to-noise ratio is needed for successful operation, noise may well be the factor which limits system performance. Noise originates both in the receiver of a communications system and externally to the receiver. Concepts concerning noise are basic to any quantitative treatment of telecommunication and remote sensing systems, and we therefore treat this topic in some depth.

Thermal noise is generated in any resistance, the thermal noise power per Hertz that is available from a resistance at the frequencies of interest being equal to kT W/Hz, where k is Boltzmann's constant, 1.38×10^{-23} J/K. T is absolute temperature, and the standard reference value of T is 290 K, which is a normal room temperature and which results in a noise power of

4×10^{-21} W/Hz. If an amplifier or receiver has a resistive input impedance at a temperature of 290 K, it will have a noise input of this magnitude. In addition such a receiver will develop some additional noise beyond this minimum thermal noise level, the additional noise being developed in its semiconductor or vacuum-tube amplifiers, mixers, etc. The amount of additional noise generated can be specified by a quantity known as the noise figure F, which is commonly expressed in decibels. Values in dB are not used in the equations of this chapter, however, unless so indicated by use of a dB subscript (e.g., F_{dB}) or unless otherwise clearly stated. A receiver having a noise figure of 12 dB, for example, has an overall output noise power about 16 times as great as that due to thermal noise at the input alone, as $F_{\mathrm{dB}} = 10 \log F$ and if $F_{\mathrm{dB}} = 12$, $F \simeq 16$.

An alternative way of expressing the additional noise generated by a receiver is in terms of an effective input noise temperature. To illustrate the meaning of noise figure and effective input noise temperature and to show the relation between them, consider the noise power W that is available at the output of an amplifier or receiver having a power gain g in a bandwidth Δf. The output noise power is given by

$$W = gkT_0\Delta f + W_i \tag{2-1}$$

where W_i is the noise generated internally in the amplifier and T_0 is the standard temperature of 290 K. The noise figure F is given by

$$F = \frac{W}{gkT_0\,\Delta f} \tag{2-2}$$

and the effective input noise temperature T_R is given by

$$T_R = \frac{W_i}{gk\,\Delta f} \tag{2-3}$$

resulting in

$$W = gk(T_0 + T_R)\,\Delta f \tag{2-4}$$

Equating the two expressions for W, we obtain

$$FgkT_0\,\Delta f = gk(T_0 + T_R) \tag{2-5}$$

or

$$F = 1 + \frac{T_R}{T_0} \tag{2-6}$$

and

$$T_R = T_0(F - 1) \tag{2-7}$$

The concept of noise temperature is illustrated in Fig. 2-1. In Fig. 2-1(a), an actual noisy amplifier or network, having input noise equivalent to that from a resistance at a temperature T_{in} of T_0, a noise figure of F, and a noise power output of W, is shown. The same effect is achieved in Fig. 2-1(b) by considering a noiseless amplifier or network which has an input resistance at

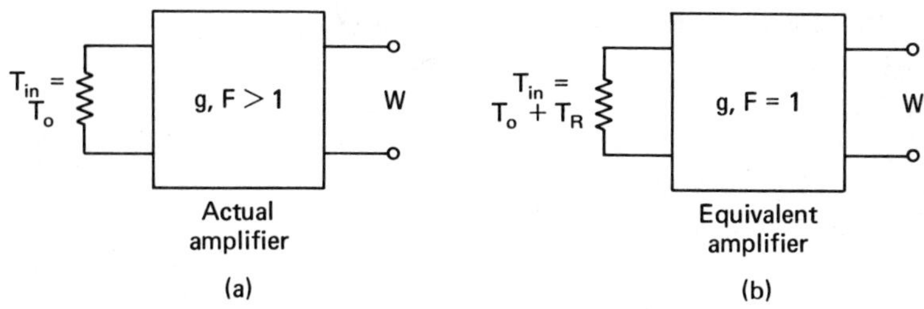

FIGURE 2-1. Alternative representations of noise performance of amplifier.

the temperature of $T_0 + T_R$ instead of T_0. Thus T_R is a measure of the additional noise added by the amplifier and has the advantage that it is additive with respect to T_{in}.

Next consider the effective input noise temperature $T_{R_{12}}$ of two amplifiers or networks connected in series. In this case

$$T_{R_{12}} = \frac{g_2 k(g_1 T_{R_1} + T_{R_2})\,\Delta f}{g_1 g_2 k\,\Delta f} \tag{2-8}$$

where g_1 is the gain of the first stage and g_2 is the gain of the second, from which

$$T_{R_{12}} = T_{R_1} + \frac{T_{R_2}}{g_1} \tag{2-9}$$

This relation can be generalized to any number of stages, and it applies whether the "stages" are amplifiers or attenuators or composites of amplifiers and other elements. It is only necessary to know the effective input temperatures of what are defined as the first and second stages and the gain of the first stage. In terms of noise figure the corresponding relation is

$$F_{12} = F_1 + \frac{F_2 - 1}{g_1} \tag{2-10}$$

Consider next the effect of an attenuator, in the form of a series resistance, that is inserted between a noise source and a point of observation as in Fig. 2-2. The noise power output of the attenuator per unit bandwidth is $W = kT_{out}$, but in such cases we shall simply show an effective noise tempera-

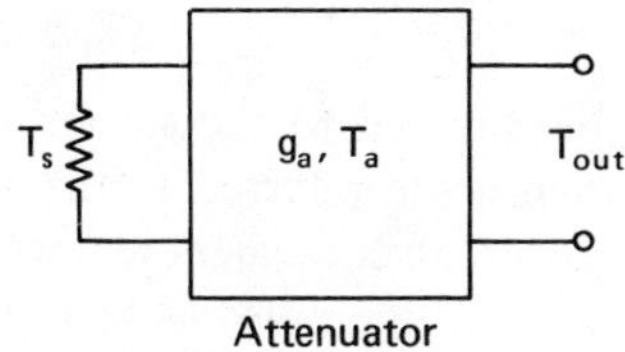

FIGURE 2-2. Attenuator placed between noise source and point of observation.

ture T_{out}. Following the treatment given in *Transmission Systems for Communication* [3] and elsewhere, note that at unity gain (no gain, no loss) $T_{out} = T_s$. On the other hand, if the gain is zero (or the loss is infinite), $T_{out} = T_a$. If $T_a = T_s$, then $T_{out} = T_s$ regardless of the attenuator gain setting. The difference $T_s - T_a$ can be regarded as a "signal" produced by the source, and this signal is attenuated so that

$$T_{out} = (T_s - T_a)g_a + T_a \tag{2-11}$$

where g_a is the gain of the attenuator. Upon rearrangement,

$$T_{out} = T_s g_a + T_a(1 - g_a) \tag{2-12}$$

(In Sec. 5-2 a comparable result is obtained for the case of attenuation in the earth's atmosphere. The factor $e^{-\tau}$ is used for the gain, or attenuation factor, in that section.) If $T_s = 0$ (or there is no source), the output temperature T_{out} of the attenuator is given by $T_{out} = T_a(1 - g_a)$. This situation is depicted in Fig. 2-3(a). It is useful, however, to consider a theoretically noiseless attenuator having the equivalent of an input resistive noise source at a temperature T_{in} such that

$$T_{in} g_a = T_{out} \tag{2-13}$$

or

$$T_{in} = (l_a - 1)T_a \tag{2-14}$$

where $l_a = 1/g_a$. T_{in} is the effective input noise temperature of the attenuator. If a noise source of temperature T_s is connected to the input of the attenuator, viewing the attenuator as in Fig. 2-3(b), then $T_{in} = T_s + (l_a - 1)T_a$ and $T_{out} = T_{in} g_a$, consistent with Eq. (2-12).

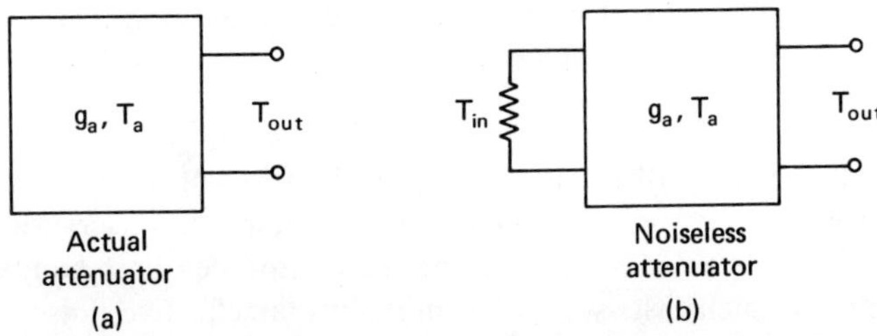

FIGURE 2-3. Equivalent representations of noise performance of attenuator.

Attention will now be directed to the system noise temperature T_{sys}, which represents total system noise referred to the antenna terminals as in Fig. 2-4. Consider first, however, an equivalent temperature, T_{RT}, that acounts for the receiver and transmission line noise as seen from the antenna terminals. T_a in Fig. 2-4 refers to the transmission line, which functions as an attenuator insofar as present considerations are concerned. An expression

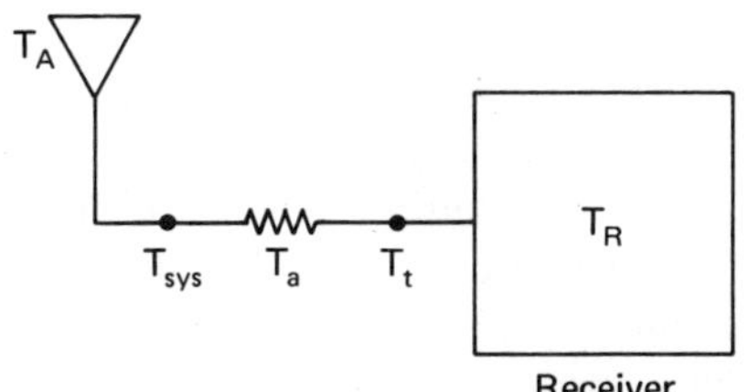

FIGURE 2-4. Locations where T_{sys} and T_t are defined.

for T_{RT} can be obtained by substituting $(l_a - 1)T_a$ for T_{R_1} and T_R, the receiver noise temperature, for T_{R_2} in Eq. (2-9). Thus

$$T_{RT} = (l_a - 1)T_a + l_a T_R \tag{2-15}$$

The system noise temperature is now defined as $T_A + T_{RT}$, so that

$$T_{sys} = T_A + (l_a - 1)T_a + l_a T_R \tag{2-16}$$

T_{sys} is commonly used in the literature on radio astronomy and satellite communications, especially in consideration of system sensitivity. It is possible, however, to make use of a corresponding temperature, T_t, at the receiver terminals, the expression for T_t being

$$T_t = T_A g_a + T_a(1 - g_a) + T_R \tag{2-17}$$

Either T_{sys} or T_t can be used, the important consideration being that signal and noise must be compared at the same place. Note that $T_t = T_{sys} g_a$ and that Eq. (2-16) includes the expression for T_{in} for an attenuator, whereas Eq. (2-17) includes the expression for T_{out} for an attenuator (as the quantities T_{in} and T_{out} are used in Fig. 2-3). Note also that if $g_a = 0$, $T_t = T_a + T_R$, and if $g_a = 1$, $T_t = T_A + T_R$. Returning to consideration of T_{sys}, representative figures for a satellite receiving system are as follows [12]:

$$T_{sys} = 40 + (1.028 - 1)290 + 1.028(20) = 68.6 \text{ K}$$

Note that $g_a = 1/l_a = 1/1.028 = 0.973$, corresponding to an attenuation of only 0.12 dB. [An attenuation of 1 dB corresponds to a g_a of 0.794 and an l_a of 1.26. If one expresses the ratio of output to input for an attenuator in dB notation, the dB value is negative. It is common practice, however, to let the sign (positive or negative) be understood and to refer specifically only to magnitude. The magnitude of $(g_a)_{dB}$ is the same as the magnitude of $(1/g_a)_{dB}$.] The noise temperature of the receiver in the above example is 20 K, corresponding to a noise figure of 1.069 or 0.29 dB. Such a noise figure can obviously not be achieved by a conventional type of amplifier. Low-noise receivers utilize masers or parametric amplifiers as their input stages, and these devices operate on principles different from those of conventional amplifiers. Furthermore, masers and some parametric amplifiers are cooled, by liquid helium in the case of masers.

Attention will now be directed to noise of external origin, which contributes to T_A in the expression for system noise temperature. In certain frequency ranges and for certain applications, external noise may be much greater than normal internal receiver noise. In such cases it is the external noise that is the principal limiting factor, and it is pointless to use a low-noise receiver. For satellite operations and in radio astronomy, however, the external noise may be rather low, and it may be advantageous to use low-noise receivers. The susceptibility of the system to external noise is increased by using a low-noise receiver, which may thus be a mixed blessing in some circumstances. External radio noise may be of natural origin or man-made and is a very important systems parameter, especially from the environmental viewpoint.

The principal varieties of naturally occurring radio noise are atmospheric noise (generated by lightning discharges), cosmic noise, thermal radiation from the atmosphere and nearby terrain, and radio noise from the sun. Atmospheric noise usually predominates at frequencies below about 20 MHz. It is the most important noise source, for example, in the VLF range, and Watt [56] gives attention to it in his treatment of VLF engineering. Such noise is especially severe when thunderstorms are in the near vicinity, but noise at frequencies below 20 MHz can propagate for long distances. In the absence of nearby thunderstorms, noise can be recorded at temperate and high latitudes from thunderstorms in major storm centers of South America, southeast Asia, and Africa. Atmospheric noise is highly variable and not readily predictable. Nevertheless, data in *CCIR Report* 322 [9] showing atmospheric noise as a function of geographic location, time of day, season, and frequency are highly valuable. (Some of the charts from the report are produced elsewhere as in Watt's text of VLF engineering.) The data in the CCIR report are given in terms of mean hourly noise factor F_{am}, which is the ratio of T_A to T_0 in dB, where T_A is the antenna temperature and T_0 is reference temperature. F_{am} values shown on world maps for a frequency of 1 MHz range as high as 100 in tropical areas and as low as 10 in the Antarctic. In the United States values near 50 are typical, but values as high as 90 may be encountered on summer afternoons.

An interesting phenomenon contributing to atmospheric noise in some locations is that of whistlers [23]. These are signals which are derived from lightning discharges and which propagate back and forth between the hemispheres along magnetic field lines. They are audio-frequency electromagnetic waves characterized by descending frequencies, the spread in frequencies resulting from the fact that the frequencies in the original lightning impulse propagate with different velocities. VLF emissions, including dawn chorus and hiss, occur in the same general frequency range as whistlers but are generated in the earth's outer atmosphere rather than by lightning.

Atmospheric noise drops off with increasing frequency, and above about 20 MHz cosmic noise tends to predominate. Jansky, an engineer with the Bell Telephone Laboratories, first identified radio noise of extraterrestrial origin while studying atmospheric noise from thunderstorms [28]. In a 1935 paper he pointed out that cosmic noise, or star static as he called it, puts a definite limit on the signal strength that can be received from a given direction at a given time "and, when a receiver is good enough to receive that minimum signal, it is a waste of money to spend any more on improving the receiver." Another pioneer was Reber, who followed up on Jansky's work and published the first maps of the radio sky. At a frequency of 250 MHz cosmic noise may cause an antenna temperature as high as 1000 K for an antenna having a sufficiently narrow beamwidth. Cosmic noise consists of radiation from discrete sources or radio stars and background radiation, the latter most intense toward the galactic center. As a function of frequency, the radiation consists of both continuous and discrete emissions. The important mechanisms for continuum radiation (covering a broad frequency range) are thermal blackbody radiation (treated in Chap. 5) and thermal emission from ionized hydrogen. The latter radiation comes from free electrons undergoing acceleration as when they pass near a proton. Line emission (emission over a narrow characteristic frequency range) takes place from neutral hydrogen at a frequency of 1420 MHz, corresponding to a wavelength of 21 cm. Cosmic radiation may be noise to the communication engineer, but it is a subject of research of the radio astronomer and has yielded important information about the structure of the universe. Cosmic radiation has also proved to be useful for studying the earth's ionosphere. In particular, records of variations in the absorption of cosmic noise in the lower E and D regions during auroral activity and polar-cap absorption events have increased the understanding of these regions. An instrument known as a riometer [38] has been designed to record such variations.

Above about 1000 MHz, atmospheric thermal noise tends to be the most important type of radio noise. All matter is continually emitting thermal radiation, and the atmosphere provides no exception. The band in which atmospheric noise tends to predominate includes the frequencies used for microwave surface, satellite, and space communications. Early studies of the effective noise temperature due to atmospheric thermal radiation were made by Hogg and Mumford [25]. Antenna temperature values vary from about 3 K for a vertically pointing antenna to about 100 K for a horizontal path which passes through much more of the atmosphere.

Radio noise is also emitted by the sun, the moon, and the planets, the sun being the most important source from the viewpoint of effects on communications. Radar workers first identified radio emissions from the sun during World War II. Radars of 5-m wavelength in the south of England experienced interference which was thought at first to be due to jamming by the Germans but was later determined to be from the sun. Radio waves are emitted from even the very quiet sun by the thermal emission mechanism, and this thermal emission from time to time shows increases for periods of days. This slowly varying component, observed at wavelengths ranging from about 3 cm to over 1 m, is associated with sunspots and plages. Burst radiations of shorter duration and of a more violent nature are also observed during times of solar activity, the more intense and longer-lived of such emissions being described as noise storms. These bursts are designated as Type I bursts, and Type II, III, IV, and V and centimeter-wavelength bursts are observed as well.

In the vicinity of metropolitan areas man-made noise and interference may be more intense in the LF through UHF bands than natural noise. Man-made sources of noise are automobile ignitions, power lines, radio-frequency stabilized arc welders, fluorescent lighting, diathermy equipment, etc. It has been noted that automobile ignition noise often dominates other sources in urban areas at and above VHF frequencies, and power transmission lines are frequently the major source of noise below 30 MHz [51]. On the basis of observations that background noise levels on aircraft communication equipment increased as aircraft approached large cities, an investigation of noise at 137 MHz was made in the Seattle area, and a 137-MHz noise map was made [7]. An F_a value (dB/KT_0B) as high as 24 dB was observed at a height of 1500 m over Seattle. The subject of noise and interference is complex. Many detailed studies are reported in the transactions of the Professional Group on Electromagnetic Compatibility and in symposium records of this same group. An antenna may receive noise and interference in the minor lobes of its pattern even when none is present in the main beam. The role of the antenna is considered in the following section.

Another type of noise, intermodulation noise, is generated in the active electronic circuits of telecommunication systems, but this topic is not treated here, and the reader is referred to references such as those by the Bell Telephone Laboratories [3] and GTE Lenkurt [20].

RECEIVER IMPROVEMENT FACTORS. Shannon's law [49] says that the maximum capacity of a channel for transmitting information, measured in bits per second, with an arbitrarily low error rate in the presence of white noise, is given by

$$\text{maximum capacity} = B \log_2 \left(1 + \frac{S}{N}\right) \tag{2-18}$$

A bit corresponds to a binary value of 0 or 1, as discussed in Sec. 2-2 with respect to PCM. B is the bandwidth of the channel, and S/N is the signal-to-noise ratio. No actual system obeys this law perfectly or achieves the indicated capacity, which is an upper theoretical bound, but some systems approach the capacity more closely than others. When S/N is large compared to unity, Shannon's law indicates that the relation of S/N to B is an inverse exponential one. For AM systems, however, S/N varies directly inversely with B for small values of N; for conventional FM systems S/N tends to vary inversely as B^2; and for PCM systems the relation is exponential but still not in agreement with Shannon's law [37].

Even though actual systems do not obey Shannon's law, they do exhibit the characteristic that a lower S/N ratio can be accepted if bandwidth is increased. Thus in FM and PCM systems the use of a relatively large bandwidth allows operation with a relatively small value of S/N, or C/X as the carrier power-to-noise ratio is designated here. In a conventional FM system, however, there is a minimum or threshold value of C/X of about 10 dB. Phase-locked FM systems can operate with lower C/X ratios. One important implication of the inverse relation between bandwidth and S/N is that the audio output of an FM receiver will have a higher S/N ratio than the C/X ratio at the receiver input terminals, where the bandwidth occupied is considerably greater than the audio bandwidth. Thus the signal-to-noise ratio of a channel S_{ch}/N_{ch} can be expressed as being equal to $R(C/X)$, where R is an improvement factor. In the case of FM modulation, the improvement factor is commonly taken to be $3(\Delta F/f_m)^2$, where ΔF is the FM frequency deviation and f_m is the highest modulating frequency.

In general the improvement in signal-to-noise ratio may be due to several causes, and R may then be expressed as a product of factors. The FM improvement factor given above may be one of the factors and can be designated specifically as R_{FM}. When frequency-division multiplexing (FDM) is employed, the signal power for n channels is less than n times the signal power for one channel, whereas the noise power for n channels tends to be n times the noise power for one channel. The result is to cause an improvement factor R_M, which may approach 20 dB when about 240 channels or more are utilized. Panter [44] gives the expression

$$R = \frac{B}{2b}\left(\frac{\Delta F}{\bar{f}_m}\right)^2 \tag{2-19}$$

for a multichannel system employing both FDM and FM modulation, where B is receiver IF noise bandwidth, ΔF is the frequency deviation caused by a standard reference (0 dBm0) signal, b is channel bandwidth, and $\bar{f}_m$ is the mean frequency of the channel under consideration.

When equalization or preemphasis is utilized a further increase, designated by R_{PRE}, may be achieved. When FM modulation is employed the channel having the highest $\bar{f}_m$ has the highest noise level and the lowest improvement factor, as indicated by Eq. (2-19), unless preemphasis is employed. Preemphasis involves modulating over a wider frequency range for higher modulation frequencies in order to achieve a more nearly constant S/N ratio. The net result, when the original relative signal levels are retrieved by de-emphasis, is an improvement in S/N ratio for the higher channels or to a lesser degree for the entire band in the case of television. Also in the case of both television and voice signals, the effect of noise on the viewer or listener has been found to be a function of frequency. Thus, the importance of noise is weighted as a function of frequency and the effective S/N ratio may be better than would be the case if weighting were not taken into account. The improvement can be represented by a factor R_W. In the analysis of a low-cost 12-GHz satellite-television receiving system, Han et al. [22] quote an S/N ratio of 54 dB when the C/X ratio is 20 dB. Of the 34-dB increase, 15.5 dB are accounted for by R_{FM}, 9.5 dB are accounted for by the combined effect of emphasis and weighting, and 9 dB are introduced for conversion from rms to peak-to-peak definition, on the basis that the peak-to-peak rather than rms value of the television signal is what the viewer responds to.

2-4 TELECOMMUNICATION SYSTEMS

Microwave frequencies are used in a large percentage of long-distance telecommunication systems, and in this section we shall concentrate primarily on microwave systems. The fundamentals of microwave terrestrial line-of-sight and earth satellite paths are considered, excluding atmospheric effects, which are treated in Chap. 4. The concepts are often applicable to other frequency ranges and systems as well. Radar systems, for example, utilize the same types of antennas in general, and noise is calculated in much the same way for conventional telecommunication and radar systems. Thus this section serves both as preparation for later chapters and covers a subject that is important in itself. The procedure for calculating the received signal power C for a microwave path is first considered, and the calculation of the noise power X and the C/X ratio is then illustrated. A brief discussion of some aspects of lower-frequency radio systems concludes this section.

2-4-1 Calculation of *C/X* Ratio for Terrestrial Microwave Links

Consider an isotropic transmitting antenna radiating a power W_T and a receiving antenna of effective area A_R at a distance d from the transmitter. The power W_R intercepted by the receiving antenna is given by

$$W_R = \left(\frac{W_T}{4\pi d^2}\right) A_R \tag{2-20}$$

A microwave relay transmitting antenna, however, concentrates the transmitted power into a narrow beam which is pointed at the receiving antenna. The degree to which the transmitted power is so concentrated is described by a quantity known as directivity. In calculating signal power on microwave links, however, we use antenna gain, which equals directivity times an ohmic efficiency factor (see Sec. 2-4-3).

The transmitting antenna gain G_T is given by

$$G_T = \frac{4\pi A_T}{\lambda^2} \tag{2-21}$$

where A_T is the transmitting antenna effective area and λ is the wavelength. (To take into account aperture and ohmic efficiencies an effective area less than the geometric aperture area should be used.) Multiplying the transmitted power by gain gives a quantity known as equivalent isotropically radiated power (EIRP). Inserting antenna gain into Eq. (2-20) gives

$$W_R = \frac{W_T A_R}{4\pi d^2}\frac{4\pi A_T}{\lambda^2} = \frac{W_T A_R A_T}{d^2 \lambda^2} \tag{2-22}$$

This expression can be rearranged to give

$$\frac{W_T}{W_R} = \left(\frac{4\pi d}{\lambda}\right)^2 \left(\frac{\lambda^2}{4\pi A_R}\right)\left(\frac{\lambda^2}{4\pi A_T}\right) = \frac{L_{FS}}{G_R G_T} \tag{2-23}$$

where L_{FS} is called free-space loss. That is, the ratio of transmitted to received power can be expressed as a quantity, $(4\pi d/\lambda)^2$, which is known as free-space path loss, divided by the gains of the transmitting and receiving antennas. The term *free-space path loss* is somewhat of a misnomer, but the expression does give the correct ratio of W_T/W_R. Taking W_T/W_R as the loss factor L, expressing the quantities in dB such that $L_{dB} = 10 \log (W_T/W_R)$, and including a miscellaneous loss factor L_a, we obtain

$$L_{dB} = (L_{FS})_{dB} - (G_T)_{dB} - (G_R)_{dB} + (L_a)_{dB} \tag{2-24}$$

or

$$L_{dB} = 20 \log \left(\frac{4\pi d}{\lambda}\right) - 10 \log \left(\frac{4\pi A_T}{\lambda^2}\right) - 10 \log \left(\frac{4\pi A_R}{\lambda^2}\right) + (L_a)_{dB} \tag{2-25}$$

As an example, consider a path length of 28.5 miles, or 45.9 km, a frequency of 4 GHz, and transmitting and receiving antennas having gains of 38.9 dB. The so-called free-space loss can be read from charts (Fig. 2-5) or calculated and in this case has the value 137.5 dB. For L_a in this example, waveguide losses of 1.8 dB and coupling network losses of 1.5 dB for the transmitting system and the same values for the receiving system are assumed. Including these combined losses of $2(1.8 + 1.5) = 6.6$ dB in the equation for L_{dB}, we obtain

$$\begin{aligned} L_{dB} &= 137.5 - 38.9 - 38.9 + 6.6 \\ &= 66.3 \text{ dB} \end{aligned}$$

If the transmitting power is 0.5 W or 27 dBm (27 dB above a milliwatt),

$$(W_R)_{dBm} = 27 - 66.3 = -39.3 \text{ dBm}$$

Next the receiver noise can be calculated, assuming noise equivalent to thermal noise at $T_0 = 290$ K at the receiver input. The thermal noise at the receiver input is thus kT_0 W/Hz or 4×10^{-21} W/Hz. Expressed in dBm, this

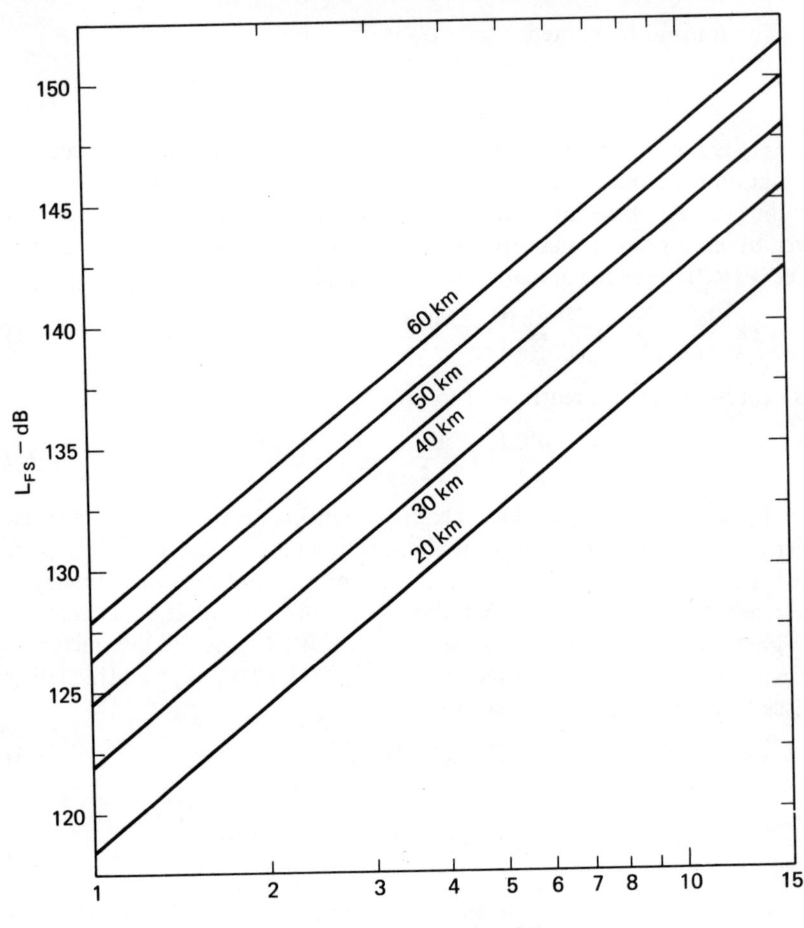

FIGURE 2-5. L_{FS} versus frequency and wavelength.

value is -174.0 dBm/Hz. Considering now a receiver bandwidth B of 15 MHz and a receiver noise figure of 13 dB, we obtain

$$X_{\text{dBm}} = -174 + 10 \log (15 \times 10^6) + 13$$
$$= -89.2$$

where X_{dBm} is noise power and the terms -174 and $10 \log (15 \times 10^6)$ represent kT_0B, the thermal noise power at the receiver input in a bandwidth B expressed in dBm. If C/X is taken as the received signal power-to-noise ratio,

$$(C/X)_{\text{dB}} = -39.3 - (-89.2) = 49.9 \text{ dB}$$

Horizontal microwave paths near the earth's surface, however, tend to be subject to signal fading, and an allowance of 35 or 40 dB is commonly allotted for fading. Using a 35-dB figure, we obtain

$$[(C/X)_{\min}]_{\mathrm{dB}} = 14.9 \text{ dB}$$

In the ordinary terrestrial microwave link in which FM modulation of the microwave signal is utilized, a minimum C/X ratio of 10 dB must be available for satisfactory operation of the FM discriminator. The calculated minimum $(C/X)_{\mathrm{dB}}$ ratio is thus adequate in this respect for a 35-dB fade but not quite adequate for a 40-dB fade. The above calculation of X_{dBm} corresponds to using a value of T_t of 290 + 5496 [see Eq. (2-17)], where $5496 = T_R$ and is the equivalent noise temperature for a 13-dB noise figure receiver. (Since waveguide and coupling network losses were taken into account in calculating C in this particular case, T_t would be the appropriate temperature to use in determining X, rather than T_{sys}.) Use of 290 for the sum of $T_A g_a + T_a(1 - g_a)$ is strictly correct only if $T_A = T_a = 290$, but this condition tends to apply for terrestrial links.

2-4-2 Satellite Communications

The above treatment can be taken as representative for surface microwave links. Microwave paths between earth stations and satellites are longer than terrestrial microwave paths but are less subject to fading because they pass through the lower atmosphere at a rather large angle from the horizontal. (A 4-dB margin has been considered sufficient for some systems.) Earth transmitting stations must have high transmitted power because of the length of the path and because satellite system noise temperatures tend to be rather high due to the fact that their receiving antennas are pointed at the earth. Satellite power outputs must be restricted for economic reasons, and earth receiving stations must therefore have high sensitivity. A low system noise temperature can be achieved by using sophisticated receiving equipment, as the ground receiving antenna looks into the largely empty sky, and the antenna temperature can therefore be small.

The INTELSAT IV satellite system uses the 5.925- to 6.425-GHz band for the up path and the 3.7- to 4.2-GHz band for the down path [1]. It provides about 5000 telephone circuits or more per satellite, utilizing 12 transponders each having a usable bandwidth of 36 MHz. The total bandwidth is 500 MHz. Satellite paths can be described in the same terms as surface microwave paths, but it is common practice to use the notation EIRP (effective isotropic radiated power) for the product of transmitter power and gain. Also the ratio of gain to system noise temperature (G/T_{sys}) is frequently taken as a figure of merit or sensitivity factor for the receiving system. In addition the quantity C/T_{sys} is sometimes utilized.

In terms of these quantities,

$$(C/T_{\mathrm{sys}})_{\mathrm{dBW}} = (\mathrm{EIRP})_{\mathrm{dBW}} - (L_{\mathrm{FS}})_{\mathrm{dB}} + (G/T_{\mathrm{sys}})_{\mathrm{dB}} \qquad (2\text{-}26)$$

where EIRP is effective transmitted power and C is received carrier power, all quantities being expressed in dB notation. The $(C/X)_{dB}$ ratio, as used previously for the case of surface paths, can be obtained by subtracting k (Boltzmann's constant) and B (the bandwidth), both expressed in dB, since $X = kT_{sys}B$.

For the earth-to-satellite path, the EIRP for the INTELSAT system may be as high as 95 dBW based on an actual power of 33 dBW and an antenna gain of 62 dB. The path loss L_{FS} is about 200 dB, and the G/T_{sys} ratio for the satellite receiving system is about -18 dB. Thus

$$(C/T_{sys})_{dBW} \simeq 95 - 200 - 18 \simeq -123$$

and, as $X = kT_{sys}B$,

$$(C/X)_{dB} \simeq -123 - (-229) - 87 \simeq 19 \text{ dB}$$

where -229 is $10 \log k$ and 87 is $10 \log B$, with $B = 5 \times 10^8$. (Unitwise, both T_{sys} and B are treated here as if they were nondimensional quantities or merely ratios, while in place of k, Boltzmann's constant alone, we actually use the product of k, a 1-K temperature interval, and a 1-Hz bandwidth. Thus the resultant value of -229 is considered to be a dBW value.)

Another ratio considered to be of fundamental importance [48] is $(C/kT_{sys})_{dB}$, which is a carrier-to-noise density ratio or the ratio of carrier power to noise power per hertz. For a single INTELSAT transponder, utilizing a single carrier, the $(C/kT_{sys})_{dB}$ ratio is quoted as 94.6 for a global antenna, whereas the $(C/X)_{dB}$ value for the 36-MHz bandwidth of a single transponder is 19 dB, the same value given above for the entire 500-MHz bandwidth. [$94.6 - 10 \log (3.6 \times 10^7) = 19.0$.] The transponders have multiple access capability (more than one carrier can be accommodated at a time by one transponder). When multiple carriers are received, the corresponding values for $(C/kT_{sys})_{dB}$ and $(C/X)_{dB}$ are quoted as 88.1 and 13 dB.

In this case of INTELSAT IV, the satellite receiving antennas are global-coverage antennas having beamwidths of 17°. For the satellite-to-earth paths, two global-coverage antennas having beamwidths of 17° and two spot-coverage antennas having beamwidths of 4.5° are available. Design EIRP values for the two antennas [1] are 22 dBW for the global antennas and 33.7 dBW for the spot antennas. The satellite-to-earth path loss is 196.9 dB. Specified minimum values for the gain of the receiving antenna and G/T_{sys} are 57.7 and 40.7 dB, respectively. These requirements are met, with a small margin of safety, by use of a 29.5-m or 97-ft antenna having an efficiency factor of 70% and a T_{sys} value of 78 K. If an EIRP value for the satellite of 33.7 dBW is used, the $(C/T_{sys})_{dBW}$ value at the earth receiving station for spot coverage is given by

$$(C/T_{sys})_{dBW} = 33.7 - 196.9 + 40.7 \simeq -123$$

which happens to be the same as that calculated above for the earth-to-satellite link. The corresponding $(C/X)_{dB}$ value would also be about the same as for the earth-to-satellite path.

The INTELSAT satellites utilize FDM-FM (frequency-division multiplexing, frequency modulation), and the signal-to-noise ratio in a given voice channel will be considerably higher than the $(C/X)_{dB}$ ratio of 19 dB, because of the receiver improvement factor R, which will be given roughly by an expression like that of Eq. (2-19) but with R_{PRE} and R_W multiplying factors (preemphasis and weighting factors) as well. It is of interest to note that 5000 telephone circuits or voice channels occupy only 20 MHz if 4 kHz is allotted per channel. Guard bands and channels for control and monitoring occupy some of the additional available bandwidth, but a large part of the difference between 20 and 500 MHz is utilized for the process of frequency modulation, as satellite systems employ large frequency deviations. Satellite links do not have to contend with the large fading allowance of terrestrial links, but, rather to the surprise of everyone, signal scintillation of ionospheric origin has proved troublesome in equatorial and auroral areas. This topic is discussed further in Sec. 4-3-5.

The G/T_{sys} value determines the available bandwidth or traffic capacity for a given EIRP, frequency, and path length. G/T_{sys} values smaller than 40.7 dB can be used if bandwidths smaller than 500 MHz are employed. For serving oases in Algeria, for example, an 11-m antenna is used, and the G/T_{sys} value is correspondingly smaller.

In the case of the ATS-F system, a satellite EIRP of 52 dBW is utilized in the S band (about 2.6 GHz), allowing reception by ground stations using a 3-m antenna and having a G/T_{sys} of 7 dB [46]. The higher EIRP is the crucial factor which permits a small, low-cost receiving installation, which can accommodate one color television channel and four voice channels. The Denver ATS-F earth station transmitted an EIRP of 84 dBW at 6 GHz and had a G/T_{sys} ratio of 28.9 dB at 4 GHz. Its antenna was 9.75 m in diameter, and its transmitted power was 3 kW. Another example of a low-cost satellite receiving system is discussed by Han et al. [22].

Atmospheric effects on microwave links are considered in Chap. 4. For a widely used treatment of the practical aspects of surface microwave systems, including factors not treated here, see the reference by GTE Lenkurt [20].

2-4-3 Microwave Antennas

A common type of microwave antenna is that in which a paraboloidal reflector is illuminated by a microwave horn or other type of feed radiator as shown in Fig. 2-6. The feed radiator is located at the focal point of the paraboloidal reflecting surface, with the result that radiation reflected from the surface is nearly uniform in phase over the area of the aperture which

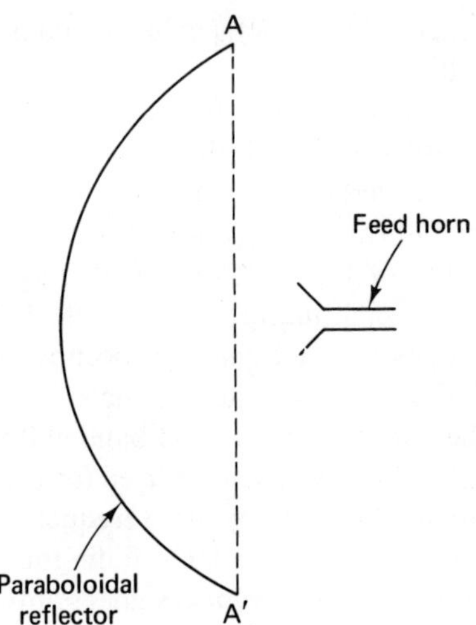

FIGURE 2-6. Paraboloidal-reflector antenna with waveguide-horn feed at focus.

lies in the plane AA'. It is obvious that the antenna concentrates radiation in a given direction rather than radiating energy uniformly in all directions as would be the case for an isotropic radiator. The degree to which an antenna concentrates energy in a given direction is indicated by a quantity known as the directivity D. By definition an antenna that concentrates energy uniformly into a solid angle of Ω_A steradians has a directivity given by

$$D = \frac{4\pi}{\Omega_A} \tag{2-27}$$

(The total solid angle surrounding a point is 4π sr or square radians. The solid angle Ω subtended at the center of a spherical surface of radius R by an area A on the surface is given by $\Omega = A/R^2$.) In the case of radiation from an aperture it can be shown that the directivity is also given by

$$D = \frac{4\pi A'_{\text{eff}}}{\lambda^2} \tag{2-28}$$

$A'_{\text{eff}} = \kappa_A A$, where κ_A is an aperture efficiency factor and is less than 1. A'_{eff} is less than the aperture area A for various reasons, including the facts that the feed radiator blocks part of the aperture and all of the radiation from the feed does not actually emanate from the focal point of the paraboloidal surface. Also A'_{eff} may be intentionally reduced with respect to A by tapering the illumination such that it is lower near the edge of the reflector than at the center. Such tapering can result in a decrease in antenna side-lobe levels. In attempting to apply Eq. (2-27) to an actual antenna for which energy is not uniformly distributed over the beamwidth, Ω_A can be assumed to be given approximately by

$$\Omega_A = \tfrac{4}{3}\theta_{\mathrm{HP}}\phi_{\mathrm{HP}} \tag{2-29}$$

where θ_{HP} and ϕ_{HP} are the half-power beamwidths in the two orthogonal directions. In the case of a lossless antenna, the received power density at a distance d from a transmitter could be calculated by multiplying $W_T/4\pi d^2$ by the directivity D, where W_T is the transmitted power. An actual antenna, however, has ohmic losses, and the power density is given instead by $(W_T/4\pi d^2)G$, where G is the antenna gain and

$$G = \kappa_0 D \tag{2-30}$$

where κ_0 is the ohmic efficiency factor. G can be calculated by using an expression identical to Eq. (2-28) but with an effective area A_{eff} that is less than A'_{eff}. Thus

$$G = \frac{4\pi A_{\mathrm{eff}}}{\lambda^2}$$

which is merely Eq. (2-21) rewritten with different notation. A_{eff} is related to A by

$$A_{\mathrm{eff}} = \kappa_{\mathrm{ANT}} A \tag{2-31}$$

where κ_{ANT} is overall antenna efficiency and is the product of κ_A and κ_0. A commonly used value of κ_{ANT} for a conventional antenna is 0.54. (The same terminology for antenna efficiencies is not used by everyone; κ_{ANT} is sometimes called aperture efficiency.)

A limitation of most directional antennas is the occurrence of minor lobes in the antenna pattern. All of the energy radiated is not concentrated into the main beam, but some appears in minor lobes having lower intensities than the main beam. Some energy is even radiated in the backward direction. A plot of the antenna pattern shows the position, shape, and intensity of the minor lobes. Figure 2-7, for example, shows the theoretical radiation pattern of a circular aperture illuminated by uniform plane wave; an actual paraboloidal-reflector antenna pattern would be similar. $J_1(ka \sin\theta)$ is the first-order Bessel function of $ka \sin\theta$.

The minor lobes are a potential source of interference and extraneous, unwanted signals and noise. If a low-noise receiving antenna, or a low value of T_A, is desired, an effort must be made to minimize the reception of radiation from undesired directions, especially radiation from the ground in the case of an antenna that points into space. The reduction in side lobes that can be accomplished by tapering the illumination was mentioned earlier. Spillover (radiation from the feed radiator that misses the paraboloidal reflector) is in an undesired direction and should be minimized; tapering helps in this respect. (And if there is spillover on transmission, a corresponding possibility for unwanted radiation exists for reception.) Other possible measures for reducing the large-angle side lobes and back lobes include mounting a

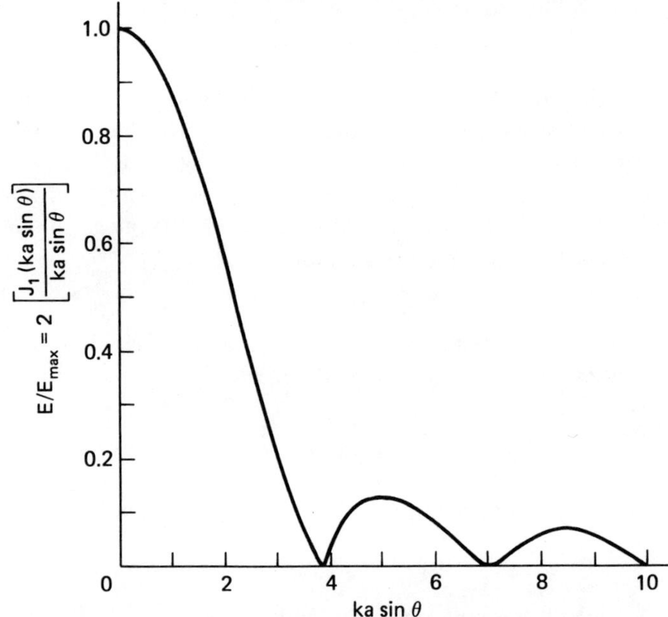

FIGURE 2-7. Radiation pattern of circular aperture illuminated by a uniform plane wave. k is the propagation constant, $2\pi/\lambda$, and a is the radius of the circular aperture.

cylinder or shroud of absorbing material so that it extends outward from the edge of the aperture, in the case of a circular paraboloidal antenna. Horn-type antennas also produce improved performance, including reduced side and back lobes, with respect to the conventional paraboloidal antenna. The construction of a horn antenna is suggested in Fig. 2-8. One of the advantages

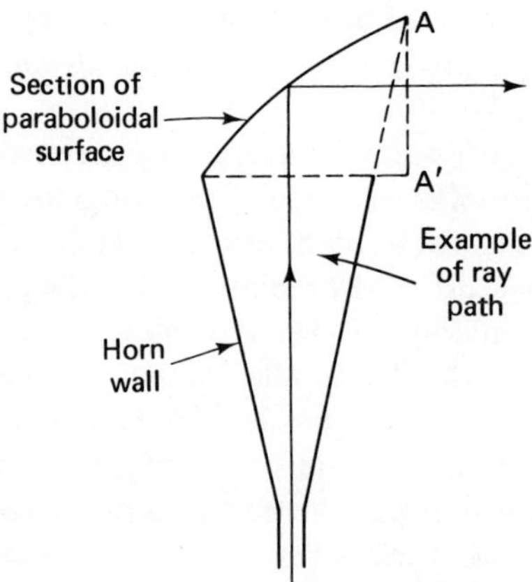

FIGURE 2-8. Microwave horn antenna. AA' represents a plane of constant phase.

of this antenna is the fact that a preamplifier can be placed immediately at the output of the horn, rather than some distance away along a lossy transmission line or waveguide. The result of so placing a preamplifier is a lower system noise temperature (see Prob. 2-4). The Cassegrain antenna utilizes a paraboloidal reflector, as in the basic antenna of Fig. 2-6, but features a hyperboloidal subreflector which is illuminated by a feed line that approaches the main paraboloidal reflector from the rear. The diagram of Fig. 2-9 illustrates the principle of the antenna, and Fig. 2-10 shows an actual Cassegrain antenna. Another antenna configuration found in some microwave systems consists of a vertically pointing antenna and a plane reflector which directs the energy into the desired path.

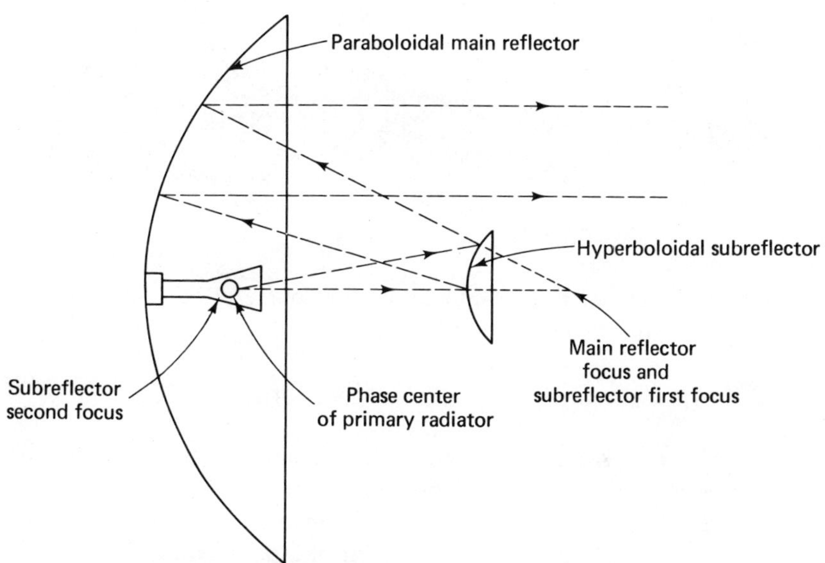

FIGURE 2-9. Geometry of Cassegrain antenna. (Courtesy of Scientific-Atlanta, Inc.)

At a particular microwave site, one half of the available frequency band is used for transmission and the other half is used for reception, the two halves being known as the high and low bands. Interference can arise most easily between two signals which are both in the same high or low band. Such interference can take place in the case of more or less parallel routes, when two or more paths join or their repeater locations are close to each other (as on the same hilltop), or when a signal from one transmitter overreaches directly to the third following repeater as well as to the next repeater site which it is intended to reach. The same antenna features that minimize noise of natural origin also tend to minimize interfering signals. Interference can also arise between terrestrial and satellite systems, and radars can interfere with microwave communication systems. In the latter case, harmonics of

FIGURE 2-10. Ten-meter Cassegrain antenna. (Courtesy of Scientific-Atlanta, Inc.)

the fundamental radar frequency may be the source of the difficulty if the harmonics are not attenuated by filtering.

2-4-4 Path Clearance and Reflections

Microwave systems require not only clearance for the direct, straight path from transmitter to receiver but for 0.5 or more of the first Fresnel zone as well. The actual required clearance in terms of distance from the direct path is also a function of the amount of atmospheric refraction; that aspect is treated in Chap. 4, and we only introduce the concept of Fresnel zones here.

Consider two paths as shown in Fig. 2-11. *TPR* is the direct path, and *TSR* is longer than *TPR*. If $TSR = TPR + \lambda/2$, where λ is the microwave wavelength, the region within the radius r of the direct path, at the distance of d_T from T and d_R from R, is defined as the first Fresnel zone. The particular value of r in this case can be designated as F_1, the first Fresnel zone radius. The concept can be extended to the case when $TSR = TPR + n\lambda/2$, for which

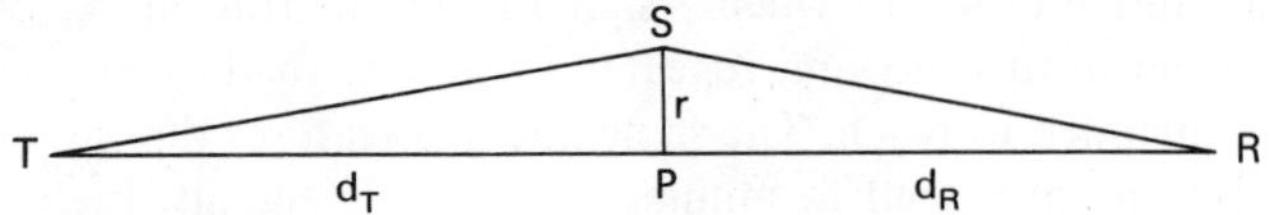

FIGURE 2-11. Geometry for consideration of Fresnel zones.

the corresponding Fresnel zone radius can be designated as F_n. Radiation passing through the first Fresnel zone tends to add in phase; radiation passing through the second Fresnel zone (between F_1 and F_2) tends to interfere destructively with radiation passing through the first Fresnel zone, etc. The tendency for constructive or destructive interference can be understood in terms of Huygen's principle, which states that every point on a wave front can be regarded as a source of secondary spherical wavelets. It can be shown that the radius of the first Fresnel zone F_1 is given, with all lengths in meters, by

$$F_1 = \sqrt{\frac{\lambda d_T d_R}{d}} \tag{2-32}$$

where $d = d_T + d_R$, or

$$F_1 = 17.3\sqrt{\frac{d_T d_R}{fd}} \tag{2-33}$$

if distances are measured in km, f is measured in GHz, and F_1 is measured in m.

Another factor which needs to be considered when energy reflected from a surface reaches a receiving antenna, in addition to energy traveling through the atmosphere, is the fact that when reflection takes place at close to grazing incidence a shift in phase of close to 180° takes place upon reflection for both horizontal and vertical polarization. Thus a signal which reaches the receiving antenna by having been reflected by a surface at a distance F_1 from the direct path will add in phase with a signal that traveled over a direct path. The two phase shifts of 180°, one due to reflection and the other due to an increase in path length, cancel out and reinforcement results. For a path length difference of λ, however, corresponding to F_2, destructive interference takes place.

The case of interference between the direct and reflected rays can be handled without reference to Fresnel zones as such. Consider the case depicted in Fig. 2-12, where a transmitter at height h_T and a receiver at height h_R are shown above a flat earth. The difference in phase between the signal following the direct path of length r_1 from T to R and the reflected ray is due

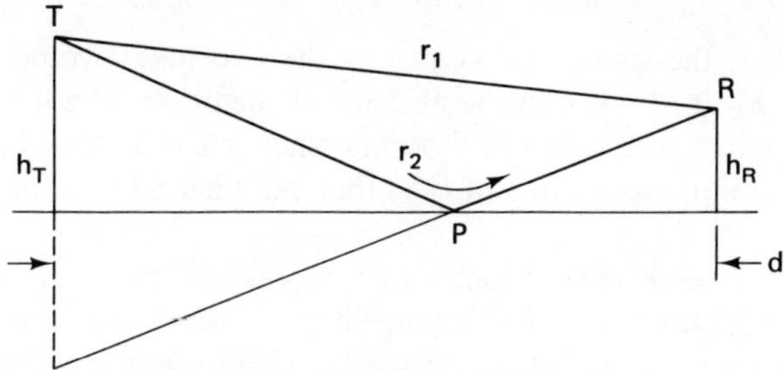

FIGURE 2-12. Direct and reflected rays.

to two factors, the difference in path length and the change in phase on reflection. The relations between r_1, r_2, d, h_T, and h_R are

$$r_1^2 = (h_T - h_R)^2 + d^2 \quad \text{and} \quad r_2^2 = (h_T + h_R)^2 + d^2$$

As $h_T + h_R$ and $h_T - h_R$ are small compared to d,

$$r_1 = d + \frac{(h_T - h_R)^2}{2d} \tag{2-34}$$

$$r_2 = d + \frac{(h_T + h_R)^2}{2d} \tag{2-35}$$

from which

$$r_1 - r_2 = \frac{2h_T h_R}{d} \tag{2-36}$$

The corresponding phase difference

$$\theta = \frac{2\pi}{\lambda}(r_1 - r_2) = \frac{4\pi h_T h_R}{\lambda d} \tag{2-37}$$

But the phase of the signal traveling the length r_2 is reversed on reflection, so the total phase difference ϕ is given by $\phi = \theta + \pi$. The phasors representing two signals C, corresponding to r_1, and D, corresponding to r_2 but without change of phase, are shown in Fig. 2-13 in one possible relative position.

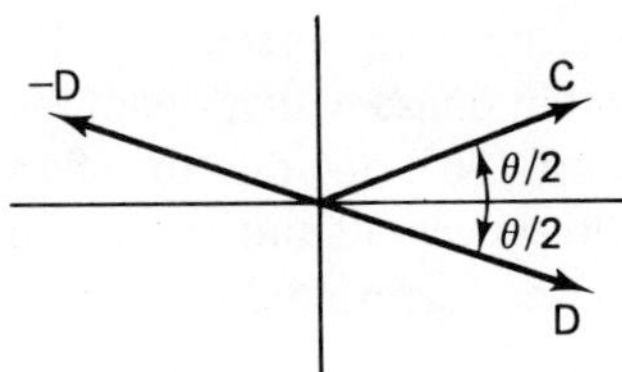

FIGURE 2-13. Phase relation between direct and reflected rays.

Because of the reversal of polarity on reflection, however, C and $-D$ are to be summed. In terms of the angle θ, the summation gives

$$E = \frac{2E_0}{|d|} \sin\left(\frac{2\pi h_T h_R}{\lambda d}\right) = \frac{2E_0}{|d|} \sin\left(\frac{\theta}{2}\right) \tag{2-38}$$

for the case that the signals arriving over the two paths would each cause field intensity E_0 at $r = 1$ m in the absence of the other. The first maximum in the expression as a function of θ occurs when $\theta/2 = \pi/2$, or $h_T h_R = \lambda d/4$, which gives the minimum value of $h_T h_R$ that can be used to obtain the maximum signal.

In the case of a curved earth this expression can be modified with the aid of Fig. 2-14, where MPN is tangent to the earth's surface. The antennas are at true heights h_T and h_R above the surface but at reduced heights h'_T and h'_R above M and N, respectively, where $h'_T = h_T - h$ and $h'_R = h_R - h$, assuming $h_T = h_R$. To determine h, make use of the relation $(r_0 + h)^2 = r_0^2$

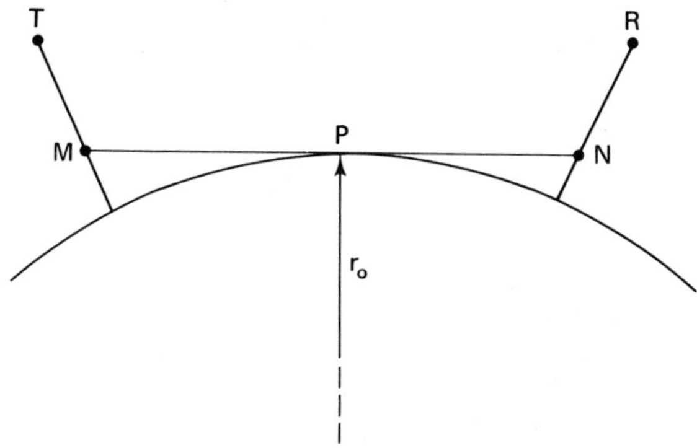

FIGURE 2-14. Ray path over curved earth.

$+ l^2$, where $l = d/2$ and d is the distance between T and R. If h^2 is neglected with respect to the other terms,

$$h = \frac{l^2}{2r_0} = \frac{d^2}{8r_0} \tag{2-39}$$

The flat earth type of analysis can now be used if h_T is replaced by $h_T - (d^2/8r_0)$ and h_R is replaced by $h_R - (d^2/8r_0)$, resulting in

$$\theta = \frac{4\pi h_T h_R}{\lambda d}\left(1 - \frac{d^2}{8r_0 h_T}\right)\left(1 - \frac{d^2}{8r_0 h_R}\right) \tag{2-40}$$

This equation applies strictly only for equal values of h_T and h_R, but if the values are not identical and not drastically different either, θ is given approximately by

$$\theta = \frac{4\pi h_T h_R}{\lambda d}\left[1 - \frac{d^2}{4r_0(h_T + h_R)}\right]^2 \tag{2-41}$$

The effects which arise when atmospheric variations are taken into account are considered in Chap. 4.

In the above analysis the reflection coefficient ρ was taken to be -1, as is the case for reflection from a smooth perfect conductor. Actually the reflection coefficient varies as a function of the nature of the surface, the electrical parameters σ (conductivity) and K (relative dielectric constant) of the earth, and the frequency and is different for horizontally and vertically polarized waves. The applicable expressions for a smooth earth for ρ_h (horizontal polarization) and ρ_v (vertical polarization) for θ measured from the earth's surface [33] are

$$\rho_h = \frac{\sin\theta - \sqrt{[K - j(\sigma/\omega\varepsilon_0)] - \cos^2\theta}}{\sin\theta + \sqrt{[K - j(\sigma/\omega\varepsilon_0)] - \cos^2\theta}} \tag{2-42}$$

and

$$\rho_v = \frac{[K - j(\sigma/\omega\varepsilon_0)]\sin\theta - \sqrt{[K - j(\sigma/\omega\varepsilon_0)] - \cos^2\theta}}{[K - j(\sigma/\omega\varepsilon_0)]\sin\theta + \sqrt{[K - j(\sigma/\omega\varepsilon_0)] - \cos^2\theta}} \tag{2-43}$$

Representative values for K and σ are shown in Table 2-1.

Table 2-1 Electrical Properties of Media.

Material	σ (*mhos/m*)	*K*
Silver	6.17×10^7	—
Copper	5.80×10^7	—
Seawater	5	80
Fresh water	8×10^{-3}	80
Rich agricultural land	1×10^{-3}	15
Rocky land, steep hills	2×10^{-3}	10

2-4-5 Lower-Frequency Systems

Attention has been concentrated primarily on microwave systems in this section, but some brief comments will be made concerning systems that operate at frequencies lower than microwave frequencies. Equation (2-24) for calculating L, the path loss, and the procedure for calculating the C/X ratio are applicable for any frequency or system, but for the lower frequencies the antenna gains may be experimentally determined or available as numerical values and are not normally calculated in terms of aperture area as in Eq. (2-25). Also the factors contributing to L_a are a function of type and location of system and operating frequency and may be difficult to determine. In this respect, note that when factors contributing to a reduction in signal level are variable and difficult to determine accurately, as in the case of fading on microwave links, for example, the practice may be not to attempt to determine a value for L_a that includes the variable factor but to utilize a fading allowance or factor of safety based on experience. Noise sources and levels also vary with location and frequency. The internal receiver noise is readily calculated, but it may be necessary to allow a margin for more highly variable, less predictable sources of noise.

HF frequencies are propagated by ionospheric reflection, sometimes by more than one hop so that the signals are alternately reflected by the ionosphere and the ground. Attenuation is experienced at each reflection. Actually the reflection process in the ionosphere is really a refraction process, and attenuation is encountered along the length of the path in the ionosphere. Data concerning the reflection and other loss factors contributing to L_a, the recommended fade margins for HF systems, and noise in the HF band, together with sample calculations, are given by Davies ([13], Chaps. 5 and 7). The theory of propagation in the ionosphere is treated in Sec. 4-3.

A large variety of antenna types is utilized at VLF-MF-HF-VHF-UHF frequencies [6,35]. Only a few types, the dipole, rhombic, and log-periodic antennas, are singled out for mention here.

A basic antenna type is the electric dipole antenna, commonly $\lambda/2$ in length and fed at the center as in Fig. 2-15. The radiated electric field inten-

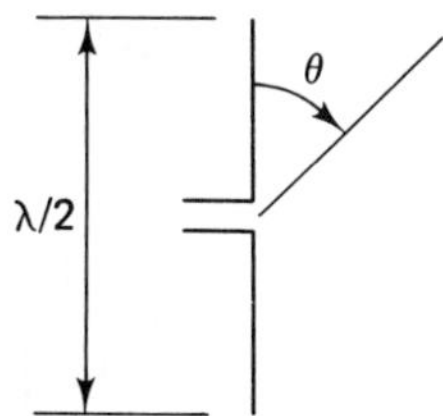

FIGURE 2-15. Half-wavelength dipole antenna.

sity at a distance r in the far field of the antenna in free space is in the θ direction and is given by

$$|E_\theta| = \frac{60I_m}{r}\frac{\cos[(\pi/2)\cos\theta]}{\sin\theta} \tag{2-44}$$

where I_m is the maximum (input) current of the dipole. The dipole is a simple example of an antenna constructed from relatively thin wires or rods. It has a broad beamwidth in the θ direction and is omnidirectional (constant signal intensity) in the ϕ direction (referring to a spherical coordinate system with θ measured from the axis of the dipole). The directivity of a single antenna is only 1.64, but dipole antennas can be arranged in linear or two-dimensional arrays.

For use of HF frequencies for long-distance commercial communications, rhombic and log-periodic antennas have been commonly used. The rhombic antenna can be regarded as an array of four horizontal long-wire antennas, arranged as in Fig. 2-16. The presence of the terminating impedance

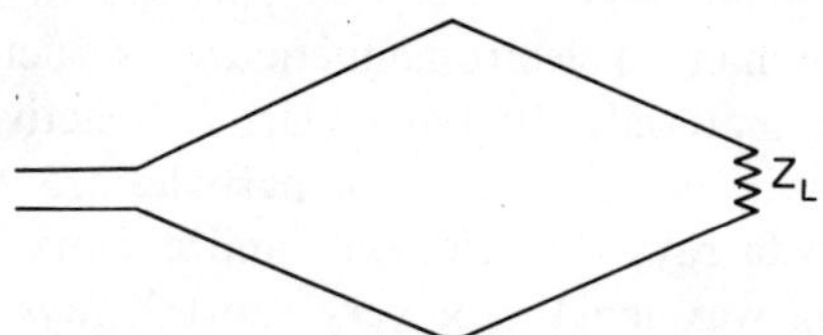

FIGURE 2-16. Plan view showing arrangement of rhombic antenna elements. The four sections are all of the same length and may be several wavelengths long.

Z_L causes the antenna to have maximum radiation in the forward direction (to the right in Fig. 2-16) at an angle above the horizontal which is a function of the height-to-wavelength ratio of the antenna. Large rhombic antenna farms on the Atlantic and Pacific coasts of the United States formerly handled a large fraction of overseas traffic, before the advent of satellites.

The log-periodic antenna has the advantage of being able to operate over an especially wide frequency range. There is a variety of forms for log-periodic antennas; one form is shown in Fig. 2-17. Such an antenna is fed at the apex (at the left in Fig. 2-17), and a wave propagates from the feed

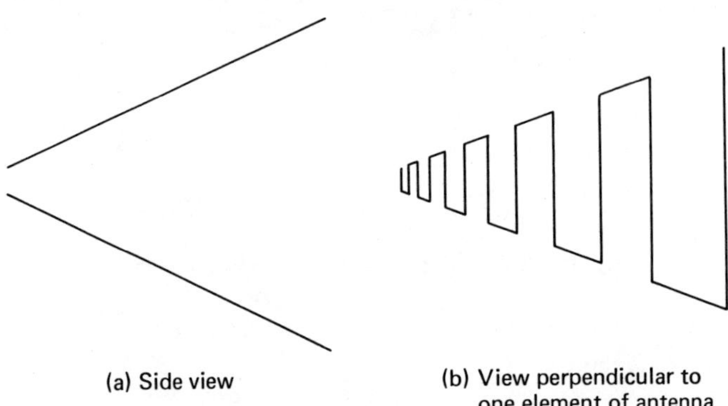

FIGURE 2-17. Trapezoidal form of log-periodic antenna. The spacings and lengths of the elements vary by the same ratio.

point to the right until a resonant $\lambda/2$ condition is encountered. Reflection then takes place, and radiation from the antenna is to the left.

2-5 BIOLOGICAL EFFECTS OF RADIATION

Radiation can be classified as ionizing or nonionizing, involving particles or waves, due to radioactivity or other causes, and natural or man-made. References to potentially harmful radiation commonly refer to ionizing radiation involving highly energetic particles or extremely high-frequency (X-ray or gamma ray) electromagnetic waves, such radiation being emitted by radioactive materials. In particular, radioactive elements emit alpha, beta, and gamma rays. Alpha rays or particles are doubly positively charged helium ions, beta rays are electrons, and gamma rays are electromagnetic waves, having wavelengths of very short X-rays and less. Alpha particles have weak penetrating power, beta particles are more penetrating than alpha particles, and gamma rays are highly penetrating. Most natural radioactive elements radiate either alpha or beta particles, but a few elements radiate both. Gamma rays often accompany the alpha and beta particles. The velocities of the particles and the wavelengths of the gamma rays vary with the element from which the radiation originates. In accordance with the concept of wave-particle duality, electromagnetic waves can be considered to consist of particles called photons, and particles also exhibit characteristics of waves.

In addition to radiation from naturally occurring radioactive materials, harmful radiation may emanate from X-ray machines, high-voltage power supplies, particle accelerators, nuclear reactors, and artificially produced radioactive materials. Neutrons emitted in reactor processes and high-energy protons and other nuclei produced by accelerators can produce harmful biological effects. X-ray burn and dermatitis were noticed on skin in 1896 within 6 months of the discovery of X-rays.

Ionizing radiation, in particular electromagnetic radiation capable of ionizing biological matter, refers to photons of energy 12 eV (electron volts) or greater. An electron volt is the energy gained by an electron in falling through a potential difference of 1 V and is 1.602×10^{-19} J. The energy E of a photon is given by

$$E = h\nu \tag{2-45}$$

where h is Planck's constant, 6.26×10^{-34} J$\cdot$s, and ν is frequency. In terms of wavelength, 12 eV corresponds to 103 nm (0.103 μm). Only shorter wavelength radiation can ionize biological matter. The visible wavelength range is 380–750 nm, and 100 nm falls in the ultraviolet wavelength range. X-rays and gamma rays have wavelengths less than 100 nm and are capable of ionizing biological matter.

It is of interest to consider the level of naturally occurring radiation, but the units with which radiation is measured must first be given. Exposure to radiation is expressed in rems, standing for roentgen equivalent man. The roentgen, named for the discoverer of X-rays, is the quantity of X-rays which will produce a certain amount of ionization (one electrostatic unit or 3.34×10^{-10} C) in 1 g of dry air. Since the production of an ion pair in air requires about 34 eV, the roentgen corresponds to the liberation of 88 ergs or thereabouts per gram of air. (An erg equals 10^{-7} J.) In tissue, however, the figure is more like 100 ergs/g. The rad is a unit which uses this amount of energy in its definition, rather than being based on the amount of ionization produced. In biological matter, however, it is necessary to consider the effectiveness of the radiation. This concept leads to the rem, where the dose equivalent (DE) in rems is given by

$$\mathrm{DE_{rems}} = (\mathrm{dose_{rads}})(\mathrm{QF}) \tag{2-46}$$

where QF is a quality factor or effectiveness [18].

Natural radiation is due primarily to cosmic rays and naturally occurring radioactivity. Values for radiation exposure given by J. M. Smith [52] will be utilized in the following discussion. The highly energetic cosmic rays impinge on the earth from outer space and produce a variety of secondary particles and effects. Although the nucleonic component (neutrons and protons) comprises only a few percent of the total particle flux near sea level, neutron monitors are one useful means of observing cosmic rays. The effects of cosmic rays in discharging gold-leaf electroscopes were first observed in about 1900, although it was not until about the 1930's that the work of Compton removed all doubt about the origin of cosmic rays [45]. Still later it was determined that highly energetic protons, now referred to as solar cosmic rays, are emitted by the sun at times of solar flares. Some of the solar proton events are sufficiently intense to be recorded by ground-level neutron monitors. The solar cosmic rays pose a potential hazard for astronauts, as a dose as large as 1500 rad behind a 1 g/cm^2 shield is possible [45]. The earth's magnetic field and atmosphere provide a degree of shielding against cosmic rays, and exposure to cosmic rays increases with altitude and with latitude up to about 50° geomagnetic latitude. The average cosmic ray exposure to a resident of the United States is 50 millirems/year, but at the altitude and latitude of Denver, for example, the exposure is about 150 millirems/year. Naturally occurring radioactive materials, in particular uranium and thorium compounds, radium, and potassium-40, cause radiation of varying amount, the level again being high in the Front Range area of Colorado [43]. Smith quotes dosages due to radiation from the ground as 15 millirems/year, assuming $\frac{1}{4}$ time is spent outdoors, and from buildings as 45 millirems/year, assuming $\frac{3}{4}$ time is spent indoors. It is stated that wooden buildings may cause a dosage rate of 50 millirems/year, while the figure for brick buildings may be as high as 100 millirems/year. (Use of uranium tailings for fill in Grand Junction, Colorado has caused a problem of excessive exposure for persons living in homes built on the tailings.) Adding 5 millirems for radiation from the air and 25 for radiation from food gives a total of 140 millirems/year from "natural" causes. To this must be added the dosage from X-ray medical and dental diagnosis and treatment and any dosage from X-rays from TV sets, etc. Smith quotes a figure of 200 millirems as an average dosage per person per year from all sources. If this dosage is received in the course of normal life, one can ask what additional incremental dosage would be acceptable for a person living in the vicinity of a nuclear plant or otherwise exposed to incremental radiation. According to one view, radiation comparable with the natural dosage should be acceptable, but, according to a divergent view, radiation in any amount is bad and any increase in dosage is bad. An additional dosage of 5 millirems/year has been adopted as an acceptable value, but Smith states that the expected value of the additional dosage near a nuclear plant may be on the order of 1 millirem/year. Two types of effects are possible from radiation—somatic and genetic. Somatic effects are experienced directly by the exposed person, while genetic effects become apparent only in subsequent generations. Orbital electrons are dislodged and molecular bands are broken by radiation, and the functioning of cells is impaired. Damage to DNA (deoxyribonucleic acid), which plays a major role in controlling genetic and functional activities, may result.

Turning our attention to nonionizing radiation or to effects of electromagnetic waves of optical and lower frequencies, including microwaves, we see that the most obvious effect is a thermal effect. The heating of biological material by microwaves can be highly useful, as in microwave ovens and in diathermy treatments. Beyond a certain point, however, heating due to microwaves and other frequencies may be harmful. Furthermore, the effects of heating are not uniform in all parts of the body. The eyes have inefficient mechanisms for dissipating heat and are especially vulnerable to damage from radiation. A value of 10 mW/cm^2 is taken as the maximum safe power density for exposure to microwaves (actually for the range from 10 to 10^5 MHz) for periods of 0.1 h or more in the United States. This corresponds to a maximum energy density of 1 mWh/cm^2 during any 0.1-h period. The U.S.S.R., however, has used a value of 0.01 mW/cm^2 for a workday but with 1 mW/cm^2 allowed for a 15- to 20-min exposure. Much discussion has taken place as to whether or not there is justification for the Soviet values, but the United States has retained the 10-mW/cm^2 value. It appears that this value is a reasonable one insofar as thermal effects are concerned. The situation is considerably less clear where nonthermal effects are concerned.

Effects of radiation differ in the different frequency ranges, and some brief comments are now made on possible effects in the various ranges. The remarks are based largely on a paper by Michaelson [41].

Nonionizing ultraviolet radiation can damage the skin and eyes, causing sunburn, skin aging, skin cancer, snow blindness, and damage to the cornea of the eye. In medical terms, ultraviolet radiation can cause erythema (a term referring to a range from slight reddening to severe blistering of the skin) and keratitis, or inflammation of the cornea. The cornea is the outer covering of the eye (Fig. 2-18), which is transparent to visible radiation but not to ultraviolet radiation. The nonionizing portion of the ultraviolet spectrum can be separated into the blacklight region (400–300 nm), erythemal region (320–250 nm), germicidal region (280–220 nm), and ozone production region (230–170 nm). Solar radiation below about 290 or 300 nm is filtered out by ozone in the atmosphere, and it is the 290- to 310-nm range of solar radiation that is responsible for sunburn and skin cancer. The ozone layer is considered further in Sec. 3-3-4. For the most part the blacklight region has rather little biological effect. Potentially harmful artificial sources of ultraviolet radiation arc electric arc welding processes and germicidal lamps. Goggles and sun creams can provide protection against ultraviolet radiation.

The eye is well adapted to visible radiation (380–750 nm) but can be damaged by excessively intense visible radiation. The cornea being transparent to visible wavelengths, damage tends to occur instead to the retina, the inner surface of the eye where the rods and cones transduce light into neuroelectrical phenomena. Irreversible injury to the retina and transient loss of visual function (flashblindness) can result from looking directly at the sun and from any extremely bright source. Injury commonly results from looking at the sun at times of solar eclipses.

The infrared wavelength range extends from 750 to 10^6 nm. Effects of infrared radiation are felt by the skin and eyes. Erythema and blistering of the skin, including that of the eyelids, can result from exposure. The cornea

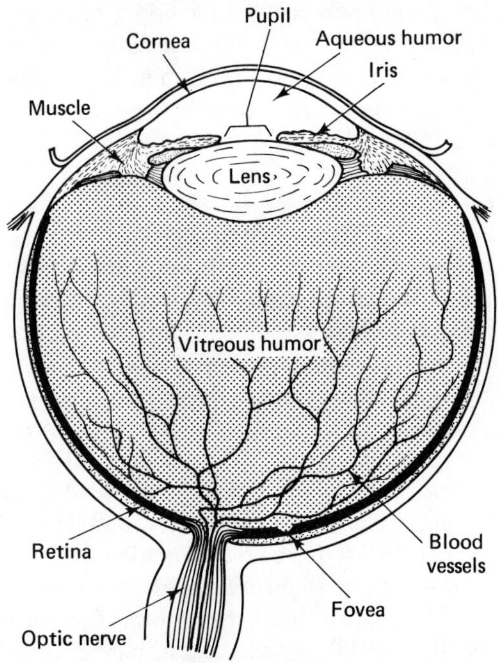

FIGURE 2-18. Structure of human eye. (From *Human Physiology—The Mechanisms of Body Function*, 2nd ed., by A. J. Vander, J. H. Sherman, and D. S. Luciano. Copyright © 1970, 1975 by McGraw-Hill, Inc. Used with permission of McGraw-Hill Book Company.)

of the eye is highly transparent to wavelengths between 750 and 1300 nm and becomes opaque to wavelengths above 2000 nm. The cornea is susceptible to damage from infrared radiation, especially for the less transparent wavelengths, and injury to the iris can also result in extreme cases. Another danger from excessive exposure to infrared radiation is the production of cataracts or opacities of the lens of the eye. Infrared radiation constitutes primarily an occupational rather than environmental hazard. For example, glassworkers, steel mill workers, and kiln operators are subject to being close to hot luminous matter and are therefore subject to hazards from infrared radiation.

Lasers can operate in the ultraviolet, optical, and infrared frequency ranges and thus tend to have the greatest effects on the cornea, retina, and lens of the eye, respectively. Safe levels of laser radiation depend on the pulse length. The smallest amount of radiation (energy density) can be tolerated in the case of the *Q*-switched laser having pulse lengths from 1 ns to 1 μs. Estimated tolerable levels are on the order of 1×10^{-8} to 1×10^{-7} J/cm^2.

For non-Q-switched lasers with pulse lengths from 1 μs to 0.1 s, the tolerable level may be an order of magnitude greater, and perhaps another order of magnitude greater for CW lasers. Adequate precautions must be taken in laser operations to protect the eyes of the personnel involved.

Microwave radiation can also cause cataracts of the eye. A level of 100 mW/cm^2 may have an injurious effect on the eye, and the figure of 10 mW/cm^2 mentioned earlier is the accepted safe value, except for short exposures where for densities up to 100 mW/cm^2 the following expression may give an estimated safe exposure time in minutes:

$$T = \frac{6000}{X^2} \tag{2-47}$$

where X is the exposure in mW/cm^2. Threshold for pain reaction for exposure of skin to microwave radiation has been found to vary from 830 mW/cm^2 for exposures longer than 3 min to 3.1 W/cm^2 for a 20-s exposure. Precautions to observe in working with microwaves are to determine and give warnings by signs and other means of areas of unsafe power densities and not to allow personnel to work in such areas, to discharge high-power generators into dummy loads rather than the surrounding area wherever practical, and to point antennas away from personnel or inhabited structures whenever radiation into space is necessary. In addition to effects from microwaves themselves, X-rays from high-voltage power supplies can be a hazard to radar personnel, for example, if shielding is removed without first turning off the high-voltage power.

Considerable controversy and confusion have arisen concerning possible nonthermal effects of microwave radiation. Several Soviet investigators have suggested that microwaves can cause effects on the central nervous system without significant heating being involved. Much interest and controversy concerning possible biological effects of low-frequency radiation have been stimulated also by the U.S. Navy's Project Seafarer, formerly Project Sanguine. This project involves the use of ELF fields (30–300 Hz) for a system of obviously limited bandwidth and information capacity for communicating with submerged submarines [5,56]. Evidence of nonthermal effects due to weak electric and magnetic fields at 45, 60, and 75 Hz on slime molds has been observed by Greenebaum et al. [19]. The process of cell division or mitosis is slowed by the presence of the fields. Possible effects of ELF fields on serum triglyceride level, blood pressure, and the central nervous system and on the weight gain and the migration and orientation of birds have been suggested and are being investigated further. A rather well-established nonthermal effect is the tendency for microscopic particles such as blood cells to align themselves in the electric field direction to form chains of particles, under the influence of radio-frequency fields in the 1- to 700-MHz range [29].

The subject of biological effects of radiation has caused much interest in recent years in meetings of the U.S. National Committee of URSI and elsewhere.

PROBLEMS

2-1 A constant-temperature, low-temperature source followed by a variable attenuator can be used as a variable-temperature noise source. Calculate the noise temperature provided by such a combination, if the low temperature is 60 K and the attenuator temperature is 290 K, for attenuator settings of 0, 1, 3, 6, and 12 dB.

2-2 What is the noise temperature of a receiver having a first stage with a noise temperature of 50 K and a gain of 10 dB, a second stage with a noise temperature of 160 K and a gain of 10 dB, and a noise temperature of 750 K for the remainder (for the following stages) of the receiver?

2-3 Find T_{sys} for a receiving system having an antenna temperature of 60 K, a transmission line with 1 dB of attenuation, and a receiver with a noise figure of 2 dB. What is the additional noise contribution of the transmission line as compared to the case of zero transmission losses?

2-4 Find T_{sys} if a preamplifier having a noise figure of 1 dB and a gain of 16 dB is placed ahead of the receiver of the previous problem. Find T_{sys} if the same preamplifier is placed at the antenna terminals.

2-5 A 1.5-m paraboloidal antenna has a gain of 42.1 dB (including the effect of the radome) at a frequency of 11.2 GHz. What is the ratio of A_{eff} to A, where A is the aperture area and A_{eff} is the effective area of the antenna? This antenna and a transmitter power of 0.1 W are used on a 40-km microwave path. The receiving station has an identical antenna and a noise figure of 9 dB, and the effective noise bandwidth of the receiver is 20 MHz. Assume a waveguide loss of 3 dB at the transmitter and the same value at the receiver. Calculate the C/X ratio, assuming $T_A = T_a = 290$ K.

2-6 Paraboloidal antennas 1.83 m in diameter, transmitting tubes having powers of 0.5 W at $f = 3000$ MHz, and receivers of 7.5 dB noise figure are available. Would this equipment be capable of providing satisfactory service for a channel of 5-MHz bandwidth over a 64-km path if conventional FM modulation is employed and if a fade margin of 40 dB must be provided? Assume that $T_A = T_a = 290$ K, use $A_{eff} = 0.54A$, and consider waveguide losses of 3 dB at each end.

2-7 In two nearly parallel microwave routes both using antennas having gains of 34 dB in the desired directions, radiation from a transmitter in one route reaches a receiver in the other route or system. The antenna gain of the transmitter in this unwanted direction is 20 dB down from the gain in the desired direction, and the gain of the receiving antenna is 16 dB down from the gain in its desired direction. The spacing between the transmitter in one

$$G = \frac{4\pi A_{\text{eff}}}{\lambda^2}$$

$$\mathrm{DE}_{\mathrm{rems}} = (\mathrm{dose}_{\mathrm{rads}})(\mathrm{QF}) \tag{2-46}$$

route and the receiver in the other system is 8 km. Compare in dB the unwanted signal magnitude and the wanted signal magnitude, assuming that all transmitters have the same power and frequency and that the path length for the wanted signal is 40 km.

2-8 Consider an FM receiving system in which an audio-frequency signal of 15-kHz bandwidth is provided at the output of an FM discriminator. Assume that the minimum acceptable S/N ratio for the high-fidelity audio signal is 45 dB. If the corresponding FM signal has a total frequency deviation of 150 kHz, what minimum S/N ratio could be tolerated at the FM receiver input if the FM and AM signals convey the same amount of information and if this information is as indicated by Shannon's law [namely $B \log_2 (1 + S/N)$]? Could a conventional FM discriminator operate with this S/N ratio? This calculation is unrealistic for several reasons but is of some theoretical interest. A more realistic approach is to consider that the improvement factor for an FM system is $3(F/f_m)^2$, where F is the FM frequency deviation and f_m is the corresponding audio-frequency bandwidth. Assuming the same audio-frequency S/N ratio of 45 dB, calculate the necessary minimum C/X ratio at the FM receiver input if the above factor applies.

2-9 If a C/X ratio of 20 dB is desired for a synchronous satellite-earth receiving station, what bandwidths can be accommodated for G/T_{sys} values of 20 and 30 dB, assuming the usual frequency band for commercial satellite-to-earth transmissions and assuming a satellite EIRP value of 33 dBW? Repeat for a satellite EIRP of 43 dBW.

2-10 A proposed 12-GHz synchronous satellite system, having low-cost earth receiving terminals, would have a satellite transmitter power of 200 W. An earth receiving antenna 1 m in diameter has been selected, and the receiving system noise temperature is expected to be 400 K. A minimum C/X ratio of 10 dB is needed, and it is planned to have a margin of 4 dB. Thus the design C/X ratio is 14 dB. The bandwidth to be accommodated is 6 MHz. What diameter of satellite circular-aperture antenna is needed if $A_{eff} = 0.54A$ for the earth antenna and $A_{eff} = 0.64A$ for the satellite antenna?

2-11 What are the relative advantages of receivers having 5- and 15-dB noise figures for

a. The HF frequency range at a time when atmospheric noise is such as to cause an antenna temperature 40 dB above the standard reference temperature?

b. A surface microwave link for which the antenna temperature can be taken to be the standard reference temperature?

(Compute the system noise temperatures in both cases.)

2-12 Derive the expression for the first Fresnel zone radius F_1. [See Eq. (2-32).]

2-13 The top of a building is known to fall at a height 18 m below a direct, straight, 48-km microwave path. The building is located halfway along the length of the path. What clearance does the path have, in terms of the first Fresnel zone radius, if the wavelength is 10 cm?

2-14 On an overwater path, 15 km in length and operating at a wavelength of

10 cm, the land at both ends of the path is flat and at a height of 30 m above mean sea level. If the transmitting antenna is 10 m above the ground level and the height h_T of the antenna above mean sea level is thus 40 m, find the optimum height h_R of the receiving antenna for mean sea-level conditions. Using this receiver height, calculate the received field intensity as a fraction of the maximum value when the sea level decreases by 4 m at a time of low tide. (Assume vertical cliffs, between the land and sea surfaces, such that the receiving antenna can only be placed above the top of the cliff on flat land. Also assume perfect reflection from the water surface and equal amplitudes for the direct and reflected rays.)

2-15 a. The value of 10 mW/cm^2 is taken as a safe level for exposure to microwave radiation. Considering this to be a safe average power density for radar operations, calculate the minimum safe distance for a person exposed continuously to a radar beam in the case of a transmitter having a peak power of 20 kW, using a 1.83-m paraboloidal antenna at a frequency of 3 GHz, and transmitting 1-μs pulses at a rate of 1000 pulses/s. Take $A_{\text{eff}} = 0.6A$.

b. Repeat for the case of a peak power of 2 MW, other parameters remaining the same.

2-16 A level of 240,000 lm/ft^2 of visible radiation can cause retinal burns. Assuming monochromatic plane-wave radiation at 555 nm, calculate the corresponding electric and magnetic field intensities in volts per meter and amperes per meter. (One lumen equals 0.00147 W.) Compare these values with the fields associated with solar radiation of 1353 W/m^2, treating the solar radiation as if it were also monochromatic. (Solar radiation actually covers a range of wavelengths, and most artificial sources do as well.)

2-17 Calculate the energy in electron volts for photons corresponding to wavelengths of 400 nm (0.4 μm) and 50 nm.

REFERENCES

[1] Bargellini, P. L. (ed.) "The INTELSAT IV communication system," *COMSAT Technical Review*, vol. 2, pp. 437–572, Fall 1972.

[2] Bekey, I., and H. Mayer, "1980–2000, Raising our sights for advanced space systems," *Astronautics and Aeronautics*, vol. 14, pp. 34–63, July/Aug., 1976.

[3] Bell Telephone Laboratories, *Transmission Systems for Communications*. Winston-Salem, NC, Western Electric: 1964.

[4] Bennett, W. R., *Introduction to Signal Transmission*. New York: McGraw-Hill, 1970.

[5] Bernstein, S. L., M. L. Burrows, J. E. Evans, A. S. Griffiths, D. A. McNeill, C. W. Niessen, I. Richer, D. P. White, and D. K. Willim, "Long-range communications at extremely low frequencies," *Proc. IEEE*, vol. 62, pp. 292–312, March 1974.

[6] BLAKE, L. V., *Antennas*. New York: Wiley, 1966.

[7] BUEHLER, W. E., C. H. KING, and C. D. LUNDEN, "VHF city noise," IEEE Electromagnetic Compatibility Symposium Record, *IEEE 68C12-ECMC*, pp. 113–118. New York: IEEE, 1968.

[8] CAMPBELL, W. H., "Geomagnetic pulsations," *Physics of Geomagnetic Phenomena*, Vol. II, pp. 821–909 (S. MATUSHITA and W. H. CAMPBELL, eds.). New York: Academic Press, 1967.

[9] *CCIR Report 322*, "World distribution and characteristics of atmospheric radio noise." Geneva: CCIR, 1963.

[10] CODDING, G. A., *The International Telecommunications Union: an Experiment in International Cooperation*. Leiden: E. J. BRILL, N.V., 1952.

[11] CROMBIE, D. D. (ed.), "Lowering barriers to telecommunications growth," *OT Special Publication 76-9*, Office of Telecommunications. Washington, D.C.: Supt. of Documents, U.S. Government Printing Office, Nov. 1976.

[12] CUCCIA, C. L., W. J. GILL, and L. H. WILSON, "Sensitivity of microwave earth stations for analog and digital communications," *Microwave Journal*, Part I, vol. 12, pp. 47–54, Jan. 1969; Part II, vol. 12, pp. 71–77, Feb. 1969.

[13] DAVIES, K., *Ionospheric Radio Propagation*, National Bureau of Standards Monograph 80. Washington, D.C.: U.S. Government Printing Office, 1965.

[14] FALK, H., "Picturephone and beyond," *Spectrum*, vol. 10, pp. 45, 49, Nov. 1973.

[15] FELDMAN, N., and S. J. DUDZINSKY, "A new approach to millimeter-wave communications," *R-1936-RC*. Santa Monica, CA: The Rand Corporation, April 1977.

[16] FICCHI, R. F., *Practical Design for Electromagnetic Compatibility*. New York: Hayden, 1971.

[17] GIERHART, G. W., R. W. HUBBARD, and D. V. GLEN, "Electrospace planning and engineering for the air traffic environment," *Report No. FAA-RD-70-71*, Office of Telecommunications, Boulder, CO. Washington, D.C.: Supt. of Documents, U.S. Government Printing Office, Dec. 1970.

[18] GLASSTONE, S., *Sourcebook on Atomic Energy*. New York: Van Nostrand Reinhold, 1967.

[19] GREENEBAUM, B., E. M. GOODMAN, and M. T. MARION, "Long-term effects of weak 45–75 Hz electromagnetic fields on the slime mold, *Physarum polycephalum*," 1975 Annual Meeting, U.S. National Committee, URSI, Boulder, CO, Oct. 20–23, 1975.

[20] GTE LENKURT, *Engineering Considerations for Microwave Communications Systems*. San Carlos, CA: GTE Lenkurt, Inc., 1972.

[21] HALLIWELL, B. J. (ed.), *Advanced Communication Systems*. London: Newnes-Butterworth's, 1974.

[22] HAN, C. C., J. ALBERNAZ, K. OHKUBO, J. M. JANKY, B. B. LUSIGNAN, "Optimization in the design of a 12 GHz low cost ground receiving system for broadcast satellites," IEEE Int. Conf. on Communications, *IEEE 73CHO 744-3-CSCB*, pp. 36-8 to 36-14. New York: IEEE, 1973.

[23] HELLIWELL, R. A., *Whistlers and Related Ionospheric Phenomena.* Stanford, CA: Stanford University Press, 1965.

[24] HINCHMAN, W. R., "Use and management of the electrospace; a new concept of the radio resource," IEEE Int. Conf. on Communications, *IEEE 69C29-COM*, pp. 13-1 to 13-5. New York: IEEE, 1969.

[25] HOGG, D. C., and W. W. MUMFORD, "The effective noise temperature of the sky," *Microwave Journal*, vol. 3, p. 80, 1960.

[26] ITT, *Reference Data for Radio Engineers*, 5th ed. Indianapolis: Howard W. Sams & Co., Inc., 1968.

[27] JACOBS, J. A., *Geomagnetic Micropulsations.* New York: Springer-Verlag, 1970.

[28] JANSKY, K. G., "Electrical disturbances apparently of extraterrestrial origin," *Proc. IRE*, vol. 21, pp. 1387–1398, Oct. 1933.

[29] JOHNSON, C. C., and A. W. GUY, "Nonionizing electromagnetic wave effects in biological materials and systems," *Proc. IEEE*, vol. 60, pp. 692–718, June 1972.

[30] Joint Technical Advisory Committee, *Radio Spectrum Utilization.* New York: IEEE, 1964.

[31] Joint Technical Advisory Committee, *Spectrum Engineering—the Key to Progress.* New York: IEEE, 1968.

[32] Joint Technical Advisory Committee, *Radio Spectrum Utilization in Space.* New York: IEEE, 1970.

[33] JORDAN, E. C., and K. G. BALMAIN, *Electromagnetic Waves and Radiating Systems.* Englewood Cliffs, NJ: Prentice-Hall, 1968.

[34] KELLER, G. V., and F. C. FRISCHKNECHT, *Electrical Methods in Geophysical Prospecting.* Elmsford, NY: Pergamon, 1966.

[35] KRAUS, J. D., *Antennas.* New York: McGraw-Hill, 1950.

[36] LARGE, D. B., L. BALL, and A. J. FARSTAD, "Radio transmission to and from underground coal mines—theory and measurement," *IEEE Trans. Communications*, vol. 3, pp. 194–202, March 1973.

[37] LATHI, B. P., *Random Signals and Communication Theory.* Scranton, PA: International Textbook Company, 1968.

[38] LITTLE, C. G., and H. LEINBACH, "The riometer—a device for the continuous measurement of ionospheric absorption," *Proc. IRE*, vol. 47, pp. 315–322, Feb. 1959.

[39] MARTIN, J., *Telecommunications and the Computer*, 2nd ed. Englewood Cliffs, NJ: Prentice-Hall, 1976.

[40] MARTIN, J., *Future Developments in Telecommunications*, 2nd ed. Englewood Cliffs, NJ: Prentice-Hall, 1977.

[41] MICHAELSON, S. M., "Human exposure to nonionizing radiant energy—potential hazards and safety standards," *Proc. IEEE*, vol. 60, pp. 389–421, April 1972.

[42] MILLER, S. E., E. A. J. MARCATILI, and T. LI, "Research toward optical fiber transmission systems," *Proc. IEEE*, vol. 61, pp. 1703–1751, Dec. 1973.

[43] OAKLEY, D., *Natural Radiation Exposure in the United States.* Washington, D.C.: U.S. Environmental Protection Agency, Office of Radiation Programs, Surveillance and Inspection Division, June 1972.

[44] PANTER, P. F., *Communication Systems Design: Line-of-Sight and Troposcatter Systems.* New York: McGraw-Hill, 1972.

[45] POMERANTZ, M. A., *Cosmic Rays.* New York: Van Nostrand Reinhold, 1971.

[46] POTTER, J. G., and J. M. JANKY, "The ATS-F health-education technology communications systems," IEEE Int. Conf. on Communications, *IEEE 73CHO 744-3-CSCB*, pp. 36-20 to 36-26. New York: IEEE, 1973.

[47] *President's Task Force on Communications Policy, Final Report.* Washington, D.C.: U.S. Government Printing Office, 1968.

[48] PRITCHARD, W. L., "Satellite communications—an overview of the problems and program," *Proc. IEEE*, vol. 65, pp. 294–307, March 1977.

[49] RAISBECK, G., *Information Theory.* Cambridge, MA: M.I.T. Press, 1964.

[50] RAMO, S., J. R. WHINNERY, and T. VAN Duzer, *Fields and Waves in Communication Electronics.* New York: Wiley, 1965.

[51] SKOMAL E. N., "The dimensions of radio noise," IEEE Electromagnetic Compatibility Symposium Record, *IEEE 69C3-EMC*, pp. 18–28. New York: IEEE, 1969.

[52] SMITH, J. M., Jr., *Natural Background Radiation and the Significance of Radiation Exposure.* San Jose, CA: General Electric Co., 1975.

[53] SMITH, R. L., *The Wired Nation.* New York: Harper & Row, 1972.

[54] THIEL, F. L., and W. B. BIELAWSKI, "Optical waveguides look brighter than ever," *Electronics*, vol. 47, pp. 89–96, March 21, 1974.

[55] Wait, J. R. (ed.), *Special Issue on Extremely Low Frequency (ELF) Communications, IEEE Trans. Communications*, vol. COM-22, April 1974.

[56] WATT, A. D., *VLF Radio Engineering.* Elmsford, NY: Pergamon, 1967.

3 THE ATMOSPHERE

3-1 INTRODUCTION

The atmosphere is an obviously important part of the environment, and pollution of the lower atmosphere is a subject of very serious concern. In addition, many of the important applications of electromagnetic waves for communications and other purposes involve effects of the earth's atmosphere. The characteristics of the atmosphere can provide increased communication capability beyond that of a vacuum or can degrade transmissions with respect to what would take place in a vacuum.

In this chapter attention is given to the entire earth's atmosphere, including the upper atmosphere. The earth's ionosphere, having peak electron densities in the 200- to 400-km altitude range, is important to the subject of communications. Proposed or completed technological programs have involved the flight of supersonic aircraft, the deposition of tiny metallic needles for communication purposes, high-altitude nuclear bursts, and the large number of space vehicles and associated debris. All of these phenomena take place in the upper atmosphere.

The structure and physical characteristics of the atmosphere are described in this chapter, whereas in Chap. 4 we shall describe the effect of the atmosphere on the propagation of electromagnetic waves, with emphasis on radio waves. In Chap. 5 we shall present the blackbody radiation laws and shall discuss thermal radiation from the sun and earth and the interaction between this radiation and the atmosphere.

The first section following this introduction contains a brief treatment of the structure of the earth's atmosphere, starting with parameters of the neutral atmosphere and proceeding to the *D*, *E*, and *F* regions of the ionosphere. Short discussions of the earth's magnetic field and solar-terrestrial relations then follow. Sections on air pollution and biogeochemical cycles complete the chapter.

3-2 STRUCTURE OF THE EARTH'S ATMOSPHERE AND MAGNETIC FIELD

3-2-1 Pressure, Density, and Temperature of Neutral Atmosphere

Several efforts have been made to determine model atmospheres, which represent the best available estimates of the average values of pressure, density, temperature, and other parameters. Two such model atmospheres are the *U.S. Standard Atmosphere, 1976* [45] and the CIRA [15] reference atmosphere. In the upper atmosphere above about 110 km all of the quantities vary with the sunspot cycle, and the CIRA reference atmosphere and also Johnson [24] give values for low-density, medium-density, and high-density atmospheres, the low- and high-density cases referring to nighttime near sunspot minimum and daytime near sunspot maximum, respectively. The pressure and density fall off continuously with increasing altitude, as would be expected, but the variation of temperature is somewhat more interesting. The 11-year sunspot cycle is discussed further in Sec. 3-2-6. Figure 3-1 shows the variation of temperature with altitude. Table 3-1 includes a very abbreviated set of values from the *U.S. Standard Atmosphere, 1976*.

The temperature falls off with increasing altitude in the troposphere, which extends to about 10 km over the poles and 16 km over the equator. Inversion layers near the surface of the earth are exceptions to this general characteristic. Discussion of these inversion layers and other points about the troposphere is found in Sec. 3-3. Above the troposphere is the stratosphere. Temperatures in this region rise with altitude, to a maximum at about 50 km, because of the absorption of solar ultraviolet radiation by ozone. Above the stratosphere and extending to about 85 km is the mesosphere, which is a region of decreasing temperature with height. Above 85 km is the thermosphere in which the temperature again increases with height, in this case because of the dissociation and ionization of the atmospheric gases by solar ultraviolet radiation. Above 300 km the temperatures change little with height. Below about 100 km the temperature varies little with time, but temperatures above about 120 km vary by nearly a factor of 3 to 1.

In the troposphere, data on temperature are obtained by balloons, which also measure pressure and thus allow a determination of height by use

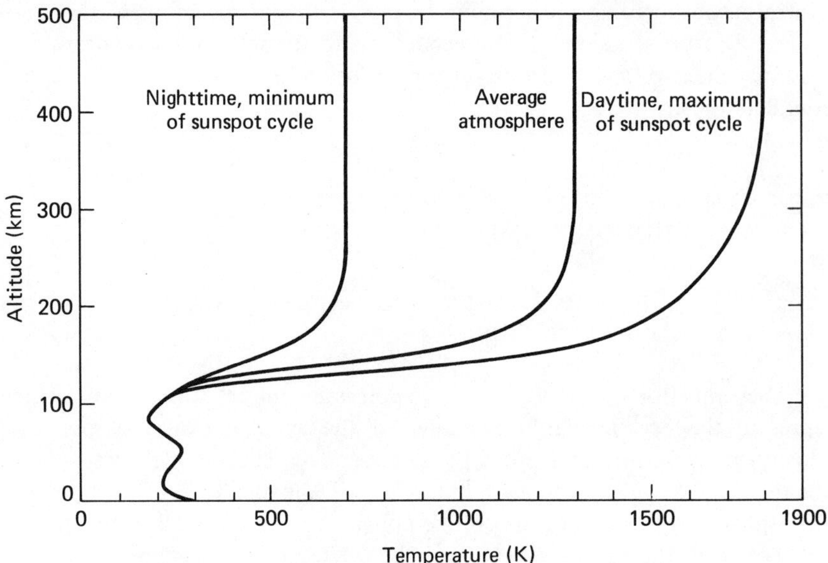

FIGURE 3-1. Atmospheric temperature distributions typical of daytime conditions near the maximum of the sunspot cycle, nighttime conditions near the minimum of the sunspot conditions, and an average in-between situation. (Reprinted from *Satellite Environment Handbook*, 2nd ed., edited by Francis S. Johnson, with permission of the publishers, Stanford University Press [24]. Copyright © 1965 by the Board of Trustees of the Leland Stanford Junior University.) The *U.S. Standard Atmosphere, 1976* [45] temperatures are lower than the values assumed in earlier years and lower than the temperature for the average atmosphere shown above. A virtue of this figure, however, is that it shows the large variation that has been believed to occur in temperatures for altitudes above 100 km.

of the hydrostatic equation [Eq. (3-1)]. Before the advent of rockets and satellites, observations of the propagation of sound to large distances and of meteors provided information about the temperature and density of the middle atmosphere [30]. More recently, rockets have been used to measure pressure and to determine temperature from velocity of sound measurements made by using rocket-launched grenade sources. Higher in the atmosphere, the drag on satellites has been used to infer atmospheric density and temperature [19].

In considering the properties of the neutral atmosphere, it is helpful to make use of the hydrostatic equation. In equilibrium the gravitational force on an element of volume of the atmosphere is balanced by the gradient of pressure force, as shown in Fig. 3-2. Thus

$$-\nabla p + \rho \mathbf{g} = 0 \tag{3-1a}$$

Table 3-1 Properties of the *U.S. Standard Atmosphere*, 1976 [45].*

Geom. Alt. (km)	*Temp. (K)*	*Pressure (mb)*	*Density (kg/m³)*	*Particle Speed (m/s)*	*Collision Frequency (per s)*
0	288.150	1.01325 +3	1.2250 +0	458.94	6.9189 +9
0.5	284.900	9.5461 +2	1.1673	456.35	6.5555
1	281.651	8.9876	1.1117	453.74	6.2075
2	275.154	7.9501	1.0066	448.48	5.5554
3	268.659	7.0121	9.0925 −1	443.15	4.9588
4	262.166	6.1660	8.1935	437.76	4.4141
5	255.676	5.4048	7.3643	432.31	3.9180
6	249.187	4.7217	6.6011	426.79	3.4671
7	242.700	4.1105	5.9002	421.20	3.0584
8	236.215	3.5651	5.2579	415.53	2.6888
9	229.733	3.0800	4.6706	409.79	2.3555
10	223.252	2.6499	4.1351	403.97	2.0558
11	216.774	2.2699	3.6480	398.07	1.7871
12	216.650	1.9399	3.1194	397.95	1.5277
13	216.650	1.6579	2.6660	397.95	1.3056
14	216.650	1.4170	2.2786	397.95	1.1159
15	216.650	1.2111	1.9476	397.95	9.5380 +8
16	216.650	1.0352	1.6647	397.95	8.1528
17	216.650	8.8497 +1	1.4230	397.95	6.9691
18	216.650	7.5652	1.2165	397.95	5.9576
19	216.650	6.4674	1.0400	397.95	5.0931
20	216.650	5.5293	8.8919 −2	397.95	4.3543
21	217.581	4.7289	7.5715	398.81	3.7161
22	218.574	4.0475	6.4510	399.72	3.1733
24	220.560	2.9717	4.6938	401.53	2.3194
26	222.544	2.1883	3.4257	403.33	1.7004
28	224.527	1.6161	2.5076	405.12	1.2502
30	226.509	1.1970	1.8410	406.91	9.2192 +7
40	250.350	2.8714 +0	3.9957 −3	427.78	2.1036
50	270.650	7.9779 −1	1.0269	444.79	5.6210 +6
60	247.021	2.1958	3.0968 −4	424.93	1.6195
70	219.585	5.2209 −2	8.2829 −5	400.64	4.0839 +5
80	198.639	1.0524	1.8458	381.05	8.6559 +4
90	186.87	1.8359 −3	3.416 −6	369.9	1.56
100	195.08	3.2011 −4	5.604 −7	381.4	2.68 +3
110	240.00	7.1042 −5	9.708 −8	431.7	5.48 +2
120	360.00	2.5382	2.222	539.3	1.63
130	469.27	1.2505	8.152 −9	625.0	7.1 +1
150	634.39	4.5422 −6	2.076	746.5	2.3

*Condensed from the complete tabulations. Numbers following the plus or minus sign are the power of 10 by which that entry and each following entry should be multiplied.

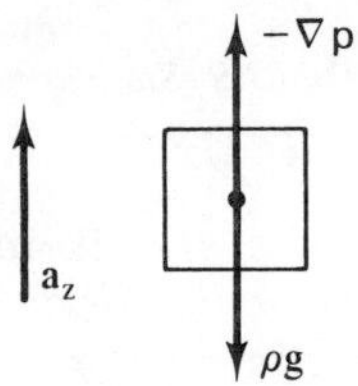

FIGURE 3-2. Forces on volume of air.

or

$$\frac{dp}{dh} + \rho g = 0 \tag{3-1b}$$

This equation is commonly written as

$$dp = -\rho g \, dh \tag{3-1c}$$

and is called the hydrostatic or barometric equation. In this equation p is pressure in newtons per square meter (N/m^2), ρ is density in kg/m^3, g is the acceleration of gravity in m/s^2 (9.8 m/s^2 at sea level), and h is distance in meters.

The equation can be put in a more useful form by making use of the perfect gas law,

$$pv = RT \tag{3-2}$$

where v is specific volume, R is the universal gas constant, and T is temperature in K. If v is taken as the volume of a kg mol, R has the value of 8.314 $\times$ 10^3 J/mol/K. The gas law can be expressed in various ways. By dividing by N_A, where N_A is Avogadro's number (6.025 $\times$ 10^{26}), taken here as the number of molecules in a kg-mol weight, $pv/N_A = RT/N_A$. But $R/N_A = k$, which is Boltzmann's constant and has the value of 1.38 $\times$ 10^{-23} J/K. Thus

$$p = \frac{N_A}{v}kT = NkT \tag{3-3}$$

where N is the number of particles per m^3. If the numerator and denominator of the right-hand side of Eq. (3-3) are multiplied by m,

$$p = \frac{NmkT}{m} = \frac{\rho kT}{m}$$

from which it can be seen that

$$\rho = \frac{pm}{kT} \tag{3-4}$$

Substituting this expression for ρ into Eq. (3-1c), we obtain

$$\frac{dp}{p} = -\frac{dh}{H} \tag{3-5}$$

where $H = kT/mg$ is called the scale height.

If H is a constant,

$$p = p_0 e^{-h/H} \tag{3-6}$$

which equation gives the pressure p at a height h above the level where the pressure is p_0. H is seen to be the distance (height interval) over which the pressure drops off by a factor of $1/e$. In general, however, H varies with altitude. When H varies with altitude a reasonable estimate of p may be obtained, if the height interval h is not too large, by using the average value of H over the interval h. To obtain the greatest accuracy in the determination of p one should theoretically use

$$p = p_0 \exp\left(-\int_0^h \frac{mg}{kT}\,dh\right) = p_0 \exp\left(-\int_0^h \frac{dh}{H}\right) \tag{3-7}$$

except that an analytical form for the variation of the parameters is seldom available and the practical procedure commonly is to replace the integration by a summation so that

$$p = p_0 \exp\left(-\sum_{i=1}^{n} \frac{h_i}{H_i}\right) \tag{3-8}$$

Use of this equation involves breaking up the total range of altitude into n layers, where n is sufficiently large, corresponding to the Δh being sufficiently small, to obtain reasonable accuracy by using average values H_i for each layer. If desired, the pressures p_i at the tops of the layers or at the levels separating the layers can also be calculated. When this is done

$$p_1 = p_0 e^{-(h_1/H_1)}, \qquad p_2 = p_1 e^{-(h_2/H_2)} \tag{3-9}$$

etc., corresponding to Fig. 3-3. One purpose in discussing the exponential

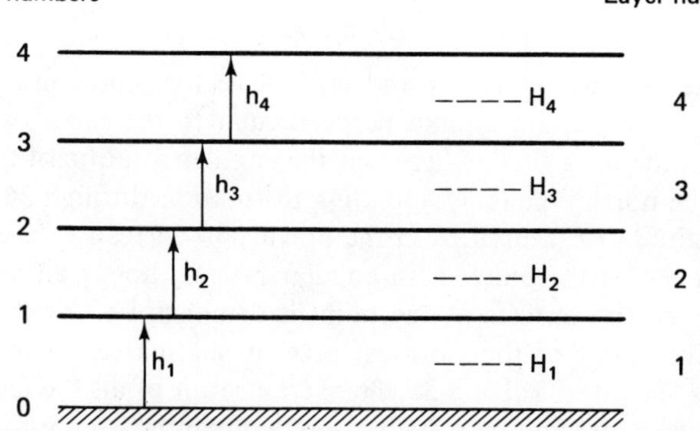

FIGURE 3-3. Levels and layers referred to in the discussion of Eq. (3-9). The H_i's are shown as being associated with dotted levels that are halfway between the numbered levels, suggesting that the H values at these halfway positions might be taken to be suitable average values.

variation of pressure with altitude in some detail here is that of illustrating how exponential functions with nonconstant coefficients can be treated in general.

In the general case m, g, and T all vary, m being average mass and g varying as

$$g = g_0 \frac{R_0^2}{(R_0 + h)^2} \tag{3-10}$$

where h is the height above the level R_0, corresponding to g_0. The scale height H is listed in some tables depicting the standard atmospheres. H has values of approximately 8.4 km at sea level and 60 km at an altitude of 300 km under average atmospheric conditions.

3-2-2 Winds and Climate

The complex subject of atmospheric circulation is beyond the scope of this volume, and the reader is referred to texts on meteorology [18, 39] for further information on this topic. Some brief remarks on wind and climate, however, may be helpful.

The surface of the earth receives a greater amount of solar radiation in equatorial regions than polar regions, and a flow of air from low latitude to high latitude tends to redistribute and equalize the heat content of the atmosphere as a function of latitude. If it were not for this flow, the difference in temperature between high and low latitudes would be much greater than it is. The low-latitude to high-latitude air currents and the associated reverse currents are modified. however, by the rotation of the earth and the consequent Coriolis "force" and by the presence of land masses, etc. The apparent Coriolis force $\boldsymbol{f}$ on a parcel of air of mass m is given by

$$\boldsymbol{f} = -2m\boldsymbol{\omega}_L \times \boldsymbol{v} \tag{3-11}$$

where $\boldsymbol{v}$ is the velocity of the air and $\boldsymbol{\omega}_L$ is a vector representing the local rotation of the earth about an axis perpendicular to the earth's surface. $\boldsymbol{\omega}_L$ has the magnitude of $\omega \sin \theta$, where ω is the angular rotation of the earth as observed at the north pole (corresponding to rotation through 360° in 24 h) and θ is colatitude (measured from the north pole instead of the equator). The fact that the earth rotates with angular rate ω_L about an axis perpendicular to the earth's surface at any point is demonstrated by the Foucault pendulum. The result of the Coriolis force on air masses in the northern hemisphere is shown in Table 3-2. The overall result of all the factors influencing air movement is that a system of predominantly west or east winds is set up over the earth, as suggested in Fig. 3-4. Near the equator the heated air rises and then moves toward the poles high in the troposphere, with the surface return flow (the trade winds) constituting an east wind. These movements conform to those of one of the three cells of the tricellular model of

Table 3-2 Effect of Coriolis Force.

Direction Wind Is from	*Direction Wind Is Blowing to*	*Deflection in Northern Hemisphere*
West	East	South
South	North	East
East	West	North
North	South	West

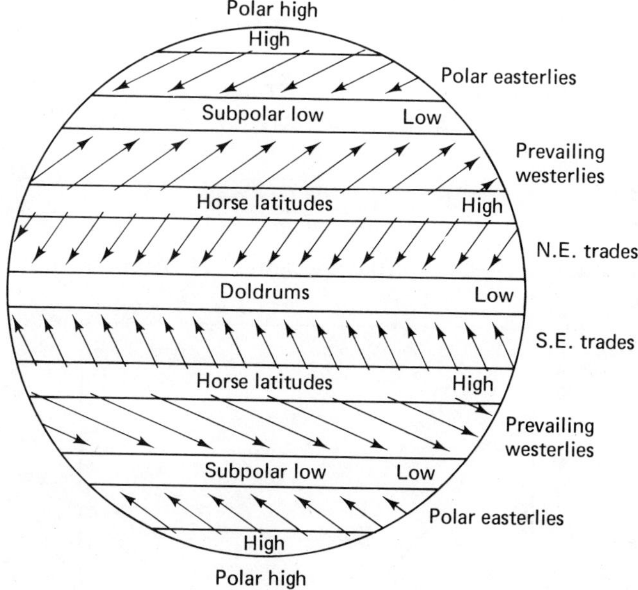

FIGURE 3-4. Ideal primary or terrestrial pressure and wind systems. (From *Meteorology*, 4th ed., by Donn [18]. Copyright © 1975, 1965 by McGraw-Hill, Inc. Used with permission of McGraw-Hill Book Company.)

atmospheric circulation. An important feature of this cell is that it involves descending air, and little horizontal movement, in what are known as the horse latitudes. Deserts tend to form on land in these regions of descending air. The region of prevailing westerlies is not characterized by steady flow, like that of the trade winds, but by the waves, surges, and disturbances that comprise the weather with which we are familiar. Limitations of the tricellular model of atmospheric circulation and the features of a more satisfactory theory of atmospheric circulation are described in modern meteorological texts including those by Donn [18] and Riehl [39]. Actually circulation near the equator, or from about 0 to 30°, where the Coriolis force is low, is much as described by the tricellular model, but the model fails more seriously farther from the equator.

It is in good measure because of the pattern of the winds that the climatic regions and biomes tend to be arranged in a regular manner also, as shown in Fig. 1-2.

3-2-3 Chemical Composition

Below about 85 km, the fractional composition of the atmosphere is the same as that at the surface. Above about 90 km dissociation of O_2 becomes important. Molecular weight as a function of altitude is given in Fig. 3-5.

Above about 105 km, diffusive equilibrium applies, and the distribution of each constituent is independent of that for the others. Each obeys a hydrostatic equation, where

$$H_i = \frac{kT}{m_i g} \tag{3-12}$$

for each constituent. Because of the appearance of the mass in the denomi-

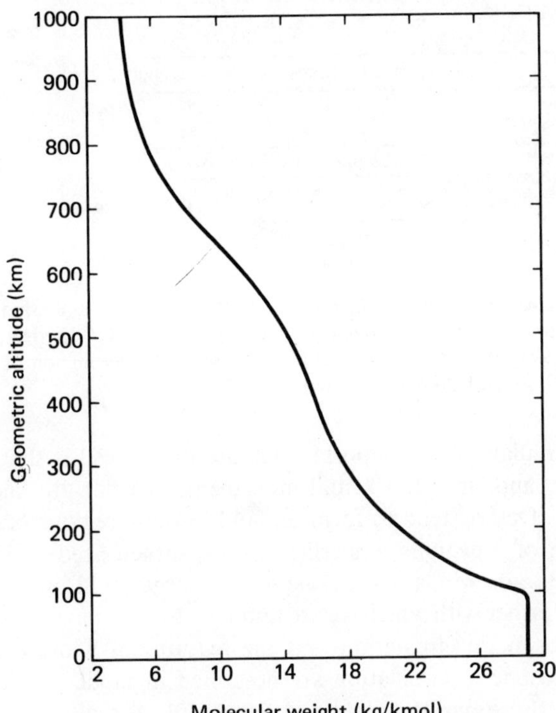

FIGURE 3-5. Mean molecular weight as a function of geometric altitude. (From *U.S. Standard Atmosphere, 1976* [45].)

nator of the expression for H_i, the values of H_i are quite different. The hydrostatic equations can be written as

$$p_i = p_{0_i} e^{-h/H_i} \tag{3-13a}$$

$$\rho_i = \rho_{0_i} e^{-h/H_i} \tag{3-13b}$$

if m, g, and T are constant. The result is that the light constituents dominate at the higher elevations. Below about 105 km turbulence tends to keep the constituents of the atmosphere mixed, and the condition of diffusive equilibrium is not encountered.

Curves showing the logarithm of the concentration of the various constituents can be found, for example, in Johnson [24]. The concentrations vary with temperature, and therefore with the sunspot cycle, in such a way that a helium layer appears above about 1500 km (above the region of oxygen atom dominance) in the daytime near the time of the maximum of the sunspot cycle, whereas the transition is directly from oxygen to hydrogen, at about 600 km, in the nighttime at sunspot minimum. In addition to nitrogen and oxygen, the lower atmosphere of the earth contains small amounts of carbon dioxide, water vapor, etc. The effects of some of these constituents are considered in later sections.

3-2-4 Ionization Processes and Formation of Ionized Layers

The ionosphere, the ionized region of the earth's atomosphere, is the region from about 60 to 500 km or more in altitude and includes the region designated as the thermosphere in Sec. 3-2-1. The region is ionized by solar radiation in the ultraviolet and X-ray frequency ranges. It is a plasma containing free electrons and positive ions such that it is electrically neutral. Only a fraction of the molecules are ionized, and large numbers of neutral particles are present also. In this section the processes by which free electrons are produced are considered. The dominant process of electron production in much of the ionosphere, especially the E and F regions, is photoionization by ultraviolet and X-ray radiation from the sun. Photodetachment of electrons is also significant in the D region at sunrise. Recombination and attachment are important electron-loss processes. Examples of the various reactions are now given. The symbol $h\nu$ stands for an incident photon, and e stands for a free electron.

1. Production of free electrons

a. Photoionization of neutral particles

$$O + h\nu \longrightarrow O^+ + e$$

$$N_2 + h\nu \longrightarrow N_2^+ + e$$

$$NO + h\nu \longrightarrow NO^+ + e$$

b. Photodetachment of electrons from negative ions

$$O^- + h\nu \longrightarrow O + e$$

$$O_2^- + h\nu \longrightarrow O_2 + e$$

c. Electron detachment of electrons from ions by the collision of the ion with a heavy particle

$$O^- + O \longrightarrow O_2 + e$$

$$O^- + N_2 \longrightarrow N_2O + e$$

$$O_2^- + O \longrightarrow O_3 + e$$

2. Exchanges of ionization

a. Charge transfer

$$N_2^+ + O \longrightarrow N_2 + O^+$$

$$N_2^+ + O_2 \longrightarrow N_2 + O_2^+$$

$$O^+ + O_2 \longrightarrow O + O_2^+$$

b. Ion-atom interchange

$$O^+ + N_2 \longrightarrow NO^+ + N$$

3. Electron-loss processes

a. Recombination

$$O^+ + e \longrightarrow O_{(\text{excited})} + h\nu \qquad \text{(radiative recombination)}$$

$$N_2^+ + e \longrightarrow N + N \qquad \text{(dissociative recombination)}$$

$$NO^+ + e \longrightarrow N + O$$

b. Attachment to neutral particles

$$O + e \longrightarrow O^- + h\nu$$

$$O_2 + e \longrightarrow O_2^- + h\nu$$

The electron concentration can be determined, theoretically, by rate equations which relate electron production and loss and the rate of change of electron concentration. If recombination is the only important loss process, one may write

$$\frac{dN(e)}{dt} = q - \alpha N(e)N(A^+) \tag{3-14a}$$

where $N(e)$ is electron concentration, q represents electron production (number/m^3/s), α is the recombination coefficient, and $N(A^+)$ is the ion concentration. If $N(e) \simeq N(A^+)$, the equation takes the form

$$\frac{dN}{dt} = q - \alpha N^2 \tag{3-14b}$$

where N stands for electron concentration, it being no longer necessary to use $N(e)$. When attachment to neutral particles is the dominant loss process

$$\frac{dN(e)}{dt} = q - KN(e)N(A) \qquad \text{(3-15a)}$$

where K is the appropriate coefficient and $N(A)$ is the neutral particle concentration. When $N(A) \gg N(e)$, which is the case for a weakly ionized gas,

$$\frac{dN}{dt} = q - \beta N \qquad \text{(3-15b)}$$

where β is the attachment coefficient. Of course recombination and attachment may also take place simultaneously.

Attention is now directed to the theory of formation of a Chapman layer. This theory was developed by Chapman [11, 40] in a pioneering paper of great interest and importance. Chapman assumed photoionization to be caused by monochromatic solar radiation, an isothermal atmosphere having a constant value of H, a horizontally stratified plane atmosphere, and equilibrium electron density such that $dN/dt = 0$ in Eq. (3-14b). He first calculated q, the electron production rate, and then calculated N by setting q equal to αN^2. The resulting expression for electron density is

$$N = N_0 \exp\left[\tfrac{1}{2}(1 - z - \sec \chi e^{-z})\right] \qquad \text{(3-16)}$$

where N_0 is peak electron density, χ is zenith angle of the sun, and $z = (h - h_{m0})/H$. h_{m0} is the height of maximum electron density when $\chi = 0°$, and H is the scale height. A plot of Eq. (3-16) is given in Fig. 3-6.

Physically, a maximum in electron production arises because above the maximum the neutral air density falls off with increasing altitude, whereas below the maximum the power density of the ionizing radiation becomes low. Above the maximum there is relatively little matter for the incident solar radiation to ionize, and below the maximum the ionizing process results in attenuation of the incident solar radiation to the extent that its capability for producing further ionization is significantly lessened. None of the ionized layers of the actual atmosphere has exactly the form of the Chapman layer, but the development of the theory of the Chapman layer constituted a major contribution to the understanding of ionospheric layers.

Several different ionospheric layers are formed because different frequency, or wavelength, ranges of the incident solar radiation react with different atmospheric constituents as shown in Fig. 3-7, which shows the altitude at which radiation intensity has dropped to $1/e$ of its original intensity. The following points concerning Fig. 3-7 are of interest:

1. From 2100 to 3000 Å, absorption is entirely due to ozone and occurs at a low altitude in the ozonosphere below the level of the ionosphere itself.
2. From 1000 to 2000 Å, molecular oxygen is the primary absorber, and it dissociates in the process.

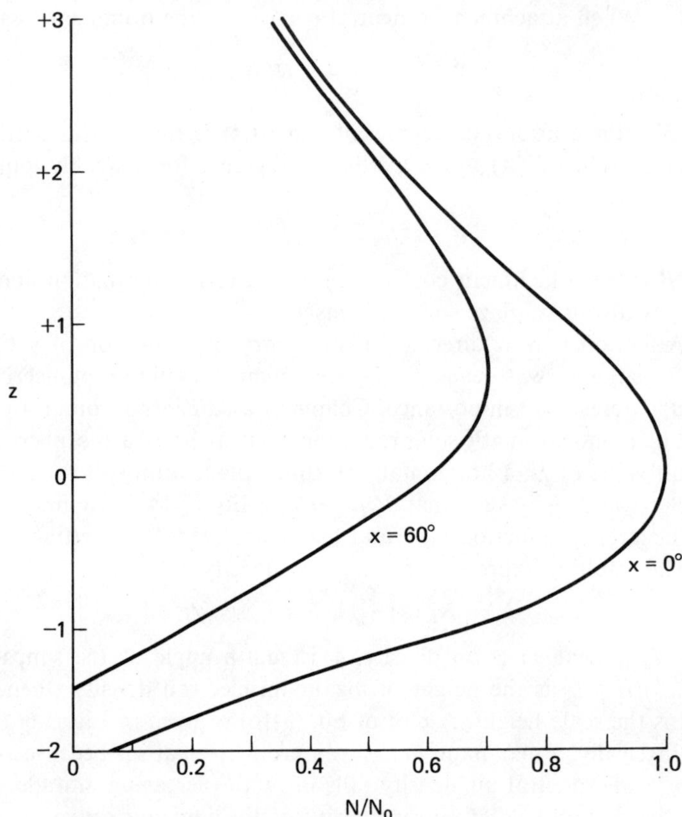

FIGURE 3-6. Form of Chapman layer for overhead sun ($\chi = 0°$) and for sun 60° from overhead ($\chi = 60°$).

3. A region of deep narrow windows is shown. One corresponds to the Lyman-α line at 1215.7 Å, which can penetrate to low altitudes near 75 km.
4. In the region from 200 to 800 Å, absorption occurs rather high in the ionosphere.
5. Penetration again increases for X-ray wavelengths in the 10- to 100-Å range.

If the development of the theory is examined in detail, it can be seen that the height of maximum electron production for $\chi = 0°$ corresponds to unit optical depth (the power density of incident wave has fallen off to $1/e$ of the original value), and for this condition

$$p = \frac{mg}{\sigma} \tag{3-17}$$

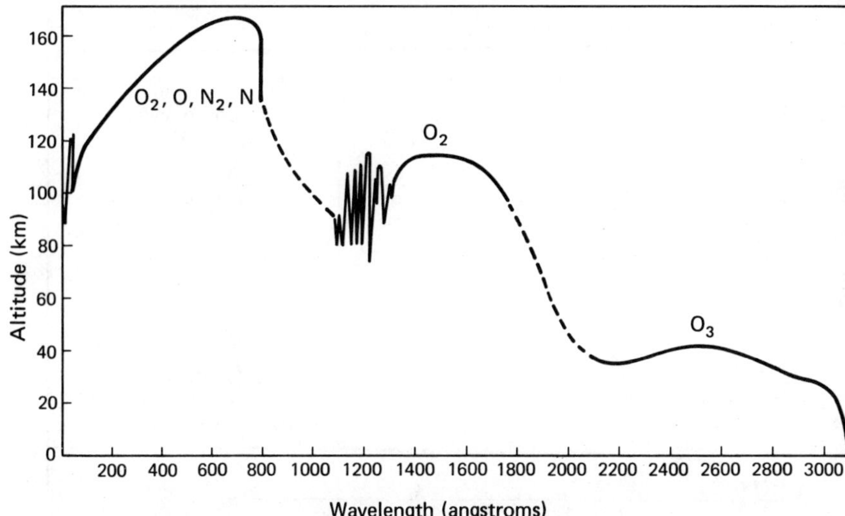

FIGURE 3-7. Penetration of solar radiation into the atmosphere. The curve indicates the level at which the intensity is reduced to $1/e$. Absorption above 2000 Å is principally due to ozone, between 850 and 2000 Å to molecular oxygen, and below 850 Å to all constituents. [From H. Friedman, "The Sun's Ionizing Radiations," in J. A. Ratcliffe (ed.), *Physics of the Upper Atmosphere*, Academic Press, Inc., 1960.] 2000 Å $= 0.2\ \mu$m.

where p is the pressure at the height of maximum production, m is the mass of the ionized particles, g is the acceleration of gravity, and σ is the cross section of the ionized particles for the frequency range in question. σ is known for the major atmospheric constituents. For example, for $\sigma = 3 \times 10^{-20}$ cm^2, the approximate cross section for the atmospheric constituents at a wavelength of $8 \times 10^{-4}\ \mu$m, $p = 1.6$ dyn/cm^2 or 0.16 N/m^2 (0.16 pascal), corresponding to a height near 90 km, as can be determined from Table 3-1.

3-2-5 The Quiet Ionosphere

Because different portions of the solar spectrum are absorbed at different altitudes, the ionosphere consists of several layers or regions [40]. The layers are not sharply defined, distinct layers, and the transition from one to the other is generally gradual with no very pronounced minimum in electron concentration in between. None of the layers correspond exactly to a Chapman layer, although the electron density of the E region is quite similar in form to the theoretical Chapman layer. The discussion begins with the lowest ionospheric region and proceeds to higher regions. Disturbed ionospheric conditions are treated in a later section. Figure 3-8 shows representative plots of electron density versus altitude.

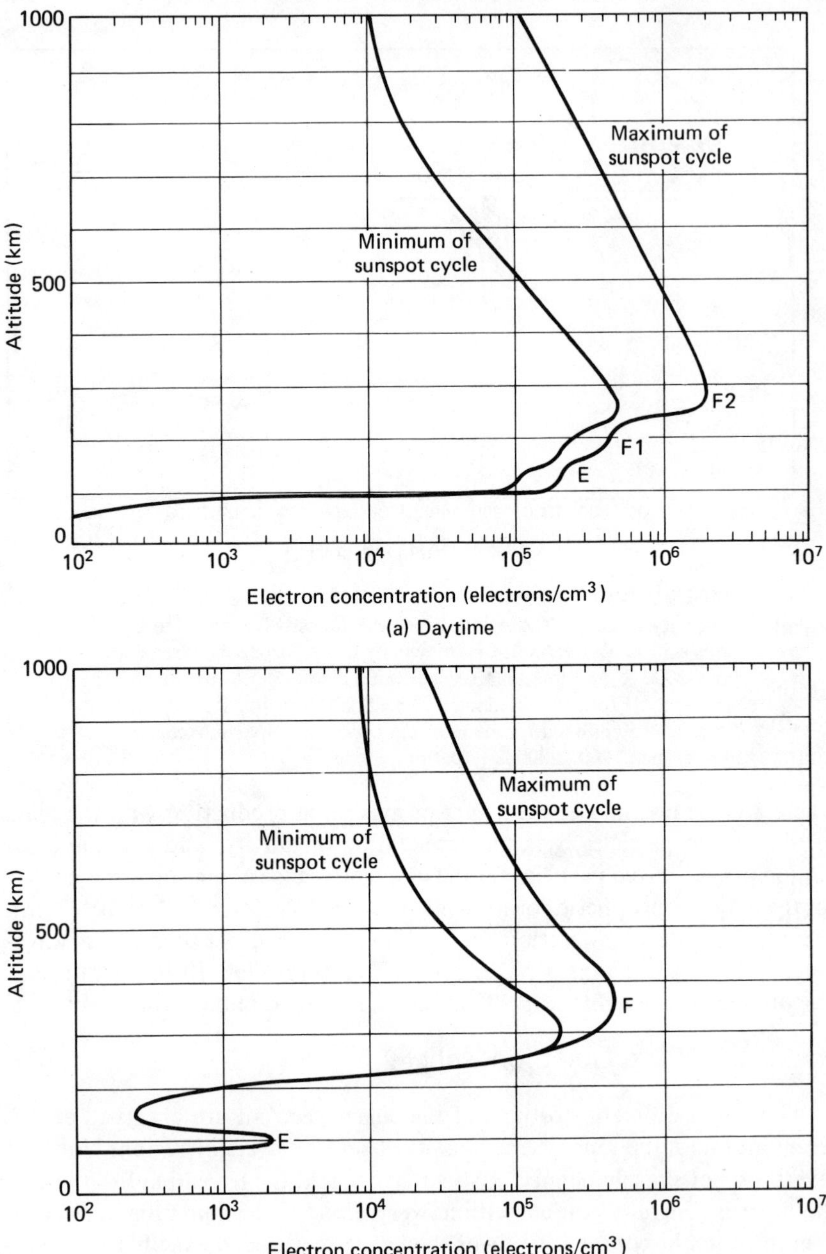

FIGURE 3-8. Normal electron distributions at the extremes of the sunspot cycle. (Reprinted from *Satellite Environment Handbook*, 2nd ed., edited by Francis S. Johnson, with permission of the publishers, Stanford University Press [24]. Copyright © 1965 by the Board of Trustees of the Leland Stanford Junior University.)

D REGION. Although the *D* region is the lowest of the ionospheric regions, it is difficult to study and is poorly understood in many ways. It covers a height range from approximately 50 to 90 km, with its maximum electron concentration of about $10^3/cm^3$ occurring between about 75 and 80 km in the daytime. The lower portion of the *D* region is believed to have a different ionizing source than the upper portion and is sometimes designated as the *C* region. Electron concentrations here may be on the order of $10^2/cm^3$. At night the electron concentration throughout the *D* region drops to vanishingly small values.

The *C* layer is believed to be caused primarily by cosmic radiation. Day-to-night variations in electron concentration are caused by the attachment at night of electrons to neutral particles to form negative ions, which are dissociated in the daytime by solar radiation.

Lyman-α radiation, at a wavelength of 1215.7 Å, penetrates deeply into the ionosphere, and the resultant ionization of NO at *D*-region heights is believed to be the important factor in forming the upper *D* region. X-ray radiation also can penetrate deeply, but it is presently believed to be more important under disturbed conditions and at the peak of the solar cycle than during quiet periods.

An important feature of the *D* region is the fact that it is at a low altitude and consequently has a high collision frequency which results in relatively high attenuation of radio waves. It is well known, for example, that radio waves in the broadcast band are highly absorbed during the daytime by the *D* region.

Among other interesting features of the *D* region is the fact that it is subject to disturbances caused by solar flares, magnetic storms, and, at high latitudes, auroral and polar-cap activity. Studies have shown that HF through VHF radio waves tend to be partially reflected at discrete heights in the *D* region. Scattering due to weak layers and/or other irregularities in the *D* region is useful for ionospheric scatter propagation. VLF waves are regularly reflected from the *D* region.

It is difficult to use ionosonde techniques to study the *D* region, although special VLF ionosondes have been constructed for the *D* region. (The ionosonde is a vertically pointing radar.) The partial reflection of HF or VHF waves and effects on VLF and LF waves have provided information about the region. Radio-wave interaction experiments have also provided some information but are very difficult to interpret. In situ rocket measurements in the region are difficult, but some useful results have been obtained by rocket techniques.

E REGION. The *E* region is comparatively well understood, and its behavior is quite similar to that of a theoretical Chapman layer. Its height extends from about 90 to 140 km, and the peak electron concentration occurs between about 100 and 110 km. At noon at the minimum of the solar cycle the peak concentration may be about $10^5/cm^3$, and the value may be about 50% greater at the maximum of the cycle. At night the concentration drops by a factor of approximately 100.

Solar electromagnetic radiation forms the E region, as is shown by observations during solar eclipses. Both X-rays and UV radiation apparently play a part in the formation of the region, and their relative importance is not clear.

The E region has a rather high electrical conductivity, and the electric currents that cause geomagnetic effects flow largely in the E region. In the equatorial and auroral regions, intense electric currents, called the equatorial and auroral electrojets, flow. The phenomenon of sporadic E, thin, somewhat sporadic or discontinuous layers of rather intense ionization, is frequent in the E region. HF radio waves may be reflected from the normal E layer, depending on the time of day, the frequency, etc., and the E layer may therefore be useful for communication purposes. Waves may also be reflected by sporadic E but not very reliably.

The daytime E layer can be studied by conventional ionosonde techniques as well as by a variety of other techniques including rocket-borne CW propagation measurements and ion-density probe measurements.

F Region. The F region of the ionosphere sometimes consists of two parts, the F_1 and F_2 layers. The F_1 layer behaves somewhat as a Chapman layer with a peak electron density of about $2.5 \times 10^5/cm^3$ at noon at the minimum of the solar cycle and a peak electron density of about $4 \times 10^5/cm^3$ at noon at the maximum of the solar cycle. The F_1 region largely disappears at night. The F_2 layer is very complicated, and its behavior is very different from that of a Chapman layer. It has the highest electron densities of any ionospheric region, and the electron densities tend to remain rather high throughout the night. It is present generally whether there is an F_1 layer or not. The peak electron density is found in the 200- to 400-km height range and ranges between about $5 \times 10^5/cm^3$ and $2 \times 10^6/cm^3$ in the daytime and between about $1 \times 10^5/cm^3$ and $4 \times 10^5/cm^3$ at night. The F_2 layer is extremely important for long-range HF propagation.

The same ionizing agent, solar radiation in the 200- to 900-Å range, is apparently effective in both the F_1 and F_2 regions, but the recombination coefficient falls off rapidly with height. Charge transport processes need to be considered and must be accounted for by a term in the rate equation, which has the form

$$\frac{dN}{dt} = q - \beta N - \nabla \cdot (vN) \tag{3-18}$$

in the F region. The loss process is recombination, but the recombination process is inversely proportional to electron concentration so that an attachment-like term appears. The $\nabla \cdot (\mathbf{v}N)$ term takes charge transport into account. The charge transport may be of three kinds. The neutral air may be in motion carrying electrons with it along field lines, the presence of an electric field may result in a drift of electrons perpendicular to both the electric and magnetic fields, and electrons may diffuse under the influence of gravity and pressure gradients.

The F region can be studied by ground-based and top-side ionosondes, by the incoherent scatter technique, by recording the amplitude, phase, and Doppler frequency of HF CW transmissions, etc. The phenomenon of spread F is frequent in the F region and at temperate and high latitudes is correlated with magnetic activity. During spread F conditions ionosonde traces lose their relatively clean, sharp condition and assume the appearance of a multiplicity of weak, fuzzy overlapping echoes.

The F region is characterized by a number of so-called anomalies. One is the diurnal anomaly, namely that the maximum electron concentrations often occur after noon, perhaps far removed from noon. A Chapman layer has the highest concentration at noon. The winter or seasonal anomaly refers to the fact that high values of electron concentration occur in winter at moderate to high latitudes. The December anomaly is the fact that high electron concentrations tend to occur in and near December over a wide range of latitudes. A double-humped distribution is found near the equator below about 600 km, a minimum of electron concentration being found at the equator itself with the two humps occurring at latitudes of about 10–15° on either side of the equator. Satellite studies have shown the existence of a narrow trough, or minimum in electron concentration, at a latitude of about 45°.

3-2-6 The Geomagnetic Field and the Magnetosphere

The study of the earth's magnetic field has a long history, one important milestone being the publishing of *De Magnete* by William Gilbert in 1600. For comprehensive treatments of geomagnetism, including historical summaries, see Chapman and Bartels ([13], Volumes I and II) and Matsushita and Campbell ([29], Volumes I and II). Some elementary points about the earth's magnetic field are mentioned in this section.

Potential functions are frequently useful for the calculation of electric and magnetic fields. For magnetic fields the vector magnetic potential **A** is a useful function that is applicable generally. For static magnetic fields in source-free regions, however, a scalar magnetic potential can be used. It is common practice for such a scalar potential to be used in treating the earth's magnetic field. By using either a vector or scalar magnetic potential it can be shown that the magnetic flux density caused by a magnetic dipole of magnetic moment M is described by the equations

$$B_\theta = \frac{\mu_0 M \sin\theta}{4\pi r^3} \tag{3-19}$$

$$B_r = \frac{\mu_0 M \cos\theta}{2\pi r^3} \tag{3-20}$$

where the dipole is oriented along the polar axis of a spherical coordinate system and θ is the polar angle. B is magnetic flux density in Wb/m^2, μ_0 is magnetic permittivity ($4\pi \times 10^{-7}$ H/m), M is magnetic moment (A m^2), and r is distance from M in meters. For the case of a current I flowing in a loop of area A, $M = IA$. To a first approximation these equations can be used to describe the earth's magnetic field if θ is measured from the magnetic polar axis. If the value of B at the earth's surface at the magnetic equator at $r = a$ is taken to be B_0, B_θ and B_r can be expressed as

$$B_\theta = \frac{B_0 a^3 \sin\theta}{r^3} \tag{3-21}$$

$$B_r = \frac{2B_0 a^3 \cos\theta}{r^3} \tag{3-22}$$

where

$$B_0 = \frac{\mu_0 M}{4\pi a^3} \tag{3-23}$$

Thus B falls off inversely as the cube of the distance from the center of the earth.

A more nearly accurate description of the earth's magnetic field can be obtained by making measurements of the components of **B** and/or the total scalar value of **B** over the surface of the earth and by then expressing the results in terms of a spherical harmonic analysis. Although **B** and its components are the measured quantities, the results are commonly expressed in terms of a scalar magnetic potential V. **B** is the negative of the gradient of V; that is, $\mathbf{B} = -\nabla V$. V can be expressed as the sum of $V^i + V^e$, where V^i is of internal origin and V^e is of external origin (external to the solid earth). The external sources, primarily ionospheric currents, will be considered next.

Superimposed on the very slowly varying earth's magnetic field of internal origin are variable components caused by ionospheric currents and hydromagnetic waves. These variations can be separated into regular variations and those associated with ionospheric disturbances. The regular variations are discussed in this section. They are caused by horizontal movements of the ionosphere that result from solar and lunar tidal effects [14] and heating by the sun. The movements result in the generation of electric fields by the dynamo effect, and the electric fields drive ionospheric currents. The electric field intensity generated by the dynamo effect is given by $\mathbf{E} = \mathbf{v} \times \mathbf{B}$, $\mathbf{v}$ being the velocity of the moving medium. ($\mathbf{v} \times \mathbf{B}$ is a vector product, and **E** is perpendicular to both **v** and **B**.) It was first suggested by Balfour Stewart in 1882 that daily magnetic variations were caused by currents flowing in a conducting region of the upper atmosphere.

The atmospheric oscillations have periods near 24 and 12 h. A diurnal oscillation (24-h period) is caused by heating by the sun. Approximately semidiurnal components are caused by solar and lunar tides, and the waveform of the heat flux from the sun also has a 12-h component. The response of the earth-atmosphere system is greatest to the solar 12-h input.

The "quiet day" ionospheric current system that results from solar heating and tides is called the S_q current system. The form of the S_q system is suggested by the sketch in Fig. 3-9. It is seen that the currents of the northern and southern hemisphere systems are in the same direction at and near the equator. Somewhat similar current systems exist in the nighttime hemisphere except that the magnitudes of the currents are smaller and the directions are reversed. The S_q current system results in a daily variation of about 50 γ in the observed magnetic field at the surface of the earth away from the magnetic equator. ($10^5\ \gamma = 1$ G, 10^4 G $= 1$ Wb/m^2.)

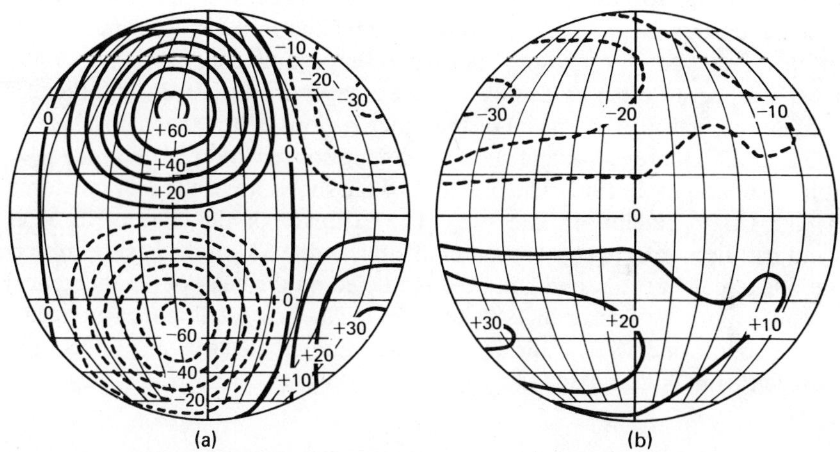

FIGURE 3-9. S_q current system at the equinoxes at sunspot minimum at a height of 100 km over the sunlit hemisphere (a) and the night hemisphere (b). A current of 10,000 A flows between adjacent lines. (From Chapman and Bartels, *Geomagnetism*, Vol. II, Oxford University Press, 1940 [13].)

An *L* current system, caused by the lunar tide, also exists, but lunar current densities are small, and it is difficult to measure its effect.

THE MAGNETOSPHERE. It is now known that the space between the earth and the sun is not a perfect vacuum but is occupied by a neutral plasma with a density of about 10 ions/cm³. This plasma flows outward from the sun at speeds of about 300–500 km/s, and the flow of plasma is commonly referred to as the solar wind. The streaming plasma confines the earth's magnetic field inside an elongated cavity. The boundary of the cavity is known as the magnetopause, and the region inside the boundary is known as the magnetosphere. Sometimes the latter is defined as the region in which the effect of the geomagnetic field dominates the motion of charged particles, as contrasted to the lower ionosphere where motions are dominated by collisions. On this basis the magnetosphere is the region above the E layer of the ionosphere. In practice the term seems to be more generally applied only to the region above the F layer, or simply to the region above the ionosphere.

The magnetosphere extends to a distance of about 10 earth radii from the center of the earth in the direction toward the sun. In the direction away from the sun the magnetosphere has a very long tail and extends to about the moon's orbit or 60 earth radii or farther. See Fig. 3-10. The velocity of the plasma streaming toward the earth from the sun is much higher than the velocities of acoustic and hydromagnetic waves in the region, and a detached shock wave is therefore produced several earth radii upstream from the magnetopause. Although Chapman and Ferraro [12] had suggested earlier that the earth's magnetic field should be confined in a cavity, confirmation of this fact and information about the shape and characteristics of the magnetosphere were provided by satellite measurements [10, 21, 32]. Pioneer I made the first measurements of the outer regions of the magnetosphere in 1958, and Explorer 10 made the first definite observation of the magnetospheric boundary in 1961. The presence of the shock wave was first measured by the IMP-1 satellite in late 1963. The existence of the continuous solar wind had been inferred by Biermann in 1951 and 1952, on the basis of observations of comet trails, and had been predicted on the basis of the theoretical considerations by Parker in 1958 and 1960. The first satellite measurements of the solar wind were made by Gringauz and his colleagues in 1960 and 1961 with Lunik 2 and 3 [28].

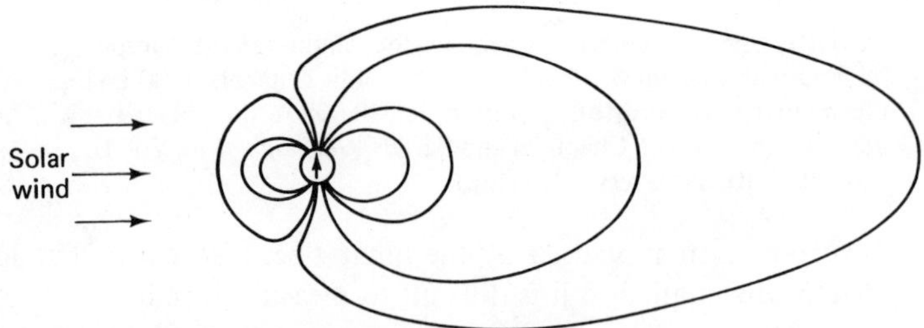

FIGURE 3-10. A section of the magnetosphere, showing the low-latitude field lines as having a roughly dipole shape and the high-latitude lines swept round over the poles to form a geomagnetic tail. [From C. O. Hines, I. Paghis, T. R. Hartz, and J. A. Fejer (eds.), *Physics of the Earth's Upper Atmosphere*, Prentice-Hall, Inc., 1965.]

A calculation of the shape of the magnetosphere is very difficult, but one feature, the boundary distance of about 10 earth radii in the solar direction, can be made plausible by a simple calculation. When a plasma and a magnetic field are in equilibrium the effective pressure of the magnetic field is equal to its energy density. The earth's magnetic field drops off as the cube of distance and can be expressed as B_0/r^3 if r is measured in earth radii. Because the magnetic dipole of the earth is imaged at the plasma boundary, however, the magnetic field at the boundary is $2B_0/r^3$. Equating the magnetic pressure and the pressure of the streaming plasma, we obtain

$$2mnv^2 = \frac{4B_0^2}{2\mu_0 r^6} \tag{3-24}$$

from which r is found to be about 9.5 earth radii if a velocity of 400 km/s, a number density of $10/\text{cm}^3$, and ionized hydrogen are considered for the impinging plasma.

The magnetosphere is the location of the earth's magnetic belts, where charged particles are trapped by the earth's magnetic field. The particles have three components of motion. They spiral around magnetic field lines, bounce back and forth from near one end of a field line to near the other end, and at the same time drift slowly around the earth. Because the particle density in the region is low, the collision rate is low, and charged particles may remain trapped for long periods. To quote a numerical example of a particle motion from Hess [20], a 50-keV electron, at a height of 2000 km near the equator, will complete a rotation around a field line in 2.5×10^{-6} s, have a bounce period of 0.25 s, and rotate around the earth in 690 min, neglecting the possibility of collisions during these periods.

Some of the trapped charged particles in the magnetosphere have very high energies, and the term penetrating radiation is used to refer to these particles. The earth's surface is protected from these charged particles by the earth's atmosphere. Only protons of energies of 1 GeV or greater penetrate significantly to sea level. The naturally occurring radiation consists of both protons and electrons. The existence of the radiation belts was not suspected until discovered by Van Allen in 1958 by means of Explorer 1. His original measurements indicated two separate radiation belts, but that result was obtained because he observed only the more energetic particles. Electron measurements have now been made between about 1 keV and 5 MeV, and proton measurements have been made in the 100-keV to 700-MeV range. (One eV corresponds to 1.6×10^{-12} erg or to 12,395 Å or to 11,600 K.)

MOTION IN STATIC MAGNETIC FIELD. The equation for the force on a charged particle in a magnetic field sufficies to determine the motion of the particle. This equation is

$$\mathbf{f} = q(\mathbf{v} \times \mathbf{B}) \tag{3-25}$$

where q is charge in coulombs, $\mathbf{v}$ is velocity in m/s, $\mathbf{B}$ is magnetic flux density in Wb/m^2, and **f** is force in newtons (N). As reference to the theory of particle dynamics indicates and as discussed in basic courses in electronics and physics, the force is perpendicular to **v**, and the particle will move in a circular orbit when **v** is perpendicular to **B**. A good treatment of particle dynamics can be found, for example, in Page ([35], Chap. 1). If the particle has components of **v** both parallel and perpendicular to **B**, it will follow a spiral path with the axis of the spiral having the same direction as **B**.

For analyzing the circular component of motion, when **B** is directed along the z axis of a cylindrical coordinate system, one may write

$$\mathbf{f} = m\mathbf{a} = \mathbf{a}_r\left(\frac{-mv^2}{r}\right) = \mathbf{a}_r(-|q|\,\mathrm{v}\mathrm{B}) \tag{3-26}$$

where $\mathbf{a}_r(-mv^2/r)$ represents the centripetal force which is directed in the $-r$ direction. If the motion of a positive ion is considered, Eq. (3-26) states that the ion must rotate in the $-\phi$ direction in order for the force to be in the $-r$ direction as required. Letting $v = r\omega_B$ in Eq. (3-26), the magnitude of ω_B is found to be qB/m, where ω_B is the angular velocity of the charged particle in rad/s. Rotation can be represented in vector form by a vector $\boldsymbol{\omega}_B$ directed along the axis of rotation, or in the $-z$ direction in the case of a positive ion, as shown in Fig. 3-11(a). The convention employed for this vector representation is the right-hand convention, namely that rotation is in the direction to which the fingers of the right hand point if they are held as if to encircle the thumb, when the latter points in the direction of the axial vector representing the rotation. The vector equation for the angular velocity of rotation of a charged particle is

$$\boldsymbol{\omega}_B = -\frac{q\mathbf{B}}{m} \tag{3-27}$$

In some treatments the magnitude of $\boldsymbol{\omega}_B$ is determined, but its vector nature, and the fact that in certain scalar equations one must substitute $\omega_B = -qB/m$, is neglected. Note that Eq. (3-27) applies to particles of either polarity. When it is desired to consider electron motion, the substitution $q = -e$ can be made, where e is the magnitude of the electron charge and is a

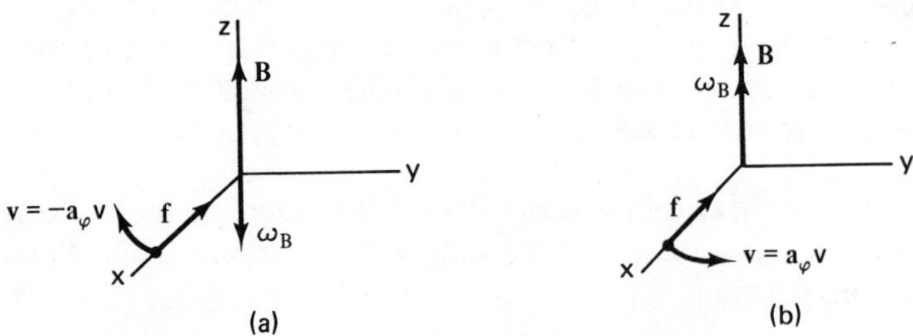

FIGURE 3-11. Forces on charged particles in a magnetic field.

positive number. The vector describing the circular motion of an electron in a magnetic field is thus $\boldsymbol{\omega}_B = e\mathbf{B}/m$, where $\boldsymbol{\omega}_B$ is in the same direction as $\mathbf{B}$, as shown in Fig. 3-11(b).

In Fig. 3-11(a), an ion is shown at the instant when it is on the x axis. Looking in the z direction, we see that the ion executes left circular (*l*c) or counterclockwise (ccw) rotation. For electrons the rotation is right circular (rc) or clockwise (cw), as shown in Fig. 3-11(b). The rotations of ions and electrons are depicted in a slightly different manner in Fig. 3-12.

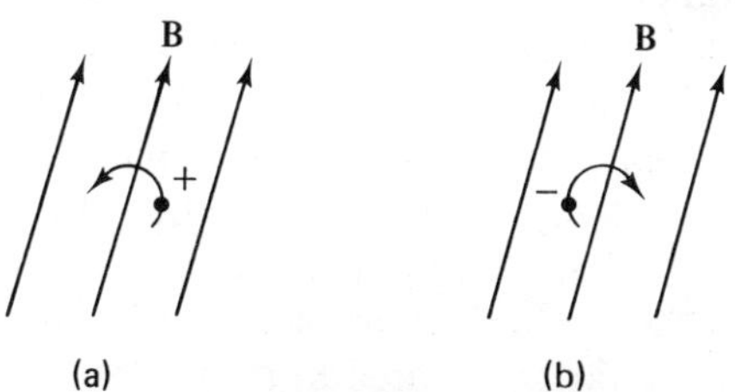

FIGURE 3-12. Motion of charged particles in a magnetic field.

3-2-7 Introduction to Solar-Terrestrial Relations and the Disturbed Ionosphere

The term *solar-terrestrial relations* is used here to refer to the effects of solar activity on the earth's upper atmosphere. The structure and surface features of the sun and the characteristics of solar activity are described briefly, and the atmospheric effects are then considered. Although the radiation from the sun in the frequency range which provides most of our heat and light is very nearly constant, radiation in the frequency range which ionizes the upper atmosphere varies with sunspot cycle and is subject to rather drastic variations when centers of activity develop on the sun.

STRUCTURE AND SURFACE FEATURES OF THE SUN. The interior of the sun, at the high temperature of about 20,000,000 K with a pressure of about 10^9 times the earth's sea-level pressure, is entirely opaque. The light and heat which the earth receives come from the photosphere, the thin surface layer of the sun at a temperature of approximately 5800 K. Sunspots occur in the photosphere.

Immediately above the photosphere is the chromosphere, which is very important from the viewpoint of solar-terrestrial relations, as flares occur there. The chromosphere is sometimes divided into three regions at heights of from 0 to 1000 km, 1000 to 4000 km, and 4000 to 12,000 km above the photosphere. The temperature increases with height in the chromosphere to temperatures in excess of 100,000 K.

The corona is the sun's outer layer and can be seen to extend for several solar radii during solar eclipses. The outer portion is now believed to extend beyond the earth. Part of the corona near the sun has temperatures on the order of 1,000,000 to 1,500,000 K. The corona is an important source of radio emissions.

The sun is characterized by centers of activity which are localized and transient regions on the sun's surface where the phenomena of flares, sunspots, faculae, plages, and prominences occur. Strong magnetic fields antedate and survive the other features. Following the appearance of magnetic fields, faculae and plages appear in conjunction. Both of these regions are hotter and brighter than their surroundings, but they occur at different levels, the faculae in the photosphere and plages in the chromosphere. As the plages and faculae brighten, sunspots may appear in the photosphere. Sunspots were the first type of solar activity to be observed and were studied by Schwabe between 1826 and 1857. Large sunspots are quite prominent in appearance. Sunspots consist of a central dark umbra surrounded by a somewhat brighter penumbra which is still dark compared to the surrounding photosphere.

The occurrence of sunspots follows the well-known 11-year solar cycle. There are also longer-period components of solar activity such that the 1957–1958 maximum during the IGY years was the largest on record. In addition there tends to be a 27-day periodicity, this figure being about the period of rotation of the sun. The Wolf number defining sunspot activity is given by

$$R = K(10g + f) \tag{3-28}$$

where g is the number of groups and f is the number of spots. Flares and other features of solar activity follow the same cycle as sunspots. Sunspots tend to occur in pairs, one preceding and one following with respect to solar rotation, and the preceding and following sunspots have magnetic fields of opposite polarity. The preceding sunspots in the northern hemisphere of the sun have one polarity, and the preceding spots in the southern hemisphere have the opposite polarity. The polarities reverse with each 11-year cycle, and the same pattern of polarities is repeated every 22 years. Thus if magnetic polarities are considered, the period of the sunspot cycle is 22 years.

Returning to the sequence of events in the solar centers of activity, note that plages and faculae continue to brighten and increase in area after sunspots appear. Then after a few days flare and prominence activity may begin, and at this time all types of activity reach a climax simultaneously. Prominences are projections from the relatively cool chromosphere into the hot corona. They may extend 30,000 or 40,000 km or even 100,000 km out from the chromosphere.

After the peak activity, the sunspots disappear first. The plages and faculae remain for weeks and then fade away. Finally the magnetic field disappears.

SUDDEN IONOSPHERIC DISTURBANCES. Solar flares can cause major, sometimes dramatic, disturbances in the earth's ionosphere. The effects can be divided into the two categories of simultaneous and delayed, the former resulting from the radiation of X-rays and the latter from charged particles emitted from the flares. The "simultaneous" effects in the ionosphere occur at essentially the same time that the flare is seen in visible light.

The simultaneous ionospheric effects of flares are commonly referred to, at the present time, under the general heading of sudden ionsopheric disturbances (SIDs). In general the X-ray radiation from the flare results in increased ionization in the lower ionosphere, the exact profile of the additional ionization depending on the intensity and spectrum of the X-rays. The additional ionization affects the amplitude, phase, and frequency of electromagnetic waves that are propagating in the ionosphere. Some of the more commonly used designations for SIDs are SWF (for shortwave fadeout, or more generally for simply a decrease in amplitude of man-made radio waves), SCNA (sudden cosmic noise absorption), SPA (sudden phase anomaly), SFD (sudden frequency deviation), SEA (sudden enhancement of atmospherics), and SFE (magnetic crochet). Actually the effects are not necessarily very sudden, and a distinction is sometimes made, for example, between S-SWF and G-SWF (sudden SWF and gradual SWF). Phase and frequency effects are related by the expression

$$df = \frac{1}{2\pi}\frac{d\phi}{dt} \tag{3-29}$$

where ϕ is phase and f is frequency. Phase has been commonly recorded at VLF and LF frequencies, and frequency deviation or Doppler frequency has usually been recorded at HF frequencies. However, a slow but perhaps large variation in phase may be best observed by recording phase directly, whereas a rapid but small variation in phase may be more readily recognized by recording frequency. At the present it is feasible to record both phase and frequency in the HF band, and there is virtue in doing so. SEAs result from the fact that the increased ionization in the D region may cause atmospherics (VLF signals of natural origin) to propagate more efficiently than before the flare. SFEs (solar flare effects) or magnetic crochets are small magnetic field variations that occur at the time of solar flares. An SFE was apparently the first SID effect to be observed. Carrington determined that a magnetic effect took place at the time of a solar flare on Sept. 1, 1859, although his observation was discounted at the time [6].

Cosmic noise absorption is studied by the use of an instrument known as a riometer (relative ionospheric opacity meter) [27]. The output of a diode in the riometer is maintained equal to the signal from the antenna, and the noise diode output level is recorded. This arrangement provides long-term stability that a simple radio receiver would not provide. Frequencies from about 10 to 50 MHz have been employed, 30 MHz being perhaps the most widely used.

UPPER ATMOSPHERIC STORMS. The delayed effects caused by solar flares are caused by particles which are ejected from the sun and which subsequently impinge upon the earth's magnetosphere. Very energetic protons or solar cosmic or subcosmic ray particles may reach the earth in from 15 min to several hours after a flare and cause polar-cap absorption (PCA). Before discussing polar-cap and auroral zone ionospheric effects, however, we turn to a consideration of temperate latitude effects and to upper atmospheric storms there which begin 20 to 40 or more hours after a flare. These storms are caused by protons and electrons which have traveled at lower velocities than the protons causing polar-cap absorption. The terms magnetic storm and ionospheric storm are more commonly used than upper atmospheric storm, but the former two designations seem to refer to different aspects of the same thing, while the latter covers all aspects. However, the term *magnetic storm* is also often used in a general sense.

MAGNETIC EFFECTS. The disturbance in the magnetic field during upper atmospheric or magnetic storms will now be treated. The disturbance component of the earth's magnetic field is designated by D. D has been separated into parts in two different ways, both methods proposed by Chapman. In his first proposal he divided D into two parts so that

$$D = Dst + DS \tag{3-30}$$

where Dst, the storm-time variation, is the same at all longitudes and DS is asymmetric. Another method of separating D is in terms of origin. On this basis

$$D = DCF + DR + DP \tag{3-31}$$

where the three components of D are caused by three current systems. The DCF current system (CF standing for corpuscular flux) is that which results when the solar particles, or the solar plasma, impinge on the magnetosphere. DCF currents are considered to flow on a thin outer layer of the magnetosphere, and changes in the magnetic field produced by the surface currents are propagated to the earth's surface as hydromagnetic waves. The resulting sudden increase in the magnetic field is designated as ssc (storm sudden commencement) or S.C. and frequently though not always marks the onset of a magnetic storm.

The DR field is believed to be developed by an effective ring current which encircles the earth at a radius less than four earth radii. The simple picture of this current is that it is due to the drift velocity of charged particles that are also spiraling around and bouncing back and forth along field lines as mentioned in Sec. 3-2-5. A ring current is now believed to flow under quiet conditions, but the current is enhanced during storms. The question of the exact nature of the ring current is something that neither theoretical analysis nor satellite measurements have completely resolved, but there seems to be such a current and the current is believed to be responsible for the decrease in the horizontal component of magnetic field that takes place during the main phase of a magnetic storm. Typically the magnetic field is increased following an S.C., and the main phase follows the S.C. by several hours.

The *DP* field is caused by the intense currents of the auroral or polar electrojet, and it has its largest amplitude in and near the auroral zone. The electrojet currents result in what are called negative bays, pronounced negative excursions of the magnetic field.

IONOSPHERIC EFFECTS OF ATMOSPHERIC STORMS. Associated with the magnetic variations during an upper ionospheric storm are variations in electron density and corresponding effects on the amplitude, phase, and frequency of electromagnetic waves that propagate through the ionosphere. The theory of electromagnetic wave propagation in the ionosphere is treated in Sec. 4-3, and a number of terms utilized in the present discussion are explained in that section.

In the typical sequence of events during a magnetic storm, the first magnetic activity is the S.C. In the general vicinity of the auroral zone, absorption of radio waves may take place briefly during an S.C., apparently because of the precipitation of electrons into the ionosphere. At middle latitudes an S.C. may cause a small variation of the frequency of HF transmissions.

In contrast to the case for SIDs for which the *D* region is affected strongly and the *F* region is little affected, ionospheric storms have a strong effect in the *F* region [36, 40]. Changes in the electron concentration in the *F* region have been analyzed in terms of $Dst + DS$ or $DCF + DR + DP$ components, which correspond to the components of the magnetic field itself. The *Dst* component tends to be characterized by a slight initial increase in N_{max} followed by a large decrease at high and middle latitudes. Near the geomagnetic equator the situation is somewhat reversed, as an initial decrease followed by a slight increase has been reported. The *DS* component is primarily diurnal in nature. In addition to the tendency for N_{max} to decrease during a magnetic storm at middle latitudes, the height of N_{max}, h_m, tends to rise. The increase in height may be on the order of 200 km or more, although caution is needed in interpreting ionosonde records during storms. The apparent increase in height may be exaggerated at times by oblique echoes, as well as being affected by changes in retardation below the reflection height. Faraday rotation measurements made by the use of satellites and by moon bounce have shown that the total electron content decreases during a magnetic storm and that the existing electrons are not merely redistributed.

These *F*-region effects are poorly understood, but thermal, chemical, and electrodynamic mechanisms have been invoked as explanations. The thermal mechanism involves heating of the ionosphere due to particle precipitation and perhaps due to current flow. A chemical theory involves the upward transport of O_2 due to storm-caused turbulence. According to this theory, the O_2 may become ionized, and the O_2^+ ion then combines with electrons in a dissociative recombination process with the result that the electron density is decreased. The electrodynamic theory supposes horizontal electric fields in the *E* region, associated with the electric currents there, and subsequent motion of electrons in the $\mathbf{E} \times \mathbf{B}$ direction. Where the magnetic field is not vertical this direction has a vertical component, and the motion should result in a redistribution of ionization in the *F* region.

Severe ionospheric storms can cause complete blackout of HF transmissions. In addition to the reduction in electron concentration and the absorption of the radio waves, certain dynamic effects take place in the middle latitude ionosphere during magnetically disturbed nights. The phase and Doppler frequency of HF transmissions commonly show large variations at such times. These variations can frequently be recognized as being caused by a traveling ionospheric disturbance followed by the condition of spread *F*. Although this section deals with the effects of storms, it should be pointed out that there is no sharp demarcation but rather a gradual transition between quiet and disturbed conditions. Magnetically disturbed conditions are often encountered at night when there is nothing that would be called a magnetic storm, or when there is only a minor magnetic storm, in progress. During such periods the nighttime phase and Doppler records show activity of the type mentioned, but the corresponding daytime records correspond essentially to those of the quiet ionosphere.

EQUATORIAL AND AURORAL IONOSPHERES. The equatorial and auroral ionospheres have features of interest which set them apart from the ionospheres of temperate latitudes. It was mentioned when discussing the S_q current system that the currents of the northern and southern hemisphere systems are in the same direction at the equator. This occurrence, plus the fact that the conductivity of the ionosphere becomes high over a restricted range of altitudes in the *E* region, results in a strong, concentrated ionospheric current known as the equatorial electrojet. Radio waves can be scattered from the irregularities in ionization that are associated with this current. Also, as mentioned further in Sec. 4-3-5, ionospheric scintillation has been observed on earth-satellite paths in the equatorial and auroral ionospheres.

Energetic particle precipitation into the auroral ionosphere causes the visible aurora, excess ionization which attenuates and scatters radio waves, and concentrated electrical currents known as auroral electrojets. The currents cause readily evident disturbances in magnetic records. The phenome-

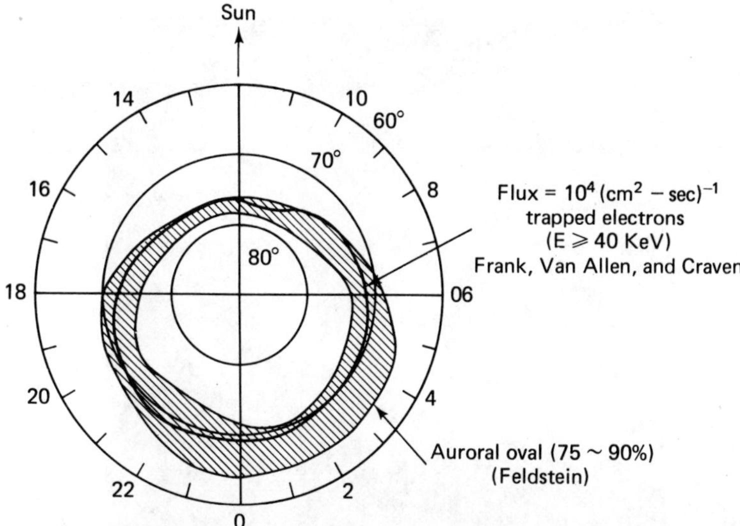

FIGURE 3-13. The auroral oval. (From Akasofu, *Polar and Magnetospheric Substorms*, Springer-Verlag New York, Inc., 1968 [3].)

non of the aurora occurs in the form of an oval around the magnetic poles, as in Fig. 3-13. The position of the oval is fixed approximately with respect to the sun, and the earth rotates beneath the oval. The auroral zone, where auroral activity occurs most frequently (essentially every night to some degree), is the locus of the midnight part of the oval. Akasofu [3] discusses activity within the auroral oval in terms of substorms, two or three or possibly even four, of which may occur in one night. Each manifestation of the aurora can be described as a substorm of its own type. Thus there is the auroral (visible) substorm, the polar magnetic substorm, the ionospheric substorm, the X-ray substorm, the proton aurora substorm, the VLF emission substorm, and the micropulsation substorm. The first auroral substorm of the evening in the auroral zone is characterized roughly by rather quiet east-west auroral arcs, which progress southward and may reach close to the zenith by 23 h local time. (See Fig. 3-14.) One or two westward traveling folds or surges in the otherwise quiet arcs may have been observed by this time. Between 23 h and 02 h, however, auroral forms become widespread and active in the sky, this phase being known as the auroral breakup (of the previously quiet arcs). After the breakup, patchy, luminous forms appear in the sky. Quiet forms may then reappear as the opening phase of a second auroral substorm.

The polar magnetic substorm tends to involve a rather weak positive bay during the period of quiet auroral arcs and a strong negative bay that

FIGURE 3-14. Auroral arcs over University of Alaska campus. (Photograph by Victor P. Hessler.)

begins near the time of breakup. The terms positive and negative bays refer to periods when the recorded magnetic field components have values greater than and less than, respectively, their presubstorm values. The ionospheric substorm involves increased ionization which results in attenuation of radio signals and cosmic noise and in the scatter of radio waves. The occurrence of such ionization can be detected by radar techniques and is commonly referred to as radar aurora or sometimes as auroral sporadic-E (Sec. 6-4-6). Evidence has accumulated that the auroral zone is an important source of traveling ionospheric disturbances (TIDs) observed in temperate latitudes. The TIDs are believed to be acoustic-gravity waves involving variations in neutral-particle density and corresponding variations in electron density. The latter affect electromagnetic waves that are incident upon or propagate through the region where the TIDs are propagating [22, 23].

X-rays are emitted when the incoming precipitating particles (energetic electrons) collide with atoms and molecules of the upper atmosphere. Protons, as well as electrons, play a role in auroral phenomena, the protons being associated with the quiet auroral arcs and also occurring equatorward of the arcs. VLF emissions are atmospheric noise that are closely associated with strong auroral activity.

The term *micropulsations* refers to extremely low-frequency electromagnetic waves (having periods of seconds or minutes). The magnetic fields of such a wave may be recorded by magnetometers of various types, or a signal proportional to electric field intensity may be recorded by using telluric current techniques (by measuring the voltage between two electrodes inserted into the earth).

During storms auroral activity occurs at lower latitudes than otherwise, and absorption and even blackout of radio waves can take place in temperate latitudes during magnetic storms whether visual aurora is seen at the same time or not. It is in and near the auroral zone itself, however, that radio-wave absorption is the most severe and frequent. Polar-cap absorption takes place over the polar cap, from approximately the auroral zone poleward, when energetic protons or solar cosmic rays precipitate into the polar ionosphere during unusually intense solar activity [38]. Transpolar propagation of radio waves may be seriously disrupted for several days. A diurnal variation in absorption occurs, when the region is not in continuous darkness or sunlight, due to the fact that at night electrons attach to neutral particles to form negative ions, whereas photodetachment of electrons from ions occurs during the daytime.

Possible Relations Between Solar Activity and Climate. Although solar-terrestrial relations were said at the beginning of this section to refer to effects on the upper atmosphere, the possibility of effects of solar activity on climate is drawing increasing attention. It has been suggested by Walter Orr Roberts that droughts occur on the Great Plains at about 22-year intervals, corresponding to the solar cycle if the polarities of the magnetic fields associated with sunspots are considered. Evidence is accumulating that a 70-year period of low temperatures before 1700 was characterized by a nearly complete lack of sunspots and auroral activity, as the latter were observed in Europe. Attention was drawn to this period by Maunder and Spörer in papers published between 1887 and 1922, and the validity of Maunder's claim has been strengthened by investigations by John Eddy. An account of Eddy's research on this subject and related papers are included in the proceedings of a workshop on *The Solar Output and Its Variation* [46]. The period of nearly zero sunspots is now known as the Maunder Minimum. An earlier period of reduced solar activity is known as the Spörer Minimum, and a period of high solar activity is believed to have occurred in the twelfth century. These periods of reduced and high solar activity correspond to periods of colder than normal and favorable weather (the Medieval Climatic Optimum) respectively. The Maunder and Spörer Minima were especially severe periods during what is known as the "Little Ice Age." The apparent validity of the minima contradicts the commonly held view that the sun is highly regular and consistent in all respects.

MONITORING OF SOLAR-TERRESTRIAL ENVIRONMENT. A real-time solar-terrestrial environment monitoring system is operated by the Space Environment Laboratory of NOAA at Boulder, Colorado in cooperation with the U.S. Air Force Air Weather Service [47]. Solar X-rays in the 0.5- to 4- and 1- to 8-Å (0.05- to 0.4- and 0.1- to 0.8-nm) wavelength ranges are monitored by SMS-2 and GOES-1 synchronous satellites at 135° west longitude and 75° west longitude, respectively. Some X-ray data are also received from the U.S. Navy's SOLRAD satellite. Solar wind data are received from the U.S. Air Force Vela Satellites, and energetic proton, electron, and alpha particle fluxes, recorded by the SMS-2, GOES-1, NOAA-4, and METEOR satellites, are received. The NOAA-4 and METEOR satellites are in polar orbit; the METEOR satellite is operated by the U.S.S.R. Other data inputs are magnetic field values at geostationary satellite altitude, total electron content from Faraday rotation measurements of geostationary satellite signals, H-α solar events, solar radio flux and bursts at a variety of frequencies and locations including Boulder (1415, 2695, and 4995 MHz) and Ottawa (2800 MHz), spectrographic solar radio events, solar calcium plage observations, coronal intensities, sunspot observations, optical auroral observations, auroral radar backscatter, ionosonde observations, high-frequency radio path signal strengths, sudden ionospheric disturbances, high-latitude riometer data, and ground-based magnetometer observations.

3-3 AIR POLLUTION AND RELATED ATMOSPHERIC PHENOMENA

3-3-1 Introduction

Air pollution is one of the major environmental problems and one that most people are familiar with in some degree. The Los Angeles Metropolitan area became notorious for its smong in the late 1940's, but the most severe episode of air pollution in terms of lives lost took place in London in 1952. It is considered that over 4000 persons died in 1 week because of severe air pollution. At present most metropolitan areas, as well as many other locations near smelters and industrial plants, experience significant air pollution. Two aspects of air pollution are discussed in this section:

1. The meteorological factors of air movement and vertical mixing
2. Air pollutants and their effects

The natural processes and cycles affecting some of the atmospheric constituents are treated in Sec. 3-4. The scattering and absorption of electromagnetic waves by particles and gases are discussed in Sec. 4-5. Remote sensing of the atmosphere is treated in Chap. 6. Possible measures for controlling atmospheric pollution are given only brief mention in these notes, and the reader interested in this topic must look to other sources, such as the series on control techniques by the U.S. Department of Health, Education and Welfare.

3-3-2 Temperature Inversions and Dispersion of Pollutants

The meteorological factors affecting air pollution are primarily the degree of horizontal movement or wind and the degree of vertical mixing.

Consider a windy city. All its pollution for a particular day is distributed over a considerable area downwind. If the average wind speed is 16 km/h, the pollution is spread out along a distance of 386 km. Next consider a city of about 16 km in diameter putting out the same pollution but having no wind. Other things being equal the city would have concentrations of pollutants about 24 times those of the windy city.

The other factor is the amount of vertical mixing. At one extreme, in the presence of thunderstorm activity, the vertical mixing may take place throughout the entire troposphere or to a height of about 9 km. At the other extreme in the presence of a temperature inversion near the surface the mixing takes place only in a shallow depth, perhaps 15 to 30 m or less. Los Angeles has the worst of conditions in that it has little wind and is in a valley or basin surrounded by mountains and typically has temperature inversions which allow vertical mixing to only about 460 m. Compared to the windy city considered earlier and one also having vertical mixing to 9 km, Los Angeles should have 480 times as much pollution, in terms of pollution concentrations.

It is the presence of temperature inversions that inhibits vertical mixing. Before discussing temperature inversions specifically, the dry adiabatic lapse rate of the atmosphere will be calculated. Consider a parcel of air which is moving vertically in the atmosphere without loss or gain of heat; that is, consider that the air is experiencing an adiabatic process. In general, heat added to a system is used to do work and to increase the internal energy or

$$Dq = du + p\,dv \tag{3-32}$$

The increase in internal energy can be expressed as $C_v\,dT$, where C_v is specific heat at constant volume, and, considering an adiabatic process for which $Dq = 0$,

$$C_v\,dT + p\,dv = 0 \tag{3-33}$$

Next consider the perfect gas law,

$$pv = RT$$

from which

$$p\,dv + v\,dp = R\,dT$$

or

$$p\,dv = R\,dT - v\,dp \tag{3-34}$$

Substituting the latter equation into that for an adiabatic process, we obtain

$$(C_v + R)T - v\,dp = 0 \tag{3-35}$$

Next substituting $C_p - C_v = R$, where C_p is specific heat at constant pressure, and using the gas law again, we obtain

$$C_p \frac{dT}{T} - R\frac{dp}{p} = 0 \tag{3-36}$$

Use of the hydrostatic equation,

$$dp = -\rho g\,dz,$$

then results in

$$C_p \frac{dT}{T} + \frac{\rho g R\,dz}{p} = 0 \tag{3-37}$$

But $\rho = M/v = Mp/RT$, where M is the molecular weight and v is specific volume, so that

$$\frac{dT}{dz} = -\frac{Mg}{C_p} \tag{3-38}$$

Inserting numbers in this equation results in

$$\frac{dT}{dz} \simeq -9.8^\circ\ \text{C/km}$$

This dry adiabatic rate does not occur commonly in the atmosphere. A figure of -6.5°C/km is used for model atmospheres. However, a parcel of air that is moved upwards or downwards will tend to vary in temperature at the dry adiabatic rate.

If the actual lapse rate of the atmosphere (rate of decrease of temperature with altitude) is 9.8°C/km, a parcel of air that is originally in equilibrium with its surroundings and which is then moved upwards or downwards will remain in equilibrium, at the same temperature as its surroundings. The parcel of air will not be subject to any restraining or accelerating force. Such a gradient of temperature is therefore called a neutral gradient. If the actual lapse rate of the atmosphere is greater in magnitude than 9.8°C/km, a rising parcel of air will tend to cool only at the adiabatic rate and be warmer than its surroundings. Thus it will be lighter than the air around it and will be accelerated still further upwards. The air in this condition is unstable. If the lapse rate is less in magnitude than 9.8°C/km, a parcel moved upwards would cool at the adiabatic rate and be cooler than its surroundings. Thus it would be subject to a restoring force that would inhibit the vertical motion. A lapse rate less than 9.8°C/km is a stable lapse rate.

In an inversion layer the temperature increases with altitude, and such a layer is highly stable. All vertical motions are strongly inhibited in an inversion layer, and pollution emitted underneath the layer tends to be confined below it. Temperature inversions may develop when the loss of heat from the surface of the earth by radiation is not compensated for by inputs of heat, the ground being a more efficient radiator than air and therefore cooling more rapidly. Surface and low-level inversions tend to develop at night and in the winter, especially under conditions of clear sky as in the desert at night and especially in the arctic and subarctic in winter and in locations such as the San Joaquin valley of California where fog forms under the inversion and prevents surface heating in the winter. Inversions may form also when warm air blows over a cool surface such as an ocean and where the cold air of a cold front extends beneath warm air, etc.

Inversions are also caused by subsiding air, and this type of inversion is of common occurrence because in developing or semipermanent anticyclones the air between about 500 and 5000 m is descending at a rate typically around 1000 m/day [43]. The Pacific coast of the United States is along the eastern edge of a semipermanent anticyclone that forms in the Pacific, and the persistent temperature inversion of the Los Angeles area is caused by subsiding air.

The formation of an inversion layer by the process of subsidence can be visualized with the aid of Fig. 3-15. A cloud layer may develop below the top of a subsidence inversion and help to maintain or strengthen the inversion. The air beneath the temperature inversion is often very moist, while that above tends to be dry. Consequently the index of refraction may decrease

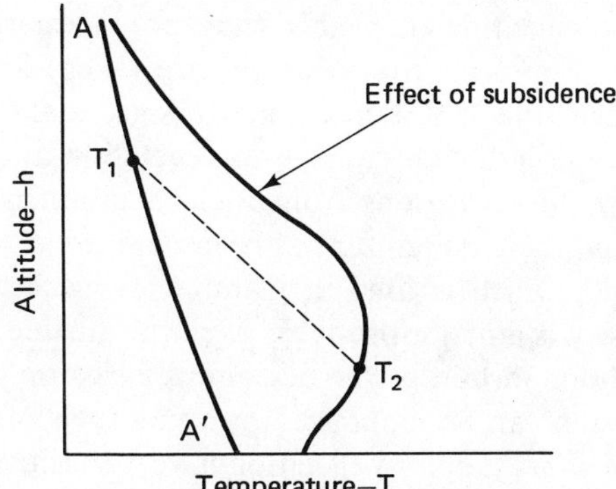

FIGURE 3-15. Temperature inversion formed by subsidence. The dotted line illustrates the heating of air from temperature T_1 to T_2 by a process of adiabatic compression. As the subsiding air cannot flow into the ground, however, the heating does not extend all the way to the ground, and a temperature inversion is formed. AA' represents an original or "normal" profile.

rapidly with height in the temperature inversion layer. This variation in index of refraction may cause severe fading on near-horizontal microwave links.

In addition to the subject of how pollutants are removed from the immediate area by winds and vertical mixing, the question of what eventually happens to the pollutants is of importance. How are they removed from the atmosphere? How do the concentrations of man-made pollutants compare with naturally occurring chemicals of the same types? Are the pollutants in the atmosphere accumulating and increasing in concentration? Some of the topics are discussed, with respect to particular pollutants, in following pages.

3-3-3 Air Pollutants and Their Effects

Much detailed information on air pollutants and their effects is available in the literature, and for that reason only a brief summary is presented here. See *Air Conservation* [1], *Cleaning Our Environment* [4], etc., for overall discussions and the extensive publications of the U.S. Department of Health, Education and Welfare such as the following series: Preliminary Air Pollution Survey of __________, Air Quality Criteria for __________, and Control Techniques for __________. Pollutants can be classified as gaseous and particulate. Among the more important or hazardous gases are sulfur dioxide, nitrogen oxides, carbon monoxide, and various organic hydrocarbons, aldehydes, mercaptans, etc. Solid and liquid particle pollutants include dusts, soots, and ashes; compounds containing lead and other metals; particles from automobile exhaust and photochemical reactions; herbicides and insecticides; and radioactive materials.

Sulfur dioxide is one of the very important air pollutants. Industrial operations, especially smelting of metals, may produce large quantities of sulfur oxides unless proper control measures are taken. Smelters at Trail, British Columbia, near the U.S. border; at Kellogg, Idaho; at Ducktown, Tennessee; etc., have done much damage to vegetation and health in their vicinities in the past, but emissions from such facilities have been greatly reduced generally, and the waste sulfur has been retained in some cases in the form of sulfuric acid, which is used in manufacturing fertilizers, etc. The electric power industry is a major producer of sulfur dioxide, and this source becomes increasingly important as the demand for electric power increases. Sulfur dioxide emission can be minimized by using fuels of low sulfur content, but these are in short supply. Vegetation is very sensitive to sulfur dioxide. The California Standards of Ambient Air Quality indicate that 1 ppm for 1 h or 0.3 ppm for 8 h can cause damage to vegetation; 5 ppm for 1 h causes bronchoconstriction in humans; and 10 ppm for 1 h causes severe distress in human subjects. Sulfur dioxide is believed to have played a major role in three air pollution disasters—the Meuse Valley in Belgium in 1930 when 63 persons died from respiratory irritation in a 5-day smog attack;

Donora, Pennsylvania in 1948 when five persons died and 43% of the population became ill in another 5-day period; and London in 1952 when 4000 persons died in a week. In such cases it is usually not sulfur dioxide alone that does all the damage but sulfur dioxide in combination with sulfuric acid aerosols, smoke, etc.

Sulfur dioxide commonly oxidizes to form sulfur trioxide which dissolves in water to form sulfuric acid. Europe, especially the Netherlands and Norway and Sweden, is becoming concerned about increasing acidity of its land and water, and industries in England and mainland Europe are suspected as the original source of the sulfur, in the form of sulfur dioxide emissions. Precipitation removes sulfuric acid and sulfate aerosols from the atmopshere and deposits them on the surface. A pH of 7.0 is neutral, with higher values increasingly alkaline and lower values increasingly acidic. pH values of 4 could be found over the Netherlands in 1962.

Hydrogen sulfide is another objectionable sulfur compound. Even 0.1 ppm can cause sensory irritation. Escape of hydrogen sulfide from a refinery in Poza Rica, Mexico in 1950 caused 22 deaths and hospitalized 320. In general, however, hydrogen sulfide is a serious pollutant on only a local scale.

The Los Angeles type of smog is a complex phenomenon and involves photochemical reactions in which nitrogen oxides play a major role. Nitrogen oxides in the atmosphere result primarily from combustion processes as in automobiles, boilers, furnaces, etc. A small amount results from lightning. The formation of NO is a nitrogen fixation process. At a temperature of 2200°C as in the spark of a combustion engine, the equilibrium concentration of NO in the reaction

$$N_2 + O_2 \longrightarrow 2NO \tag{3-39}$$

is quite high. When the gas is cooled to 20°C the equilibrium concentration is very low, but on the other hand equilibrium is not reached rapidly in this case, and a relatively high NO content can remain. NO can then take part in the generation of NO_2 by a process such as

$$RO_2 + NO = RO + NO_2 \tag{3-40}$$

where R stands for free radicals which are always present in polluted air containing hydrocarbons. NO_2 can then lead to the production of ozone by the processes

$$NO_2 + h\nu = NO + O \tag{3-41}$$

$$O + O_2 = O_3 \tag{3-42}$$

where $h\nu$ represents an incident photon of solar energy and O_3 is the ozone molecule. Ozone content is commonly taken as a measure of pollution, and values of 0.2 ppm may be encountered in light smog and of 0.4 ppm in heavy smog. The essential ingredients in producing the type of smog considered in this paragraph are seen to be nitrogen oxides, hydrocarbons, and sunlight, all present liberally in the Los Angeles basin. The results of this combination of pollutants include severe eye irritation, lowered visibility, and damage to plants.

In addition to their participation in photochemical smog, the nitrogen oxides have direct noxious effects on people. Nitrogen dioxide, which is yellow-brown in color, is especially irritating and effective in reducing visibility. The smoke from cigarettes, tobacco, and cigars contains several hundred parts per million of nitrogen dioxide. The Los Angeles Pollution Control District has set 3 ppm of oxides of nitrogen as the first alert level. The nitrogen cycle in nature is discussed in Sec. 3-4.

Carbon monoxide is almost entirely a man-made pollutant. It is tasteless and odorless but causes damage by combining with hemoglobin, the substance in the red blood cells normally responsible for carrying oxygen throughout the body. The effect of carbon monoxide in this case is to lower the oxygen-carrying capacity of the blood. It is toxic to humans at concentrations of 100 ppm, and the federal government considers that values above its standard of 9 ppm are hazardous to health. Values near Los Angeles freeways having slow heavy traffic have been as high as 54 ppm, and the level in Denver in the 1975–1976 winter reached a 8-h average concentration of 27.8 ppm on one occasion and exceeded the 9 ppm standard on at least 78 days. The automobile is a major producer of carbon monoxide. The carbon monoxide in cigarette smoke is one of its hazards. Persons driving over the high mountain passes of Colorado should be especially careful to have adequate ventilation in the car to prevent the buildup of carbon monoxide.

Carbon dioxide, which results from complete combustion as distinguished from carbon monoxide, is increasing in the earth's atmosphere as the result of the increasing combustion of fossil fuel. It is not poisonous but could possibly affect the earth's climate. The cycle of carbon dioxide is discussed in Sec. 3-4.

Particles in the earth's atmosphere are of both natural and man-made origin. A listing by sources, following generally that of *Man's Impact on the Global Environment* [42], is

1. Natural continental aerosols (using the term aerosol interchangeably with particle)
   - 1.1. From dust storms and desert areas, with size range above about 0.6 μm in diameter
   - 1.2. From photochemical gas reactions between ozone and hydrocarbons from plants, resulting in very small particles of less than 0.4-μm diameter
   - 1.3. From photochemical reactions between trace gases such as SO_2, H_2S, NH_3, and O_3 or O (influenced strongly by humidity or the presence of cloud droplets)

1.4. From volcanic eruptions which emit particles of all sizes and trace gases (especially SO_2) that subsequently can become particles in the troposphere
1.5. From forest fires of natural origin

2. Natural oceanic aerosols: From evaporation of ocean spray (salt particles with size above about 0.6 μm in diameter)

3. Man-made aerosols and fumes:
3.1. Smoke and dust from combustion and other activities of man, including burning for clearing land, forest fires caused by man, and man-induced soil erosion, etc.
3.2. From photochemical reactions between unburned or partially burned organic fuel (e.g., gasoline) and oxides of nitrogen (small particles of less than 0.4-μm diameter)
3.3. Photochemical reactions between SO_2 and O_3 or O (essentially same reactions as those acting on natural SO_2).

Industrially advanced nations tend to pollute the air most seriously with particles from industrial sources, automobiles, etc. Developing countries, which have little industry and which may not be able to carry out proper soil management practices, tend to pollute with dust from man-induced wind erosion, etc. In the 1930's the United States also experienced severe dust storms in the Great Plains dust bowl, but conditions in that area have been largely stabilized since then.

The removal of particles from the air is accomplished primarily by rain and snow, although some gravitational settling and contact with surfaces take place. Representative lifetimes in the lower troposphere range from 6 days to 2 weeks or more. In the upper troposphere 2–4 weeks may be required, in the lower stratosphere the time period may be 6 months to a year, and 3–5 years may be needed in the upper stratosphere. The number of small particles (less than 0.2 μm) falls off with altitude, but the number of larger particles (0.2–2.0 μm) has a maximum at 18 km. The larger higher particles are believed to be derived from SO_2 emitted by volcanoes, especially in the tropics. The very large particles in the lower troposphere (10 μm in diameter or more) are usually solid dirt or flyash particles which settle out of the air rapidly. Smoke and fumes may be either solid or liquid and range from 5 to 0.1 μm or smaller. Cigarette-smoke particles are about 0.2 μm in diameter.

The effects of particles are to reduce visibility, to reduce the intensity of sunlight reaching the earth, to soil and damage materials and vegetation, and to be harmful to health. (Effects on electromagnetic radiation are considered quantitatively in Sec. 4-5.) The larger particles tend to lodge in the nostrils or be collected by the mucous lining of the upper airway. The smaller particles, as in cigarette smoke, penetrate more deeply into the lungs. Harm can result from excessive intake of particles of any kind; airway resistance increases in humans when exposed to a number of inert particles. Some particles are harmful by their chemical nature. Chimney sweeps as early as 1775 were found to have developed cancer of the scrotum due to aromatic hydrocarbon carcinogens in chimney soot. Sulfuric acid mist damages plants. Rates of lung cancer are higher in metropolitan than in rural areas. Smoking is a more serious cause of lung cancer than air pollution, but the combination of air pollution in general and smoking causes a higher lung cancer incidence than smoking alone.

Lead aerosol is a common contaminant in urban areas of the United States, amounts ranging from average annual values of 1–3 $\mu g/m^3$ to as high as 25 $\mu g/m^3$ over Los Angeles freeways and 44 $\mu g/m^3$ in a vehicular tunnel for relatively short periods of time. Lead may be ingested by mouth in which case 5–10% may be retained. Lead aerosols are about 1 μm in diameter and thus enter the lung readily. Up to 50% of the inhaled lead may be retained. Lead aerosols also find their way into water and soil. Rainwater has been found in some cases to have twice the allowable standard for drinking water of 0.05 mg/l. Lead found in cigarette tobacco is about in the same proportion as arsenic and is believed to result from lead arsenate sprays. Even though such sprays have not been used for years, the soil in which the tobacco is grown retains the earlier contamination. A major source of atmospheric lead is the tetraethyl lead added to gasoline, in which case the lead is emitted as lead bromides, chlorides, etc.

The SCEP group recommends monitoring of lead and other possibly harmful metals in the environment including mercury, beryllium, cadmium, etc. These may occur as atmospheric dusts or aerosols, in some chemical form, as well as in soil, water, and food. Pesticides such as DDT and other toxic chlorinated hydrocarbons are also found in the atmosphere at least locally and briefly at the time of application.

3-3-4 The Ozone Layer

It was mentioned in Sec. 3-2-1 that temperature increases with altitude in the stratosphere because of the absorption of ultraviolet radiation by ozone, and possibly harmful effects of ultraviolet radiation were mentioned in Sec. 2-5. Concern has been expressed in recent years about a possible depletion of ozone content in the stratosphere, or in what can be called the ozone layer, due to supersonic aircraft [31] or due to fluorocarbons [41]. Supersonic aircraft would inject NO into the stratosphere, and NO can lead to destruction of ozone as indicated in the following two equations:

$$NO + O_3 \longrightarrow NO_2 + O_2 \tag{3-43}$$

$$NO_2 + O \longrightarrow NO + O_2 \tag{3-44}$$

One molecule of NO can be recycled so as to destroy a number of molecules of O_3. In the case of the fluorocarbons, which may diffuse to the stratosphere after being emitted from spray cans, pertinent reactions are

$$h\nu + CFCl_3 \longrightarrow CFCl_2 + Cl \tag{3-45}$$

$$h\nu + CFCl_2 \longrightarrow CF_2Cl + Cl \tag{3-46}$$

$$Cl + O_3 \longrightarrow ClO + O_2 \tag{3-47}$$

$$ClO + O \longrightarrow Cl + O_2 \tag{3-48}$$

$h\nu$ stands for the energy of an incident photon. Chlorine, derived from the fluorocarbons, can thus lead to destruction of ozone in much the same way as NO. It is difficult, however, to calculate very closely how much NO and Cl may be introduced into the stratosphere and what extent of ozone depletion will result from given amounts of NO or fluorocarbons. Factors that must be taken into account are that some concentration of NO occurs naturally and also that there are natural sources of chlorine in the stratosphere. The natural source of chlorine is primarily hydrogen chloride emitted during volcanic eruptions [9]. The motors of the space shuttle will also emit hydrogen chloride into the stratosphere.

Another potential mechanism for depletion of the ozone layer, though one apparently occurring in geologic history under a special combination of circumstances, has been proposed [37]. This mechanism involves the production of NO by dissociative ionization of N_2, followed by the reaction of N atoms with oxygen. The ionization of N_2 is caused by an influx of high-energy solar protons which are able to penetrate to the stratosphere at a time of reversal of the earth's magnetic field, when the magnetic field which ordinarily diverts the protons to high latitudes is near zero intensity. The earth's field has reversed on a number of occasions, and the occurrence of solar flares at such times would provide a source of solar protons, or solar cosmic rays as they are sometimes called. It has been suggested that the increase in ultraviolet radiation at the surface of the earth, resulting from the production of NO and consequent depletion of ozone, has been responsible for faunal extinctions which have been observed in fossil records, particularly of Radiolaria (single-celled marine microorganisms).

The National Academy of Sciences 1971 report on the effects of stratospheric flight gave much attention to the biological effects of ultraviolet radiation. The 290- to 320-nm wavelength range of ultraviolet radiation would be increased in intensity at the earth's surface by the depletion of the ozone layer; this range is approximately that which is classified as the UV-*B* range. Although solar radiation is necessary for life on the earth, the Academy report states that nearly all known effects of UV-*B* radiation are deleterious. It is this wavelength range which is believed to cause skin cancer. Experiments have indicated also that supplementary ultraviolet radiation, especially UV-*B*, inhibits the growth of plants and causes degenerative changes in the cell structure of plants.

3-4 BIOGEOCHEMICAL CYCLES

3-4-1 Carbon Dioxide and Oxygen

The cycles considered in this section are of interest in the study of the atmosphere, and they involve the solid earth, the oceans, and living matter as well. Ecology is among the fields that are concerned with such cycles; for example, see the texts on ecology by Odum [33,34]. Carbon dioxide and oxygen will be considered together here because of their interrelationships.

The process of photosynthesis in higher plants can be represented in simple form by

$$CO_2 + 2H_2O + h\nu \longrightarrow (CH_2O) + H_2O + O_2 \qquad (3\text{-}49)$$

The parentheses indicate that a number of units of CH_2O form the final product, which might be, for example, $C_6H_{12}O_6$. Some bacteria, namely the green and purple sulfur bacteria, carry out the same process except that hydrogen sulfide is used in place of water and free sulfur results as an end product instead of free oxygen. In this case

$$CO_2 + 2H_2S + h\nu \longrightarrow (CH_2O) + H_2O + 2S \qquad (3\text{-}50)$$

If $C_6H_{12}O_6$ is taken as the product in which carbon is stored and quantities are shown, one can write

$$CO_2 + 2H_2O + h\nu \longrightarrow C_6H_{12}O_6 + H_2O + O_2 \qquad (3\text{-}51)$$

CO_2	$2H_2O$	$h\nu$	$C_6H_{12}O_6$	H_2O	O_2
6 mol, 264 g	12 mol, 216 g	0.785 kWh	1 mol, 180 g	6 mol, 108 g	6 mol, 192 g

Energy is supplied as photons of solar radiation, and the amount of such energy required to produce 6 g mol of oxygen is about 0.785 kWh or 675 kcal.

The reverse process of respiration (a form of oxidation), carried on by both plants and animals, is

$$CH_2O + O_2 \longrightarrow CO_2 + H_2O + \text{energy} \qquad (3\text{-}52)$$

Plants utilize carbon dioxide and produce oxygen by the process of photosynthesis, which is carried out in sunlight. They also respire, a process which uses oxygen and produces carbon dioxide. Both photosynthesis and respiration take place in the daytime, but photosynthesis tends to dominate in the daytime, and only respiration takes place at night. Decay of plant and animal products, for example the decay of leaves which have fallen on the ground, can be regarded as a form of respiration. Decay takes place commonly in the soil, and the term soil respiration is applied in this case. Another less efficient process of obtaining energy is fermentation, which is carried out without oxygen, as in

$$C_6H_{12}O_6 \longrightarrow 2C_2H_5OH + 2CO_2 + \text{energy} \qquad (3\text{-}53)$$

This process is anaerobic and proceeds without air. Oxidation of glucose yields 686 kcal/g·mol, while the fermentation of glucose yields only 50 kcal/g·mol. Not all of the carbon fixed in the photosynthesis process is released quickly. Some may be stored in the body of the plant for many years, and some may be converted eventually to coal and oil (fossil fuel).

It is commonly considered that the earth's atmosphere has developed from gases emitted from the earth's interior, such as water vapor and carbon dioxide [16,25]. These gases contain no free oxygen. The latter has been formed partly from photodissociation of water vapor but mainly from photosynthesis in excess of respiration or decay of the products of photosynthesis, according to this theory. The original water vapor and carbon dioxide have been largely removed from the atmosphere to form the oceans and deposits of carbonate rocks. Nitrogen has become the principal constituent of the atmosphere because it is relatively inert and has accumulated. One of the important effects of oxygen, in the ozone form, is to screen out ultraviolet radiation. It was perhaps not until sufficient ozone had developed that extensive photosynthesis could take place at the surface of the earth, and the early forms of life probably developed at sufficient depths in water so that they were shielded by water from ultraviolet radiation. Also the very early forms of life at a time when little oxygen was available must have been anaerobic.

The average concentration of carbon dioxide in the atmosphere in 1969 was about 320 ppm, but the concentration at noon at treetop level may be as low as 305, whereas at night the value may rise to as high as 400 ppm at the ground. In addition to this diurnal variation there is an annual variation of about 6 ppm at Mauna Loa, Hawaii and about twice that in Scandinavia and Alaska, the concentration tending to diminish in the northern hemisphere in summer and increase in the winter. The average value recorded at Mauna Loa increased from about 313 ppm in 1959 to 330 ppm in 1974, as shown in Fig. 3-16. From 1958 to 1968, the average increase was 0.64 ppm/year compared with a value of 1.24 ppm/year that would have resulted from the combustion of fossil fuel if all the CO_2 so produced had remained airborne. Measurements at other locations show increases comparable to those at Mauna Loa [5].

The increase in CO_2 content of the atmosphere has caused consideration of possible climatic change, involving an increase of average temperature of the earth because of an increased greenhouse effect [5, 26, 44]. At the same time there has been thought that the average temperature may decrease because of increased particulate content of the atmosphere or because of natural variations [8]. The increase in CO_2 content has been commonly blamed primarily on man's consumption of fossil fuels, but consideration has also been given more recently to effects due to the expansion of agriculture and forestry [2,7]. The clearing of forest land for agriculture and the burning of the trees for disposal or of wood in general for fuel appear to have a dual

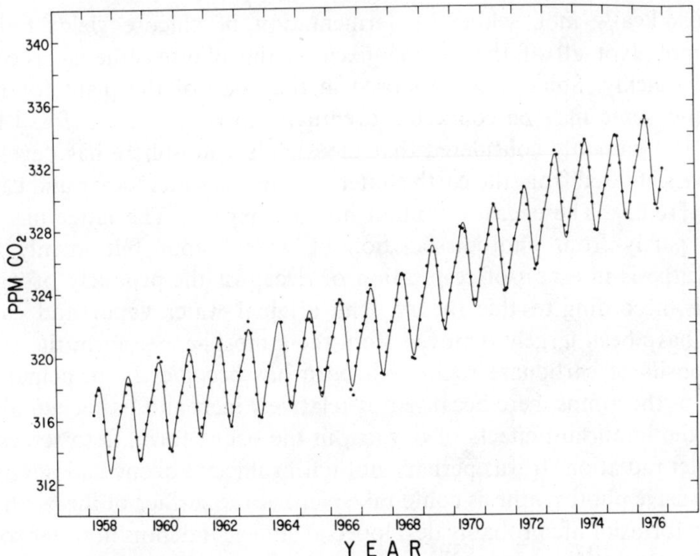

FIGURE 3-16. Monthly average values of CO_2 content in the atmosphere at Mauna Loa Observatory, Hawaii. (1958–1971 values are from C. D. Keeling et al., "Atmospheric Carbon Dioxide Variations at Mauna Loa Observatory, Hawaii," *Tellus*, vol. 28, No. 6, pp. 538–551, 1976; 1971–1976 values are from unpublished data supplied by C. D. Keeling. A companion paper in the same issue of *Tellus*, "Atmospheric Carbon Dioxide Variations at the South Pole," pp. 552–564, provides data showing smaller annual variations than in Hawaii but essentially the same buildup over the years.)

effect. Most obviously, the burning of wood adds directly to the CO_2 content of the atmosphere in the same way as does the burning of fossil fuel. Deforestation also decreases the forest area which is available for removing CO_2 from the atmosphere. Of course crops as well as forest trees use CO_2, but trees store a larger amount of CO_2 in wood and develop a larger biomass per unit area than crops.

3-4-2 Nitrogen

Nitrogen, a constituent of protein, is essential to life, as are carbon, oxygen, sulfur, etc. Although nitrogen constitutes 79% of the atmosphere, it cannot be used directly by the majority of living things. Some microorganisms, namely some bacteria and blue-green algae, are able to fix nitrogen (provide nitrogen in the form of a chemical compound which can be used by other living organsims). Some of the microorganisms are free-living, but others live in symbiotic association with fungi and higher plants. Certain lichens (those which are symbiotic associations of blue-green algae and fungi) are capable of fixing nitrogen. Among the higher plants forming symbiotic associations with nitrogen-fixing bacteria are legumes, alders, *Ceanothus*, cycads, and ginkgos. A small amount of nitrogen is also fixed by lightning and ionizing radiations. Natural humus contains nitrogen in a useful form. Naturally occurring nitrate deposits in Chile were at one time the major source of fixed nitrogen for fertilizer and explosives. They are still an important source, but variations of the Haber-Busch catalytic fixation process developed during World War I now supplies the bulk of fixed nitrogen, originally as NH_3.

The nitrogen oxides fixed by man-controlled combustion processes are an important atmospheric pollutant on a local scale. In some areas the nitrogen oxides from automobiles contribute to high nitrogen levels in the nearby land and waters. Perhaps more important, however, are the effects of the large-scale industrial fixation of nitrogen and the use of nitrogen compounds on a large scale for fertilizer. One of the major steps which have been taken to increase food supplies is the extensive use of fertilizer. Nitrogen, phosphorus, and potassium, in the form of suitable compounds, are important elements of fertilizers.

In the natural state the fixation of nitrogen by nitrogen-fixing bacteria is balanced by the action of denitrifying bacteria. At the present, however, industrial fixation supplies an amount of fixed nitrogen approximately equal to that produced naturally before the advent of modern agriculture. Legume crops produce an additional amount. As a result it is estimated [17] that fixation of nitrogen exceeds denitrification by about 10%. The increase in fixed nitrogen is especially significant in lakes, where it causes eutrophication. Also the nitrate content of some soils is said to have increased to dangerous levels in some areas.

PROBLEMS

3-1 a. Compare the pressures at a height of 40 km as given by Table 3-1 and as determined by starting with the sea-level pressure p_0 using the expression $p = p_0 e^{-h/H}$.

b. Make the same comparison as for part a (for pressures at a height of 40 km only), again starting with sea-level pressure but breaking the exponential factor into four factors corresponding to the four 10-km layers from 0 to 40 km.

For both parts a and b, use average values of H for the 40- or 10-km intervals or layers considered. (For example, for part a, where the height interval utilized is 40 km, a suitable average value can be taken to be the value for H for 20 km.) In calculating H, use values of g that are corrected for the altitudes in question in order to obtain as much accuracy as possible.

3-2 a. Determine the electron density at a given level in the ionosphere assuming an equilibrium between production and recombination and assuming a production rate q of 5000/cm^3/s and a recombination coefficient α of 10^{-7}/cm^3/s.

b. Assuming that the value for electron density calculated in part a is the electron density at $h = 110$ km and is the peak electron d nsity of a Chapman layer, plot the electron density of this theoretical layer as a function of z and h for $\chi = 0°$. (Use the value of H corresponding to 110 km.)

3-3 Determine the approximate height of maximum production of ionization due to incident solar radiation for the case of atmospheric particles having a cross section of 2×10^{-18} cm^2.

3-4 If the total earth's magnetic field at the surface of the earth is taken to be 0.5 G, calculate the angular gyrofrequency and gyrofrequency in Hz for electrons at the surface, at a height of 300 km, and at a height of one earth radius, assuming a dipole field. (The heights are heights above the earth's surface.)

3-5 List the values of f_B in Hz for electrons and H^+ ions at the earth's surface and at distances of 3, 6, and 9 earth radii from the surface (at 1, 4, 7, and 10 earth radii from the center of the earth).

3-6 Calculate the distance to the edge of the magnetosphere for a solar wind involving a velocity of 500 km/s, a number density of 50/cm^3, and ionized hydrogen atoms, assuming the value of B_0 to be 0.5 G.

REFERENCES

[1] AAAS, *Air Conservation.* Washington, D.C.: AAAS, 1969.

[2] ADAMS, J. A. S., M. S. M. MANTOVANI, and L. L. LUNDELL, "Wood versus fossil fuel as a source of excess carbon dioxide in the atmosphere: a preliminary report," *Science*, vol. 196, pp. 54–56, April 1, 1977.

[3] AKASOFU, S. I., *Polar and Magnetospheric Substorms.* New York: Springer-Verlag New York, Inc., 1968.

[4] American Chemical Society, *Cleaning Our Environment. The Chemical Basis for Action.* Washington, D.C.: American Chemical Society, 1969.

[5] BAES, C. F., Jr., H. E. GOELLER, J. S. OLSON, and R. M. ROTTY, "Carbon dioxide and climate: the uncontrolled experiment," *American Scientist*, vol. 65, pp. 310–320, May/June 1977.

[6] BARTELS, J., "Solar eruptions and their ionospheric effects—a classical observation and its new interpretation," *Terrestrial Magnetism and Atmospheric Electricity*, vol. 42, pp. 235–239, 1937.

[7] BOLIN, B., "Changes of land biota and their importance to the carbon cycle," *Science*, vol. 196, pp. 613–615, May 6, 1977.

[8] BRYSON, R. A., and T. J. MURRAY, *Climates of Hunger*. Madison, WI: University of Wisconsin Press, 1977.

[9] CADLE, R. D., C. S. KIANG, and J. F. LOUIS, "The global scale dispersion of eruption clouds from major volcanic eruptions," *Journal of Geophysical Research*, vol. 81, pp. 3125–3132, June 20, 1976.

[10] CAHILL, L. J., "The magnetosphere," *Scientific American*, vol. 212, pp. 58–68, March 1965.

[11] CHAPMAN, S., "The absorption and dissociative or ionizing effect of monochromatic radiation in an atmosphere on a rotating earth," *Proc. Physical Society* (*London*), vol. 43, I: pp. 26–45, II: pp. 483–501, 1931.

[12] CHAMPMAN, S., and V. C. A. FERRARO, "A new theory of magnetic storms, I. The initial phase," *Terrestrial Magnetism and Atmospheric Electricity*, vol. 36, pp. 77–97, 171–186, vol. 37, pp. 147–156, 421–429; "II. The main phase," vol. 38, pp. 79–96, 1931–1933.

[13] CHAPMAN, S., and J. BARTELS, *Geomagnetism*, Vols. I and II. London: Oxford University Press, 1940.

[14] CHAPMAN, S., and R. S. LINDZEN, *Atmospheric Tides*. New York: Gordon & Breach, 1970.

[15] CIRA, *COSPAR International Reference Atmosphere*. Amsterdam: North-Holland, 1965.

[16] CLOUD, P., and A. GIBOR, "The oxygen cycle." *Scientific American*, vol. 223, pp. 110–123, Sept. 1970.

[17] DELWICHE, C. C., "The nitrogen cycle," *Scientific American*, vol. 223, pp. 136–146, Sept. 1970.

[18] DONN, W. L., *Meteorology*, 4th ed. New York: McGraw-Hill, 1975.

[19] EVANS, J. V., "Ground-based measurements of atmospheric and ionospheric particle temperatures," in *Solar-Terrestrial Physics* (J. W. King and W. S. Newman, eds.), pp. 289–340. New York: Academic Press, 1967.

[20] HESS, W. N., "The earth's radiation belt," in *Introduction to Space Science* (W. N. Hess, ed.), pp. 165–203. New York: Gordon & Breach, 1965.

[21] HESS, W. N., and G. D. MEAD, "The boundary of the magnetosphere," in *Introduction to Space Science* (W. N. Hess, ed.), pp. 347–381. New York: Gordon & Breach, 1965.

[22] HINES, C. O., "Motions of the neutral atmosphere," in *Physics of the Earth's Upper Atmosphere* (C. O. Hines, I. Paghis, T. R. Hartz, and J. A. Fejer, eds.), pp. 134–156. Englewood Cliffs, NJ: Prentice-Hall, 1965.

[23] HINES, C. O., et al., *The Upper Atmosphere in Motion*, Geophysical Monograph 18. Washington, D.C.: American Geophysical Union, 1974.

[24] JOHNSON, F. S. (ed.), *Satellite Environment Handbook*, 2nd ed. Stanford, CA: Stanford University Press, 1965.

[25] JOHNSON, F. S., "The balance of atmospheric oxygen and carbon dioxide," *Biological Conservation*, vol. 2, pp. 83–89, Jan. 1970.

[26] KELLOGG, W. W., "Global influences of mankind on the climate," in *Climate Change* (J. Gribbin, ed.). Cambridge, England: Cambridge University Press.

[27] Little, C. G., and H. Leinbach, "The riometer—a device for continuous measurement of ionospheric absorption," *Proc. IRE*, vol. 47, pp. 315–320, Feb. 1959.

[28] Lüst, R., "The properties of interplanetary space," in *Solar-Terrestrial Physics* (J. W. King and W. S. Newman, eds.), pp. 1–44. New York: Academic Press, 1967.

[29] Matsushita, S., and W. H. Campbell (eds.), *Physics of Geomagnetic Phenomena.* New York: Academic Press, 1967.

[30] Mitra, S. K., *The Upper Atmosphere.* Calcutta: The Asiatic Society, 1952.

[31] National Academy of Sciences, *Environmental Impact of Stratospheric Flight.* Washington, D.C.: National Academy of Sciences, 1975.

[32] Ness, N. F., "Observations of the interaction of the solar wind with the geomagnetic field during quiet conditions," in *Solar-Terrestrial Physics* (J. W. King and W. S. Newman, eds.), pp. 57–89. New York: Academic Press, 1967.

[33] Odum, E. P., *Ecology.* New York: Holt, Rinehart and Winston, 1963.

[34] Odum, E. P., *Fundamentals of Ecology*, 3rd ed. Philadelphia: Saunders, 1971.

[35] Page, L., *Introduction to Theoretical Physics.* New York: Van Nostrand Reinhold, 1935.

[36] Ratcliffe, J. A., *An Introduction to the Ionosphere and Magnetosphere.* Cambridge, England: Cambridge University Press, 1972.

[37] Reid, G. C., "Ionospheric disturbances," in *Physics of Geomagnetic Phenomena* (S. Matsushita and W. H. Campbell, eds.), pp. 627–662. New York: Academic Press, 1967.

[38] Reid, G. C., I. S. A. Isaksen, T. E. Holzer, and P. J. Crutzen, "Influence of ancient solar-proton events on the evolution of life," *Nature*, vol. 259, pp. 177–179, Jan. 22, 1976.

[39] Riehl, H., *Introduction to the Atmosphere.* New York: McGraw-Hill, 1972.

[40] Rishbeth, H., and O. K. Garriott, *Introduction to Ionospheric Physics.* New York: Academic Press, 1969.

[41] Rowland, F. S., and M. J. Molina, "Chlorofluoromethanes in the environment," *Reviews of Geophysics and Space Physics*, vol. 13, pp. 1–36, 1975.

[42] SCEP (Study of Critical Environmental Problems), *Man's Impact on the Global Environment.* Cambridge, MA: M.I.T. Press, 1970.

[43] Scorer, R., *Air Pollution.* Elmsford, NY: Pergamon, 1968.

[44] SMIC (Study of Man's Impact on Climate), *Inadvertent Climate Modification.* Cambridge, MA: M.I.T. Press, 1971.

[45] *U.S. Standard Atmosphere, 1976*, sponsored by NOAA, NASA, USAF. Washington, D.C.: Supt. of Documents, U.S. Government Printing Office, 1976.

[46] White, O. R. (ed.), *The Solar Output and its Variation.* Boulder, CO: Colorado Associated University Press, 1977.

[47] Williams, D. J., "SELDADS: an operational real-time solar-terrestrial environment monitoring system," *NOAA Technical Report ERL 357-SEL 37.* Boulder, CO: NOAA, March 1976.

4-1 SIGNIFICANCE OF INDEX OF REFRACTION AND RELATIVE DIELECTRIC CONSTANT

Many of the important applications of electromagnetic waves for communication and scientific purposes involve transmission in the earth's atmosphere. The atmosphere may reflect or scatter electromagnetic waves, it may cause bending of waves and changes in polarization, and it may attenuate electromagnetic waves. These effects are functions of frequency. Bending, reflection, and scattering are caused by variations in index of refraction. In this section the significance of the index of refraction and how it is calculated are discussed. To do this satisfactorily requires brief reference to Maxwell's equations and the elementary theory of electromagnetic waves. In the lossless case and when the medium has no magnetic properties so that the magnetic permittivity μ equals μ_0, n, the index of refraction, is related to K, the relative dielectric constant, by the simple relation

$$n = \sqrt{K} \tag{4-1}$$

Thus any remarks made about the effects of the index of refraction can be readily interpreted in terms of relative dielectric constant and vice versa.

The index of refraction n of a particular type of wave in a given medium is the ratio of phase velocity of an electromagnetic wave in a vacuum to phase velocity of the wave in the medium. Thus

$$n = \frac{c}{v_p} \tag{4-2}$$

ATMOSPHERIC EFFECTS ON ELECTROMAGNETIC WAVES

4

and $v_p = c/n$, where c is 3×10^8 m/s, the phase velocity of an electromagnetic wave in a vacuum, and v_p is phase velocity in the medium. The propagation constant k of an electromagnetic wave is also influenced by the value of the index of refraction. k can be expressed in the following ways:

$$k = \frac{2\pi}{\lambda} = \frac{\omega}{v_p} = \omega\sqrt{\mu_0 \varepsilon_0 K} = k_0 n \tag{4-3}$$

where ω is angular frequency, λ is wavelength, μ_0 is magnetic permittivity of free space ($4\pi \times 10^{-7}$ H/m), ε_0 is electric permittivity of free space (8.854×10^{-12} F/m), and $k_0 = \omega\sqrt{\mu_0 \varepsilon_0}$. Note also that $c = 1/\sqrt{\mu_0 \varepsilon_0}$ and $\lambda = \lambda_0/n$, where λ_0 is free-space wavelength. The significance of the propagation constant k in the case of a plane electromagnetic wave propagating in the z direction, for example, is that the electric field intensity E, having units of V/m, varies as indicated by $E = E_0 e^{-jkz}$, where j is the square root of -1 and e is the natural base of logarithms. The propagation constant k can in general be considered to have a real part β and an imaginary part α, which accounts for losses, so that $k = \beta - j\alpha$. When α is zero, $k = \beta$, and the two quantities can be used interchangeably.

Snell's law is an important relation which involves n. At a plane boundary between two media, Snell's law can be expressed as

$$n_1 \cos \theta_1 = n_2 \cos \theta_2 \tag{4-4}$$

where n_1 and n_2 are the indices in the two media and θ_1 and θ_2 are the directions of the electromagnetic waves in the two media and are measured from the boundary line. For an electromagnetic wave perpendicularly incident from medium 1 onto a plane boundary and in the case that medium 2 extends to infinity,

$$\rho = \frac{n_1 - n_2}{n_1 + n_2} \tag{4-5}$$

where ρ is the reflection coefficient. These ways in which the index of refraction affects electromagnetic waves illustrate its significance to the subject of the atmosphere as a communications medium.

To show analytically how the index of refraction affects wave propagation, Maxwell's equations, the basic equations governing electromagnetic theory, are now presented. Maxwell's equations in differential and integral form are

$$\nabla \cdot \mathbf{D} = \rho \tag{4-6a}$$

$$\oint \mathbf{D} \cdot \mathbf{dS} = q \tag{4-6b}$$

$$\nabla \cdot \mathbf{B} = 0 \tag{4-7a}$$

$$\oint \mathbf{B} \cdot \mathbf{dS} = 0 \tag{4-7b}$$

$$\nabla \times \mathbf{E} = \frac{-\partial \mathbf{B}}{\partial t} \tag{4-8a}$$

$$\oint \mathbf{E} \cdot \mathbf{d}\mathbf{l} = -\frac{d}{dt}\int \mathbf{B} \cdot \mathbf{dS} \tag{4-8b}$$

$$\nabla \times \mathbf{H} = \mathbf{J} + \frac{\partial \mathbf{D}}{\partial t} \tag{4-9a}$$

$$\oint \mathbf{H} \cdot \mathbf{d}\mathbf{l} = I + \frac{d}{dt}\int \mathbf{D} \cdot \mathbf{dS} \tag{4-9b}$$

where $\mathbf{D}$ = electric flux density in C/m^2,
$\mathbf{B}$ = magnetic flux density in Wb/m^2,
$\mathbf{E}$ = electric field intensity in V/m,
$\mathbf{H}$ = magnetic field intensity in A/m,
ρ = charge density in C/m^3,
q = charge in C,
$\mathbf{J}$ = current density in A/m^2,
I = current in A,
$\mathbf{dS}$ = element of surface,
$\mathbf{d}l$ = element of length.

Vector quantities, indicated by bold type in the equations, have both a magnitude and a direction in space, in contrast to scalar quantities which have only a magnitude. Force is a familiar example of a vector quantity; temperature is an example of a scalar. We will have occasion, however, to refer to the magnitude of vectors and will not use bold type for that purpose. For example, $\mathbf{E}$ represents the vector, electric field intensity, and E represents the magnitude, or amplitude, of $\mathbf{E}$. Maxwell's equations are the basis of electromagnetic theory, and the equations and their applications are treated in a number of texts [15, 16, 26, 27, 32].

Making use of the relation $\mathbf{D} = \varepsilon_0\mathbf{E} + \mathbf{P}$, where $\mathbf{P}$ is electric dipole moment per unit volume, the $\nabla \times \mathbf{H}$ equation may be written as

$$\nabla \times \mathbf{H} = \mathbf{J} + \frac{\partial(\varepsilon_0 \mathbf{E})}{\partial t} + \frac{\partial \mathbf{P}}{\partial t} \tag{4-10}$$

The quantities in the equation vary sinusoidally with time in the situations we are interested in here. In this case, differentiation with respect to time may be replaced by multiplication by $j\omega$ and

$$\nabla \times \mathbf{H} = \mathbf{J} + j\omega\varepsilon_0\mathbf{E} + j\omega\mathbf{P} \tag{4-11}$$

where the quantities should now be understood to be complex quantities having an amplitude and phase in contrast to Eq. (4-10) where the quantities are real variables. (Sometimes a caret is used to indicate a complex quantity, as in $\hat{\mathbf{E}}$.) The terms on the right-hand side represent current densities. $\mathbf{J}$ is conduction or convection current density in A/m^2. $j\omega\varepsilon_0\mathbf{E}$ is vacuum displacement current density, and the factor j shows that vacuum displacement current is 90° out of phase with $\mathbf{E}$. $j\omega\mathbf{P}$ is polarization current density, which is also 90° out of phase with $\mathbf{E}$.

In this treatment of propagation in the troposphere and ionosphere, loss processes will be neglected initially. Thus any component of current density that is in phase with $\mathbf{E}$ is neglected. In particular, conduction current density, which is equal to $\sigma\mathbf{E}$, where σ is conductivity (mhos per meter), will be considered to be unimportant. In this case Eq. (4-11) can be written as

$$\nabla \times \mathbf{H} = j\omega\varepsilon_0 K\mathbf{E} \tag{4-12}$$

where K is relative dielectric constant and is real. Dielectric materials have a finite value of **P** which causes K to be greater than unity. For ordinary solid dielectrics K may be in the range of about 2–6. In the gaseous troposphere, however, K is only slightly greater than unity. In the case of a plasma it can be considered that a finite value of **P** exists also, but an alternative approach is to consider **P** to be zero and to calculate a convection current density **J**. In a lossless isotropic plasma, **J** will be 90° out of phase with **E** but also 180° out of phase with $j\omega\varepsilon_0\mathbf{E}$ so that K will be less than unity. In the following section K will be calculated for such a plasma by using Eqs. (4-11) and (4-12) with **P** set equal to zero in Eq. (4-11).

The wave equation for electromagnetic waves can be formed by taking the curl of Eq. (4-8a) and then substituting Eq. (4-12) for $\nabla \times \mathbf{H}$. By assuming sinusoidal time dependence the resulting wave equation has the form

$$\nabla \times (\nabla \times \mathbf{E}) - \omega^2 \mu_0 \varepsilon_0 K \mathbf{E} = 0 \tag{4-13}$$

The phase velocity of the wave is given by

$$v_p = \frac{1}{\sqrt{\mu_0 \varepsilon_0 K}} = \frac{c}{\sqrt{K}} = \frac{c}{n}$$

as stated earlier.

The above forms of Maxwell's equation and the wave equation are written in terms of the relative dielectric constant K, but as $n^2 = K$, n^2 can be substituted for K to show how the equations involve n.

The concept of index of refraction can also be applied to acoustic waves. In this case the index of refraction is the ratio of some reference velocity v_0 to the actual acoustic velocity v_s, or

$$n = \frac{v_0}{v_s} \tag{4-14}$$

Acoustic waves do not exist in a vacuum, as is the case for electromagnetic waves, but the velocity in a standard atmosphere or that for the ambient conditions can be used as the acoustic reference velocity. The index of refraction then indicates the effects of departures from these standard or ambient conditions, and it has much the same significance for acoustic waves as for electromagnetic waves. The term *dielectric constant*, however, is applied only to electromagnetic waves.

4-2 TROPOSPHERIC PROPAGATION

4-2-1 Indices of Refraction for Radio, Optical, and Acoustic Waves

The index of refraction or refractivity of the troposphere for radio waves is a function of temperature, pressure, and water vapor content. Variations in the index and the fact that it differs from unity, though only slightly so, may result in significant effects on radio waves. In considering the index of refraction of the troposphere a possible starting point is the Clausius-Mossoti expression for the relative dielectric constant of a gas, with a Debye relaxation term, $1 + j\omega\tau$, included [26]. This expression is

$$\frac{K-1}{K+2} = \frac{N_m}{3}\left[\alpha_0 + \frac{p^2}{3kT(1+j\omega\tau)}\right] \tag{4-15}$$

where K is relative dielectric constant, N_m is the number of molecules per unit volume, α_0 is the average polarizability of the molecules, p is the permanent dipole moment of the molecules, k is Boltzmann's constant, τ is a relaxation time for external-field-oriented molecules to return to their original random orientation, and ω is angular frequency. The right-hand side of Eq. (4-15) involves two effects, the distortion of molecules by the applied field and the orientation of polar molecules which have a permanent dipole moment. For sufficiently low frequencies the $j\omega\tau$ term can be dropped. Also $K + 2$ can be replaced by its approximate value of 3. Making these simplifications and recognizing that N_m is directly proportional to pressure and inversely proportional to temperature on the basis of the perfect gas law and considering the earth's atmosphere, we obtain

$$K - 1 = K_1' \frac{p_d}{T} + K_2' \frac{e}{T}\left(A + \frac{B}{T}\right) \tag{4-16}$$

where K_1', K_2', A, and B are constants, p_d is the pressure of dry nonpolar air, e is water vapor pressure, and T is absolute temperature. Water vapor molecules have a permanent dipole moment and a degree of polarizability as well. For air for which K differs only slightly from unity,

$$n = \sqrt{K} = \sqrt{1 + (K-1)}$$

and

$$n - 1 \simeq \frac{K-1}{2} \tag{4-17}$$

Using values for the constants as determined by Smith and Weintraub [30], one can write

$$N = (n-1)10^6 = \frac{77.6p_d}{T} + \frac{72e}{T} + 3.75 \times \frac{10^5 e}{T^2} \tag{4-18}$$

where large N units are used for convenience because $n - 1$ is such a small quantity. (Elsewhere N is used for particle density, but here it is a dimensionless number and a measure of index of refraction.) If the equation is expressed in terms of p, the total pressure, where $p = p_d + e$, it becomes

$$N = \frac{77.6p}{T} - \frac{5.6e}{T} + 3.75 \times \frac{10^5 e}{T^2} \tag{4-19}$$

Finally the last two terms can be combined to give, approximately,

$$N = \frac{77.6p}{T} + 3.73 \times \frac{10^5 e}{T^2} \tag{4-20}$$

or, more commonly,

$$N = \frac{77.6}{T}\left(p + \frac{4810e}{T}\right) \tag{4-21}$$

Thus index of refraction is a function of pressure, temperature, and water vapor pressure and depends strongly on the latter. In Eq. (4-21), pressures are measured in millibars (mb), where 1 mb $= 10^3$ dyn/cm^2 $= 10^2$ N/m^2 and 1 atmosphere $= 1013$ mb $= 1.013 \times 10^5$ N/m^2.

Propagation of radio waves in the troposphere is influenced by the index of refraction profile, or how the index of refraction varies with altitude. The profile varies with time and geographical location, but efforts have been made to establish standard or reference atmospheres. One such reference atmosphere is the exponential atmosphere [4], which is described by

$$N = N_s e^{-h/H} \tag{4-22}$$

where N_s is the surface refractivity, h is altitude, and H is scale height. Average values for N_s and H in the United States are stated to be 313 and 7 km, respectively. Values of N_s of 338 for a wet atmosphere and 262 for a dry atmosphere have been used by Campen and Cole [6] in a different model atmosphere. Average surface refractivities may vary, however, from as low as about 240 in dry high-altitude areas to about 400 in humid tropical areas. Temperature is commonly assumed to vary as

$$T = 288 - 6.5h$$

where h is altitude in km. Another form of the variation of index of refraction with altitude that has been used commonly is a decrease of 40 N units/km. Propagation in such an atmosphere is equivalent to propagation in one with a uniform index of refraction and an earth's radius $\frac{4}{3}$ times the actual radius. The actual index of refraction profile, which can be determined to varying degrees of accuracy by radiosondes and airborne microwave refractometers, may be quite irregular. The index may be high below an inversion layer, for example, and may drop abruptly with increasing altitude within the layer. A refractometer is a resonant microwave cavity with holes that allow air to enter. Precise measurement of the resonant frequency of the cavity provides a determination of the index of refraction of the air.

Optical Waves. Free electrons do not respond to optical frequency waves, and water molecules respond only very weakly. The net effect is that, to a useful approximation, the refractivity at optical frequencies in N units is given by

$$N = \frac{79p}{T} \tag{4-23}$$

Actually the value varies by about 10% over the optical frequency range. It also shows approximately 1% variation with humidity. Note that the major difference between the radio and optical indices is that the radio index is strongly affected by water vapor content and the optical index is only very slightly affected by water vapor.

Acoustic Waves. The velocity of acoustic sound waves is given by

$$v_s = \sqrt{\frac{p\gamma}{\rho}} \tag{4-24}$$

where p is pressure, γ is the ratio of specific heat at constant pressure to specific heat at constant volume, and ρ is density. For present purposes note that $p = nkT$, where n is particle number density, k is Boltzmann's constant, and T is absolute temperature, and that $\rho = nm$, where n is again particle density and m is average mass. Thus

$$v_s = \sqrt{\frac{\gamma kT}{m}} \tag{4-25}$$

The important consideration regarding this form of the equation is that the acoustic velocity is proportional to the square root of the temperature. Inserting numerical values, we obtain

$$v_s = 20.05T^{1/2} \text{ m/s} \tag{4-26}$$

Consider now that $T = T_0 + T'$, where T_0 is a reference or ambient temperature and T' is a perturbation in T. Then

$$v_s^2 = v_0^2\left(1 + \frac{T'}{T_0}\right) \tag{4-27}$$

where v_0 is the velocity corresponding to $T = T_0$, and if $T' \ll T_0$,

$$v_s \simeq v_0\left(1 + \frac{T'}{2T_0}\right) \tag{4-28}$$

Thus the index of refraction, $n = v_0/v_s$, is given in this case by

$$n \simeq 1 - \frac{T'}{2T_0} \tag{4-29}$$

if temperature variations only are considered.

The index of refraction for acoustic waves in fluids (gases or liquids) can also be considered to be affected by variations in fluid velocity, or by winds and turbulent motions in the case of the earth's atmosphere. If a local region is moving with a velocity w, then in that region

$$v_s = v_0 + w \cos\theta$$

where θ is the angle between the direction of the acoustic wave and the direction of the velocity w. Considering only the affect of this motion, we obtain

$$n = \frac{v_0}{v_0 + w\cos\theta} = \frac{1}{1 + (w\cos\theta/v_0)} \simeq 1 - \frac{w\cos\theta}{v_0} \tag{4-30}$$

If temperature and velocity variations occur simultaneously, one can define a quantity $n_1 = n - 1$ such that

$$n_1 = \frac{-T'}{2T_0} - \frac{w \cos\theta}{v_0} \tag{4-31}$$

Also if

$$N = (n - 1) \times 10^6$$

then

$$N = -\left(\frac{T'}{2T_0} + \frac{w \cos\theta}{v_0}\right) \times 10^6 \tag{4-32}$$

Note that the index of refraction for acoustic waves is much more sensitive to temperature variations than is the case for electromagnetic waves. The index is also sensitive to winds and turbulence. It is only slightly sensitive to water vapor content and

$$n_{\text{moist}} = \frac{n_{\text{dry}}}{1 + 0.14e/p} \tag{4-33}$$

corresponding to

$$v_{s_{\text{moist}}} = v_{s_{\text{dry}}}\left(1 + \frac{0.14e}{p}\right) \tag{4-34}$$

where e is water vapor pressure and p is total pressure.

The relative sensitivities of the three indices mentioned can be illustrated as follows [21]. A 1°C variation in temperature results in a 1700 N unit change in acoustic refractive index but only a 1 N unit change in radio refractive index. A 1-m/s variation in wind speed results in a 3000 N unit change in acoustic index and essentially no change in radio or optical indices. A 1-mb change in water vapor pressure causes a 140 N unit change in acoustic index, a 4 N unit change in radio index, and about a 0.04 N unit change in optical index. The sensitivity of the acoustic index has prompted the development of acoustic sounders for probing the lower atmosphere. These sounders are discussed further in Chap. 6.

4-2-2 Tropospheric Propagation: Refraction and Time Delay

In Sec. 2-4 the concepts of microwave systems were considered, but discussion of atmospheric effects was omitted. In this section we shall partially remedy that deficiency.

In considering the propagation of electromagnetic waves in the troposphere, it is commonly assumed that the waves travel in straight lines with time delays given by using the velocity c of 3×10^8 m/s. The index of refraction n of the troposphere, however, differs slightly from unity and varies with location and time, with the result that electromagnetic waves experience bending and follow curved rather than straight paths. Also, excess time delays, greater than predicted by using the velocity c, are encountered.

A useful simple concept concerning propagation in a region where n varies with position is that of the curvature C of a ray. Curvature has units of rad/m, or more commonly μrad/km or μrad/100 m, and, in the case of a ray having a constant curvature, the total change in direction of the ray in a distance d is simply Cd. The curvature C has the value at any point of $1/\rho$, where ρ is the radius of curvature. (These statements can be readily checked for the case of a circular path.) An expression for curvature will now be derived, following the treatment in Bean and Dutton [3].

Consider a wave front that moves from AB to $A'B'$ in time dt in Fig. 4-1, which refers to propagation in the earth's atmosphere in the case where the index of refraction n varies linearly with height h. Assuming that n decreases with height, so that dn is negative, the distance BB' is greater than AA', $B'A'$ and BA are not parallel (their projections meet at some point), and AA' and BB' are arcs of circles that have radii of ρ and $\rho + d\rho$, respectively. It follows that

$$\rho \, d\theta = v \, dt$$

$$(\rho + d\rho) \, d\theta = (v + dv) \, dt$$

where $\rho \, d\theta$ and $v \, dt$ are seen to represent the length of the arc AA'. Thus

$$\frac{v}{\rho} = \frac{v + dv}{\rho + d\rho} \tag{4-35}$$

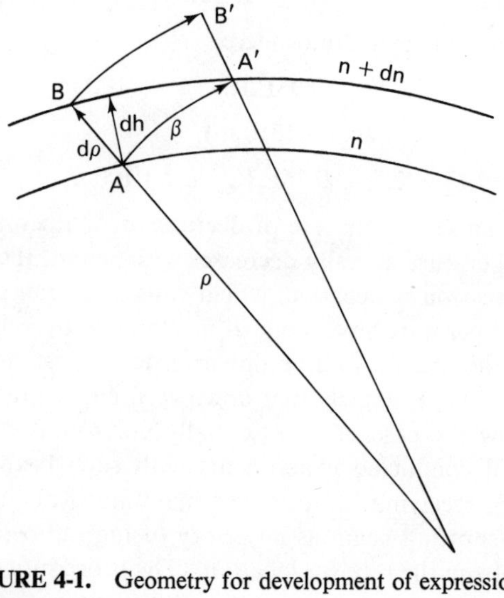

FIGURE 4-1. Geometry for development of expressions for curvature.

where v is phase velocity, which varies with h because n varies with h. Manipulation of this equation leads to

$$\frac{dv}{v} = \frac{d\rho}{\rho} \tag{4-36}$$

Also, $v = c/n$ by definition, so that

$$\frac{dv}{v} = \frac{-dn}{n} \tag{4-37}$$

Equating the two expressions for dv/v then results in

$$C = \frac{1}{\rho} = -\frac{1}{n}\frac{dn}{d\rho} \tag{4-38}$$

or as $dh = d\rho \cos \beta$,

$$C = -\frac{1}{n}\frac{dn}{dh}\cos \beta \tag{4-39}$$

In the case where $\beta \simeq 0°$ and for the troposphere for which $n \simeq 1$,

$$C = -\frac{dn}{dh} \tag{4-40}$$

It is of interest to consider the application of Eq. (4-40) at optical frequencies. As

$$N = (n - 1) \times 10^6 \simeq \frac{79p}{T}$$

for optical frequencies, it follows that in this case

$$C = -\left(\frac{79}{T}\frac{\partial p}{\partial h} - \frac{79p}{T^2}\frac{\partial T}{\partial h}\right) \times 10^{-6} \tag{4-41}$$

At sea level for a standard atmosphere,

$$C = 3.26 - 0.93\text{L.R.} \quad \mu\text{rad/100 m} \tag{4-42}$$

or at Boulder, Colorado (elevation 1610 m),

$$C = 2.8 - 0.75\text{L.R.} \quad \mu\text{rad/100 m} \tag{4-43}$$

where L.R. is the lapse rate or rate of decrease of temperature with height (K/100 m). If temperature actually decreases with height, the second term in the curvature expression is negative, which causes the ray to tend to curve upwards. In a temperature inversion, T increases with height, the second term is positive, and the rays curve downwards, consistent with the usual convention for curvature, namely that downward curvature is positive.

Consider now the case of an essentially horizontal laser beam. Ochs and Lawrence [20] conducted experiments with such beams in which the incident beam at a receiving location was maintained at the same position by tilting the transmitting beam as necessary (using radio signals sent back to the transmitter from the receiver location). The procedure can be described by reference to Fig. 4-2. Assume path A to be the direct path from transmitter

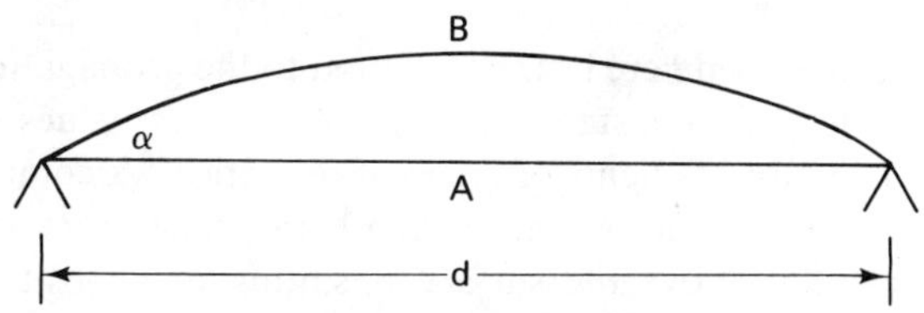

FIGURE 4-2. Curved ray path.

to receiver, but assume that path *B* is the path which must be followed in the presence of a temperature inversion. The transmitter beam must in this case be tipped up by an angle α from the direct path. Consider that path *B* has a constant downward curvature. At the center of the path the beam must be horizontal. Therefore $\alpha = C(d/2)$, and, by measurement of α, C and the average value of L.R. along the path can be determined. The procedure is feasible because of the very narrow beamwidth which can be achieved at optical frequencies and because the index of refraction at optical frequencies is insensitive to water vapor content.

In the propagation of electromagnetic waves in the troposphere, an important quantity is the difference between the curvature of the earth's surface and the ray curvature, which has been discussed in the last paragraphs. This difference is

$$\frac{1}{r_0} - C = \frac{1}{r_0} + \frac{dn}{dh} \tag{4-44}$$

where r_0 is the radius of the average earth's surface. The same relative curvature can be maintained if, instead of using the actual earth and actual ray curvature, one uses a radius of kr_0 and a ray of zero curvature as illustrated by the following equation:

$$\frac{1}{r_0} + \frac{dn}{dh} = \frac{1}{kr_0} + 0 \tag{4-45}$$

This equivalence is commonly used in analyzing tropospheric propagation, with $k = \frac{4}{3}$ corresponding to $dn/dh = -40\ N$ units/km. The procedure followed may be to use charts constructed with radii corresponding to $\frac{4}{3}$ times the true radius. Ray paths can then be drawn as straight lines.

Although a vertical gradient of $-40\ N$/km is a typical value, the actual gradient can vary over rather wide limits. In the case where the gradient is $-157\ N$/km, $k = \infty$, and a ray that is launched parallel to the earth's surface will remain parallel. Gradients near this value have been responsible for the abnormally long radar ranges which are encountered at times in some locations.

It is sometimes more convenient to use a quantity $K = 1/k$ such that

$$\frac{1}{kr_0} = \frac{K}{r_0}$$

(Note that the k and K used here bear no relation to the propagation constant k or the relative dielectric constant K.) Table 4-1 shows values of k and K corresponding to certain gradients of index of refraction. A common type of mirage results when the earth's surface is at a high temperature and the temperature drops rapidly above the surface, resulting in a positive value of dn/dh. The condition of looming can result when large negative values of dn/dh occur.

Table 4-1 k and K Values.

$\frac{dn}{dh}$ (N/km)	k	K
157	1/2	2
78	2/3	3/2
0	1	1
−40	4/3	3/4
−100	2.75	0.364
−157	∞	0
−200	−3.65	−0.274
−300	−1.09	−0.917

The curvature or bending referred to above results in an elevation angle error, for example, in the case of a radar receiving an echo from an object in or above the troposphere. The elevation angle error can be calculated by use of Eq. (4-55) once the bending has been calculated. The expression for the bending $d\tau$ in a path length of ds is, by a slight modification of Eq. (4-39),

$$d\tau = -\frac{1}{n}\frac{dn}{dr}\cos\beta\, ds \tag{4-46}$$

But $dr = \sin\beta\, ds$, so

$$d\tau = -\frac{dn}{n\tan\beta} \tag{4-47}$$

or as n is essentially unity,

$$d\tau = -\frac{dn}{\tan\beta} \tag{4-48}$$

Although the assumption of a constant gradient of n, such as that which leads to a $\frac{4}{3}$rds earth effective radius, is frequently useful, greater accuracy and generality is achieved by using a measured index of refraction profile, if such is available, or the profile of a reference atmosphere such as the exponential atmosphere [Eq. (4-22)]. In a layer of finite thickness, the total bending is given by

$$\tau_{12} \simeq -\int_{n_1,\beta_1}^{n_2,\beta_2} \frac{dn}{\tan\beta} \tag{4-49}$$

where τ_{12} is the bending experienced between levels 1 and 2, but in the case of a finite but sufficiently thin layer,

$$\tau_{12} = \frac{2(n_1 - n_2)}{\tan \beta_1 + \tan \beta_2} \tag{4-50}$$

If τ is in milliradians (mr) and the index of refraction is given in N units,

$$\tau_{12}\,(\text{mr}) = \frac{2(N_1 - N_2) \times 10^{-3}}{\tan \beta_1 + \tan \beta_2} \tag{4-51}$$

For small angles, the tangents of the angles can be replaced by the angles themselves and

$$\tau_{12}\,(\text{mr}) = \frac{2(N_1 - N_2)}{\beta_1\,(\text{mr}) + \beta_2(\text{mr})} \tag{4-52}$$

The total bending in n layers is given by a summation of terms like Eq. (4-51) or (4-52). For example,

$$\tau_n\,(\text{mr}) = \sum_{k=0}^{n-1} \frac{2(N_k - N_{k+1}) \times 10^{-3}}{\tan \beta_k + \tan \beta_{k+1}} \tag{4-53}$$

The values of N are determined from the profile used. The β's can be determined by starting with β_0 and calculating successive values of β from [3]

$$\beta_{k+1}\,(\text{mr}) = \sqrt{\beta_k^2\,(\text{mr}) + \frac{2(r_{k+1} - r_k) \times 10^6}{r_k} - 2(N_k - N_{k+1})} \tag{4-54}$$

Weisbrod and Anderson [39] have supplied charts which allow graphical determination of the β's, thus avoiding the use of Eq. (4-54). Once the total bending is calculated the elevation angle error can be determined if desired by use of

$$\delta = \frac{\tau \tan \beta - (N_0 - N) \times 10^{-6} + \tau^2/2}{\tau + \tan \beta - \tan \beta_0} \tag{4-55}$$

where β_0 is the initial angle of the ray where the index of refraction is N_0, β is the final angle where the index of refraction is N, and τ is the total bending.

The determination of range error will now be discussed briefly. For this purpose, consider the phase path length P as defined by

$$P = \int n\, dl \tag{4-56}$$

where the integral is evaluated along a path of interest. If n were exactly unity, P would equal the actual path length. Insofar as n differs from unity, P is different from the actual length. The difference ΔP between P and the actual length is given by

$$\Delta P = \int (n - 1)\, dl \tag{4-57}$$

Weisbrod and Anderson [39] give the equation

$$\Delta r_{jk} = \int_{l_j}^{l_k} \frac{N \times 10^{-6}\, d\rho}{\sin \beta} \tag{4-58}$$

for the range error induced in a particular layer, where the integration is carried out in the vertical direction. By using the same type of approximation as for determining bending, they arrive at

$$\Delta r_{jk} = \frac{(N_k + N_j)(\rho_k - \rho_j) \times 10^{-6}}{\sin \beta_k + \sin \beta_j} \tag{4-59}$$

where the average values of both N and $\sin \beta$ are used. This equation gives the range error in a particular layer, or between two particular levels, j and k. The total error along a path can be determined by a summation of such terms.

A considerable amount of effort has gone into determining radar coverage by calculating ray paths using the above or similar techniques. Radar antennas have finite beamwidths, and calculations need to be made for the range of angles included within the beam. Figure 4-3 shows an example of the coverage of an airborne radar in the presence of an elevated atmospheric layer [23]. Where the ray paths are missing or widely separated, radar coverage is lacking or minimal.

In addition to possible errors, or variations, in range and angle, the atmosphere causes variations in phase and distortion of wave fronts. Phase variations are important to systems using phase comparison for position location [37], and distortion of wave fronts is important to imaging systems, such as side-looking radars [14,36].

4-2-3 Tropospheric Effects on Microwave Links

In investigating path clearances on line-of-sight microwave links, as discussed in Sec. 2-3-4 in terms of Fresnel zones, the effects of the atmosphere need to be included. k values ranging from $\frac{2}{3}$ to ∞ may need to be taken into account. Typical clearance criteria for highest-reliability systems of $0.3F_1$ for $k = \frac{2}{3}$ and $1.0F_1$ for $k = \frac{4}{3}$, whichever is greatest, are quoted by GTE Lenkurt [11]. When less stringent reliability requirements are to be met, $0.6F_1 + 3$ m at $k = 1.0$ may be satisfactory. In determining whether the criteria are met or not, the relation, $h = l^2/2r_0$ [Eq. (2-39)] can be modified. As to units, it can be expressed as

$$h = \frac{l^2}{12.75} \tag{4-60}$$

if h is in m and l in km for $r_0 = 6370$ km, or with h in ft and l in miles the

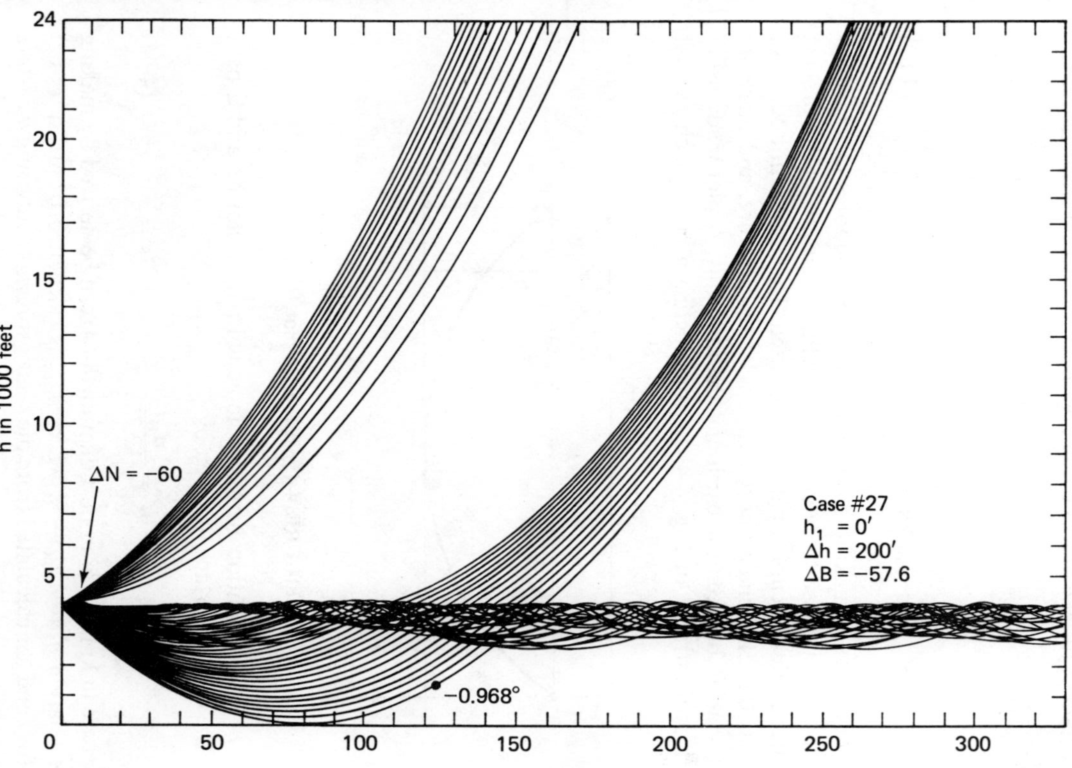

FIGURE 4-3. Ray-tracing diagram applicable to an airborne transmitter at the base of a 200-ft (61-m) layer in which the index of refraction decreases by 60 N units (by 57.6 units more than in a standard atmosphere; i.e., $\Delta B = -57.6$). The diagram depicts the occurrence of radar holes. (From Moreland, *Estimating Meteorological Effects on Radar Propagation*, Vols. I and II, Air Weather Service, U.S. Air Force, 1965 [23].)

expression becomes

$$h = \frac{l^2}{1.5} \tag{4-61}$$

More importantly, to take into account a value of k different from 1.0, use

$$h = \frac{l^2}{12.75k} \tag{4-62}$$

or

$$h = \frac{l^2}{1.5k} \tag{4-63}$$

The procedure of making plots based on an earth radius of kr_0, mentioned earlier, may be used to determine ray paths, or the alternative process of using a flat earth and curved paths may be used. The justification for both approaches is that they maintain the same relative curvature, and therefore the proper spacing between the earth and the ray paths. A plot made using a flat earth is illustrated in Fig. 4-4. In constructing a curve such as that

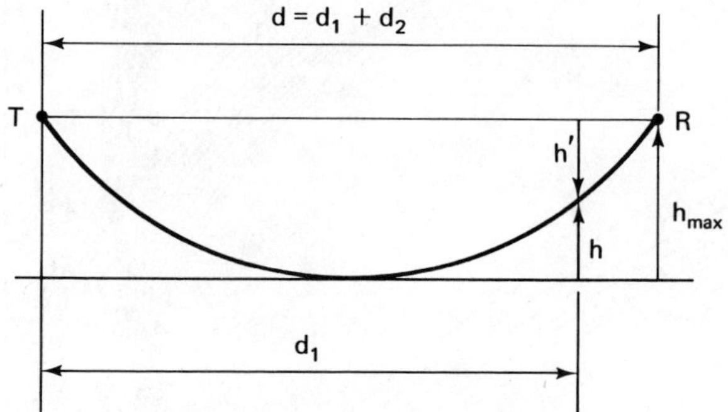

FIGURE 4-4. Flat earth plot.

illustrated, h', the separation between the actual path followed and a path parallel to the earth's surface, is given by

$$h' = \frac{d_1 d_2}{12.75k} \tag{4-64}$$

if distances are in km and h' in m. Equation (4-64) can be derived by making use of the relation $h' = h_{max} - h$, where h_{max} is calculated from Eq. (4-62) using $d/2$ for l and h is calculated from the same expression but using $d_2 - d/2$ or $d_1 - d/2$ for l.

When prepared for general-purpose use, ray paths, such as that of Fig. 4-4, are drawn for equal heights at the two ends. However, the curves can be utilized when the transmitter and receiver are at unequal heights. The path profile (or merely particular points of interest) is plotted to the same scale as the ray-path plot, and the profile is moved until the transmitter and receiver locations both fall on the ray-path curve. Care must be taken to keep the horizontal lines on both plots parallel to each other in the process. It is simple to construct ray-path curves for different k values oneself, and there is no need to rely on printed curves such as those in the useful GTE Lenkurt publication on microwave systems [11], but the latter contribute to convenience and efficiency when large numbers of paths are to be analyzed. For the unequal height case, however, the distance between transmitter and receiver corresponds to d_1 or d_2 in Fig. 4-4, and it is necessary to first solve for d in order to construct the ray path.

When a reflection from the terrain between the transmitter and receiver takes place, the effect of the reflection depends on the vertical separation r between the direct and reflected rays, among other factors. If r equals F_1, the first Fresnel zone radius, maximum reinforcement takes place, as the reflected path is then $\lambda/2$ longer than the direct path, but a reversal of phase tends to take place on reflection. The two phase reversals result in the direct and reflected rays being in phase. If r equals F_2, the second Fresnel zone radius, however, maximum destructive interference takes place. When $r = 0.6F_1$, the sum of the direct and reflected rays merely results in the same amplitude as for the direct ray alone. For r less than $0.6F_1$, a decrease in signal amplitude results. The direct path is that which applies for the particular k value of interest.

k is a variable quantity, and reflection may take place for certain k values and not for others. Also the location of the point of reflection [11] and the phase and amplitude of the reflected wave may vary as a function of k. Because of the possibility of destructive interference and variation of phase and amplitude, reflections should be avoided or minimized if possible by arranging for obstructions in the way of potentially reflecting paths or arranging for reflection to take place from rough poorly reflecting surfaces rather than from smooth, highly reflecting surfaces. Another possible procedure, if terrain permits, is to position one site at a relatively high location with respect to the adjacent site. The difference in path lengths between direct and reflected rays in such a case tends to be small, and the undesirable reception of second- and higher-order Fresnel zone energy can be more readily avoided.

It is not always possible to avoid reflections and the associated destructive interference which may occur for some values of k. Furthermore, even if surface reflections are avoided, undesirable multipath conditions and associated fading may take place when certain index of refraction profiles occur. Nearly horizontal paths through or near temperature inversion layers are subject to severe fading. Figure 4-5 shows an example of fading at a wavelength of 8.6 mm on an 18.4-km (11.4-mile) path in the Los Angeles basin in the presence of a temperature inversion, and Fig. 4-6 shows a contrasting stable condition that occurred on the same path at a time when no temperature inversion was present. As mentioned in Chap. 2, a margin of 35 or

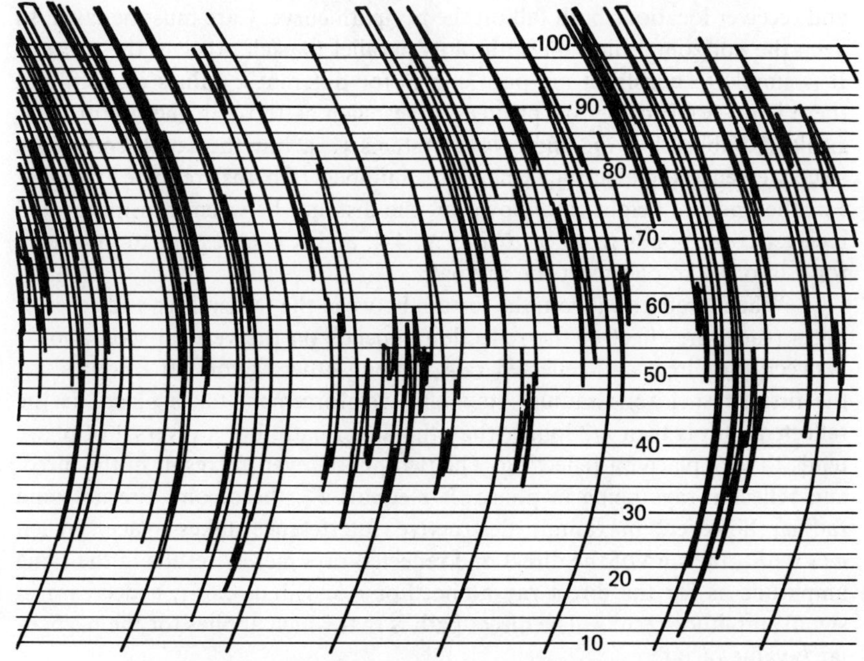

FIGURE 4-5. Signal fading at 35 GHz on 18.4-km path from UCLA campus to Los Angeles City Hall, Sept. 14, 1956, at a time of a surface temperature inversion. Horizontal scale: 1 div. = 15 min; vertical scale: 1 large div. = 2 dB. (Fades of 30 dB or more were observed on an oscilloscope, but the recorder time constant limited the displayed fading range to that shown.)

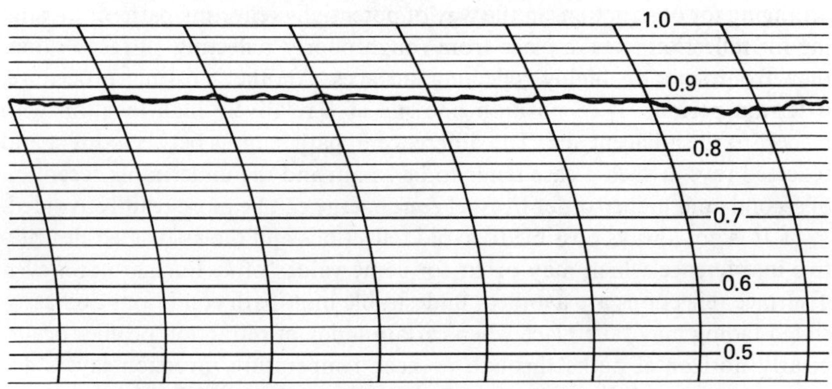

FIGURE 4-6. Steady signal at 35 GHz on 18.4-km path from UCLA campus to Los Angeles City Hall, April 18, 1956, at a time of no temperature inversion. Horizontal scale: 1 div. = 15 min; vertical scale: 1 large div. = 2 dB.

40 dB is usually allowed in order to minimize the effects of such fading. This margin may well be sufficient to give satisfactory performance, say with 99.99% reliability or better, with respect to strictly atmospheric fading and reflection-atmospheric fading [11]. If higher reliability is needed, however, as may be the case when data rather than voice signals are being transmitted, it may be necessary to utilize diversity. The latter term refers to using two frequencies (frequency diversity) or two antennas at two different heights (space diversity) or a combination of the two to obtain greater reliability. Very large negative gradients of refractivity, corresponding to negative k values (Table 4-1), may cause a severe form of fading known as blackout fading, which may not be amenable to remedy by diversity techniques. The variation of signal level on radio paths is perhaps best described in statistical terms and can be displayed as in Fig. 4-22, which refers to a troposcatter path.

The effects of various values of k are depicted in Figs. 4-7 and 4-8. It needs to be kept in mind that the actual atmosphere is highly variable with respect to both time and location. Designing a system to provide satisfactory operation over a range of k values (perhaps from $k = \frac{2}{3}$ to $\frac{4}{3}$) is normally an appropriate, practical procedure, but the actual index of refraction may not

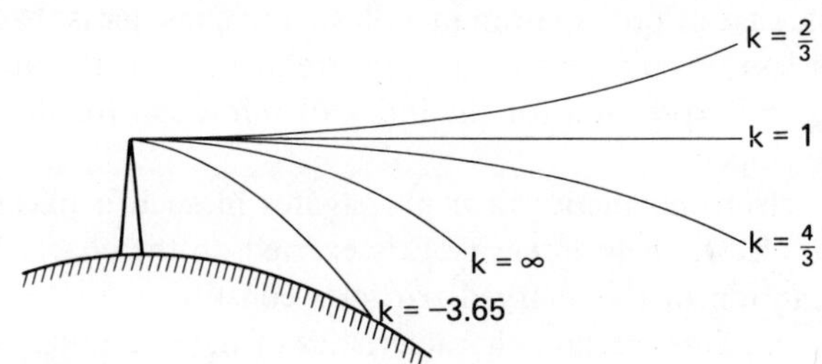

FIGURE 4-7. Effects of various values of k (exaggerated and illustrative only) for rays launched horizontally in the presence of a curved earth of radius r_0.

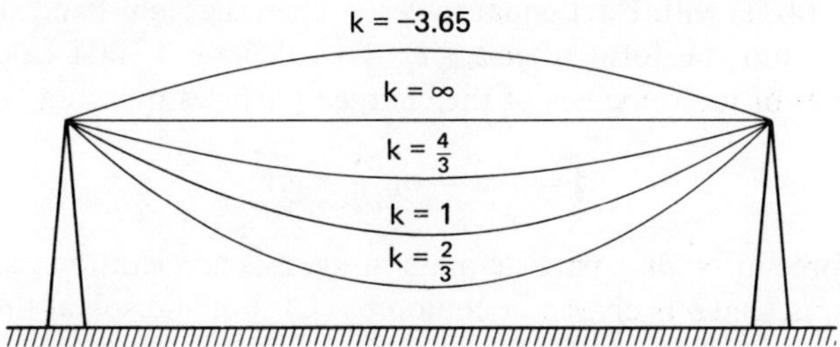

FIGURE 4-8. Effects of various values of k (exaggerated and illustrative only) for rays directed to reach a particular target, with a flat representation of the curved earth's surface.

vary linearly with height as implied by use of fixed k values. Ray tracing carried out in accordance with the concepts of Sec. 4-2-2 could provide a highly accurate description of the ray path corresponding to a given (measured or assumed) index of refraction profile, but, as the profile will vary with time and location, this procedure has its limitations also. In spite of the complexity and variability of the atmosphere, microwave systems can generally be designed to provide satisfactory results.

4-3 IONOSPHERIC PROPAGATION

4-3-1 Characteristic Waves

The ionosphere is an ionized medium or plasma, the ionization being formed as described in Sec. 3-2-4. The total amount of negative charge in a plasma is the same as the amount of positive charge, and a plasma is thus electrically neutral. In this treatment it is assumed that the negatively charged particles are free electrons and that the positively charged particles are singly charged positive ions. In the ionosphere only a fraction of the molecules are ionized, and relatively large numbers of neutral particles are present. The most simple case of propagation in a plasma to consider is that of propagation in a lossless or cold plasma at high frequencies in the absence of a magnetic field. The expression for the index of refraction for this condition will now be developed.

When an electromagnetic wave propagates in such a plasma a force proportional to electric field intensity **E** is exerted on the charged particles. The resulting motion of the charged particles constitutes electrical current, and this current modifies certain characteristics of the electromagnetic wave. A procedure to calculate the value of the index of refraction is to calculate the value of convection current density **J** and to insert this value into

$$\nabla \times \mathbf{H} = \mathbf{J} + j\omega\varepsilon_0\mathbf{E} \tag{4-65}$$

which is Eq. (4-11) with **P** set equal to zero. Then the right-hand side of Eq. (4-65) is put into the form of $j\omega\varepsilon_0 K\mathbf{E}$. To calculate **J**, first calculate the average values of the velocities of the charged particles by using

$$\mathbf{f} = m\mathbf{a} = m\frac{d\mathbf{v}}{dt} = q\mathbf{E} \tag{4-66}$$

where **f** is force in N, m is particle mass in kg, **a** is acceleration in m/s^2, **v** is velocity in m/s, and q is charge in coulombs (C). For sinusoidal time dependence the form of the equation is

$$mj\omega\mathbf{v} = q\mathbf{E} \tag{4-67}$$

from which

$$\mathbf{v} = \frac{q\mathbf{E}}{jm\omega} \tag{4-68}$$

The convection current density is given by $\mathbf{J} = N_e q_e \mathbf{v}_e + N_i q_i \mathbf{v}_i$, where N is particle density, the subscript e stands for electrons, and the subscript i stands for ions. Note that the velocities are inversely proportional to mass. As the mass of a positive ion is at least 1840 times as great as that of an electron, the effects of ions can be neglected in this treatment and only electrons need be taken into account. In general, the force on a charged particle in electric and magnetic fields is given by

$$\mathbf{f} = q[\mathbf{E} + (\mathbf{v} \times \mathbf{B})] \tag{4-69}$$

where $\mathbf{B}$ is magnetic flux density. One might wonder if a component of force due to the magnetic flux density of the wave should not be included. However, $E/B = v_p$, and as v_p, the phase velocity of the wave, is usually much greater than v, the charged particle velocity, the magnetic force due to the magnetic field of the wave itself can usually be neglected. Thus considering electrons and the force due to the electric field only, we obtain

$$\mathbf{J} = Nq\mathbf{v} = -\frac{jNq^2\mathbf{E}}{m\omega} \tag{4-70}$$

Inserting this into the right-hand side of Eq. (4-11) and forcing the right-hand side into the form $j\omega\varepsilon_0 K\mathbf{E}$ leads to

$$K = 1 - \frac{Nq^2}{m\varepsilon_0\omega^2} \tag{4-71}$$

If $Nq^2/m\varepsilon_0$ is represented by ω_p^2, the equation can be written as

$$K = 1 - \frac{\omega_p^2}{\omega^2} \tag{4-72}$$

where ω_p is called angular plasma frequency. This expression confirms the previous statement that K is less than unity for the case considered. If $K > 0$, electromagnetic wave propagation can take place in the plasma. For $K < 0$, however, wave propagation cannot take place, and only an evanescent "wave" can exist. ω must be larger than ω_p for propagation to occur, and the frequency $\omega = \omega_p$ plays the role of a cutoff frequency, as in the case of a waveguide for which f must be greater than f_c for propagation to occur. For frequencies greater than the cutoff frequency in both a plasma and a waveguide, the phase velocities are greater than c (3×10^8 m/s). The propagation constant k and the alternate propagation constant γ, where $\gamma = jk$, take the following forms in these cases. When $K > 0$, k is real and γ is imaginary or

$$k = \beta, \qquad \gamma = j\beta$$

When $K < 0$, however, k is imaginary and γ is real or

$$k = -j\alpha, \qquad \gamma = \alpha$$

Next, consider the effect of an applied external static magnetic field, in particular the earth's magnetic field. In Sec. 3-2-5 it was shown that charged particles in a magnetic field tend to rotate with an angular velocity or gyrofrequency ω_B, given by $\omega_B = -qB/m$. The gyrofrequency is much lower for positive ions than for electrons, just as in the absence of a magnetic field the ion velocity due to an applied electric field is much lower than the corresponding electron velocity. If electromagnetic waves having frequencies high compared to the ion gyrofrequency are considered, the effects of ions can be neglected just as they were in the previous section. Consider that the earth's magnetic field $\mathbf{B}_0$ is directed along the z axis as in Fig. 4-9. The vector equation of motion for electrons in this case is

$$q[\mathbf{E} + (\mathbf{v} \times \mathbf{B}_0)] = j\omega m\mathbf{v} \tag{4-73}$$

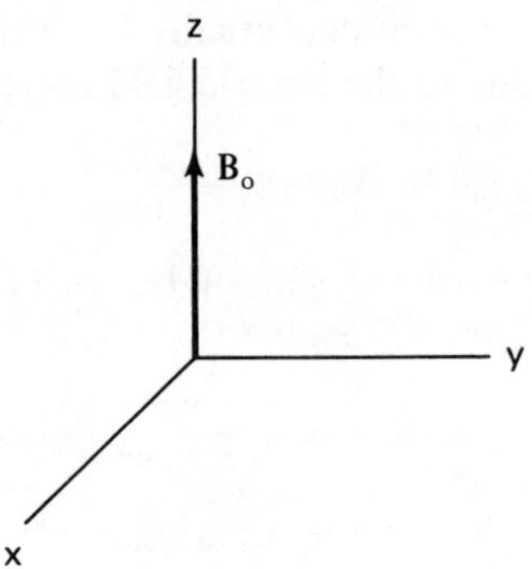

FIGURE 4-9. Coordinate system utilized for calculation of relative dielectric constant K.

The three rectangular components of this equation are

$$qE_x + qv_yB_0 = j\omega m v_x \tag{4-74}$$

$$qE_y - qv_xB_0 = j\omega m v_y \tag{4-75}$$

$$qE_z = j\omega m v_z \tag{4-76}$$

The first two equations are two equations with two unknowns, v_x and v_y, and have as solutions

$$v_x = \frac{-j\omega(q/m)E_x + \omega_B(q/m)E_y}{\omega^2 - \omega_B^2} \tag{4-77}$$

$$v_y = \frac{-\omega_B(q/m)E_x - j\omega(q/m)E_y}{\omega^2 - \omega_B^2} \tag{4-78}$$

where $\omega_B = -qB_0/m$. If the x and y components of velocity are substituted into

$$\nabla \times \mathbf{H} = Nq\mathbf{v} + j\omega\varepsilon_0\mathbf{E}$$

the two rectangular components of $\nabla \times \mathbf{H}$ can be expressed as

$$(\nabla \times \mathbf{H})_x = j\omega\varepsilon_0(K_\perp E_x - K_x E_y) \tag{4-79}$$

$$(\nabla \times \mathbf{H})_y = j\omega\varepsilon_0(K_x E_x + K_\perp E_y) \tag{4-80}$$

where

$$K_\perp = 1 - \frac{\omega_p^2}{\omega^2 - \omega_B^2}, \qquad K_x = \frac{j\omega_p^2\omega_B}{\omega(\omega^2 - \omega_B^2)}$$

An efficient way to proceed at this point is to consider that $K_\perp$, K_x, and $K_\parallel$ are terms of a second-order tensor [where $K_\parallel$ is obtained from Eq. (4-76) and has the same value as K of Eq. (4-72)] [2]. The tensor can be substituted into the wave equation, which can then be solved for the indices of refraction of the characteristic waves for waves propagating at any angle θ with respect to the magnetic field $\mathbf{B}_0$. It turns out, however, that the characteristic waves which can propagate parallel to the magnetic field in a plasma are left and right circularly polarized waves, where the term characteristic wave refers to a wave whose polarization does not change with distance as the wave propagates. By anticipating this result and considering propagation parallel to the magnetic field or in the z direction and making use of Eqs. (4-79) and (4-80), the two rectangular components of the wave equation (4-13) can be written for this case as

$$\frac{\partial^2 E_x}{\partial z^2} + \omega^2\mu_0\varepsilon_0(K_\perp E_x - K_x E_y) = 0 \tag{4-81}$$

$$\frac{\partial^2 E_y}{\partial z^2} + \omega^2\mu_0\varepsilon_0(K_x E_x + K_\perp E_y) = 0 \tag{4-82}$$

But a left circularly polarized wave is one for which $E_y = jE_x$, as a left circularly polarized wave can be generated by two electric field intensity vectors which are oriented 90° from each other in space and are 90° out of phase in time. A circular Lissajous figure can be formed on an oscilloscope in the same way. Substituting $E_y = jE_x$ into Eqs. (4-81) and (4-82) results in

$$\frac{\partial^2 E_x}{\partial z^2} + \omega^2\mu_0\varepsilon_0(K_\perp - jK_x)E_k = 0 \tag{4-83}$$

$$\frac{\partial^2 E_y}{\partial z^2} + \omega^2\mu_0\varepsilon_0(K_\perp - jK_x)E_y = 0 \tag{4-84}$$

where

$$K_\perp - jK_x = 1 - \frac{\omega_p^2}{\omega(\omega + \omega_B)}$$

Since these equations refer to a left circularly polarized wave, the quantity $K_\perp - jK_x$ can be considered to be the relative dielectric constant for a left circularly polarized wave K_l, so that

$$K_l = 1 - \frac{\omega_p^2}{\omega(\omega + \omega_B)} \tag{4-85}$$

Similarly, substituting $E_y = -jE_x$, which applies to a right circularly polarized wave, leads to

$$K_r = 1 - \frac{\omega_p^2}{\omega(\omega - \omega_B)} \tag{4-86}$$

The indices of refraction for left and right circularly polarized waves are related to the corresponding relative dielectric constants in the usual way; that is, $n_l^2 = K_l$ and $n_r^2 = K_r$.

The expressions for K_l and K_r are commonly stated in a different notation as follows:

$$K_l = 1 - \frac{X}{1 + Y} \tag{4-87}$$

$$K_r = 1 - \frac{X}{1 - Y} \tag{4-88}$$

where $X = \omega_p^2/\omega^2$ and $Y = \omega_B/\omega$. These expressions apply strictly only for propagation parallel to the magnetic field. However, it develops that if

$$4(1 - X)^2 Y_L^2 \gg Y_T^4 \tag{4-89}$$

then

$$K_l = 1 - \frac{X}{1 + Y_L} \tag{4-90}$$

and

$$K_r = 1 - \frac{X}{1 - Y_L} \tag{4-91}$$

where $Y_L = Y \cos \theta$ and $Y_T = Y \sin \theta$. These expressions apply for propagation of circularly polarized waves at an angle of θ with respect to the magnetic field under the condition that Eq. (4-89) applies. The approximation involved in this case is known as the *QL* approximation.

For propagation perpendicular to the magnetic field, the characteristic waves are no longer circularly polarized. One characteristic wave, that having only a z component of electric field intensity, is unaffected by the magnetic field. The relative dielectric constant $K_\parallel$ for this wave, commonly called the ordinary wave, is the same as if there were no B_0 and is given by Eq. (4-72). The other characteristic wave for perpendicular propagation is complicated by the fact that it has a longitudinal component of **E** (a component in the direction the wave is propagating). This wave is called the extraordinary wave and has a relative dielectric constant K_{ex} given by

$$K_{ex} = \frac{K_l K_r}{K_\perp} \tag{4-92}$$

We have concentrated here on the concepts of propagation in a lossless plasma, but electromagnetic waves in plasmas experience attenuation because of collisions of electrons with the neutral atoms and molecules of the atmosphere and sometimes in other ways as well. In the collision process, the electrons lose to the neutral particles some of the ordered energy they have gained from the incident electromagnetic wave. Thus the electromagnetic wave imparts energy to the neutral particles and is attenuated. The attenuation coefficient for a wave of angular frequency ω is proportional to the electron density N and collision frequency ν, as indicated by Eq. (4-93) for the ordinary wave,

$$\alpha = \frac{Nq^2\nu}{2m\varepsilon_0 c n_{re}(\omega^2 + \nu^2)} \tag{4-93}$$

where n_{re} is the real part of the index of refraction, which can be considered to become complex when losses are taken into account. If losses are not very great, however, n_{re} has essentially the same value as n for the lossless case. Equations for the attenuation coefficients for the left or right circularly polarized characteristic waves can be obtained by replacing ω^2 by $(\omega + \omega_B)^2$ or $(\omega - \omega_B)^2$, respectively.

4-3-2 The Ionosonde

Electromagnetic waves which are vertically incident upon the ionosphere from the earth are reflected when and if the relative dielectric constant for the particular wave type reaches zero, as the wave enters the ionosphere from below and travels upwards through a region of increasing electron density. For the case of the "ordinary" wave which is unaffected by the magnetic field, reflection occurs at the elevation at which f, the wave frequency, equals f_p, the plasma frequency. Information about the distribution of electron density with height below the peak of the F layer can be obtained by using a variable-frequency pulse radar known as an ionosonde.

The ionosonde is a type of radar which receives echoes from the overhead ionosphere over a rather wide range of frequencies. A typical ionosonde sweeps over about a 0.5- to 25-MHz frequency range in 10–15 s, transmitting pulses at a repetition rate of about 100 pps during the time interval of the sweep. In the usual recording system, the echo pulse received from the ionosphere forms a spot on a cathode-ray tube at a distance from a reference position that is proportional to the delay time of the echo. The delay time is also proportional to the virtual height h' of the ionospheric reflection point and is recorded photographically by moving a photographic film at right angles to the time base. h' can be determined by assuming that the reflected signal traveled with a velocity of c, 3×10^8 m/s. The actual velocity of propagation in the ionosphere below the reflection point is less than c, however, so the true height, h is less than h'. Procedures are available for recovering the true height from the virtual height curves. [The Environmental Data Service of the National Geophysical and Solar-Terrestrial Data Center in Boulder, Colorado, will produce N(h) profiles from ionograms (ionosonde records) as a service.] At the lowest frequencies at which echoes can be recorded, reflection occurs at a lower height than for higher frequencies. When the frequency becomes sufficiently high, the wave penetrates the ionospheric layer considered, and no reflection is received. In a simple case of only one reflecting layer, as at night when reflection is from the F_2 layer, an ionosonde record might look roughly as shown in Fig. 4-10, which shows that the ionosonde trace divides into o and x portions near the penetrating frequencies. The notations o and x stand for the ordinary and extraordinary waves.

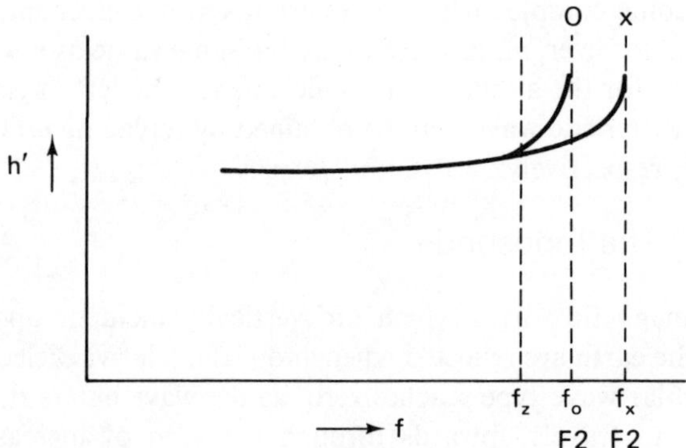

FIGURE 4-10. Ionosonde trace, showing f_o and f_x.

The ionosonde is a very useful and time-honored system for obtaining electron density profiles below the peak of the F layer of the ionosphere. Corresponding information about the electron density profile above the peak of the F layer can be obtained from top-side sounders in satellites, and the incoherent scatter technique (Sec. 4-4-5) can provide information about the complete profile. Some brief remarks concerning elementary aspects of the interpretation of ionosonde records follow. First, consider the notation of ordinary and extraordinary. These terms have been used in these notes to apply only to propagation perpendicular to the magnetic field, but they are often applied to propagation at any angle, including the case of propagation parallel to the field. In the case of parallel propagation, the lc (left-circularly polarized) wave is called ordinary and the rc (right-circularly polarized) wave is called extraordinary. There is some basis for these designations as the lc wave transforms into the o wave and the rc wave transforms into the x wave as the angle of propagation with respect to the magnetic field is increased from 0 to 90°.

Although the ionosonde usually radiates merely a linearly polarized wave, this wave consists of the two component characteristic waves, namely the ordinary and extraordinary. Each wave type is reflected from the ionosphere at normal incidence when the index of refraction becomes zero (neglecting losses and using cold plasma theory) or when cutoff for the particular wave type is encountered. The maximum frequency of reflection occurs when cutoff is not encountered in the course of the upward passage of the wave until the peak of the layer in question is reached. Near reflection, the X of Eq. (4-89) becomes large so the QL approximation mentioned following Eq. (4-89) does not apply. On the contrary, the QT approximation, which holds when the inequality signs of Eq. (4-89) point in the other direction, applies, and the two characteristic waves have dielectric constants given by Eqs. (4-72) and (4-92). Reflection occurs for the ordinary wave when $K_\parallel = 0$ and for the extraordinary wave when $K_r = 0$.

At high latitudes it develops that there can be some coupling of energy between the ordinary and extraordinary modes near the reflection height for the ordinary wave. This coupling results in the condition that some energy propagates upward in the ionosphere until the $K_l = 0$ value is reached. A third ionospheric trace, called the z trace, is then developed. This phenomenon is referred to as triple splitting.

Consider now the separation between f_o and f_x of Fig. 4-10, f_o being the critical frequency for the extraordinary wave. Let f_z be the corresponding value for the z trace, when it occurs, and let f_p be the plasma frequency at the peak of the layer. Then, corresponding to $K_\parallel = 0$, $K_l = 0$, and $K_r = 0$, one can write

$$f_p^2 = f_o^2 \tag{4-94}$$

$$f_p^2 = f_z^2 + f_z f_B \tag{4-95}$$

$$f_p^2 = f_x^2 - f_x f_B \tag{4-96}$$

The last two equations follow from Eqs. (4-85) and (4-86). From Eqs. (4-95) and (4-96), it can be determined that

$$f_x - f_z = f_B \tag{4-97}$$

where f_B is the gyrofrequency in Hz, or $\omega_B/2\pi$. Also from Eqs. (4-94) and (4-96) it develops that if $f_x \gg f_B$ so that $f_x + f_o \approx 2f_x$,

$$f_x - f_o \approx \frac{f_B}{2} \tag{4-98}$$

4-3-3 Reflection at Oblique Incidence

Knowledge of the electron densities in the E and F regions of the ionosphere, as provided by an ionosonde or otherwise, allows determining the frequencies which can be utilized for communication by reflection from the ionosphere. In communicating from one location to another, however, one would normally utilize reflection at an oblique angle of incidence on the ionosphere rather than at perpendicular incidence. An important concept about ionospheric communication is that of the maximum usable frequency at oblique incidence. This frequency can be determined for the ordinary wave, when the earth's curvature is neglected, by a simple application of Snell's law. If this law is written in terms of the angle measured from the zenith, it takes the form

$$n \sin \phi = n_0 \sin \phi_0 \tag{4-99}$$

where n_0 is the index of refraction at the location where the wave is launched at an angle ϕ_0 from the vertical. n_0 is essentially unity and will be so considered. At the highest point in the path of the wave it must travel horizontally for which condition

$$n = \sin \phi_0$$

For the ordinary wave for which $K = 1 - (f_p/f)^2$, where f_p is the plasma frequency at the height in question,

$$n^2 = K = \sin^2 \phi_0 = 1 - \left(\frac{f_p}{f}\right)^2 \tag{4-100}$$

from which $\cos \phi_0 = f_p/f$ and

$$f = f_p \sec \phi_0 \tag{4-101}$$

This expression gives the maximum frequency f which will be reflected from the height where the plasma frequency is f_p in the case of a wave launched at an angle ϕ_0 from the vertical. The frequency actually used for communication must be equal to or less than f and, as a practical matter, would normally be significantly less if f_p is the value for the peak of the layer, in order to provide a margin of reliability.

The prediction of the performance of HF systems and the choice of operating frequencies has long been of interest to the Institute of Telecommunication Sciences (ITS) and to the National Bureau of Standards (NBS), which had responsibility for such matters before the formation of ITS [8,12]. Oblique sounders, similar to ionosondes but operating with a transmitter at one location and a receiver at a distant location, are highly useful tools for obtaining both long-term and real-time data on propagation conditions. Oblique sounder terminals can be located at essentially the same positions as the terminals of a particular HF path and used to select suitable operating frequencies for the path.

Communication by ionospheric reflection at HF frequencies was formerly a principal means of long-distance communication, and it is still an important and useful method. Communication by satellite, however, has taken over a large share of long-distance communication, and ionospheric propagation is not so widely used as it was. Disadvantages of communication using HF frequencies and ionospheric reflection are the variability of the ionosphere and posssible disruptions of service due to ionospheric storms and disturbances, the low bandwidth available, and the susceptibility to mutual interference between various users. Advantages are low cost and the ability to communicate between a number of rather widely scattered locations. An example of an application that favors HF techniques is that of obtaining information from a number of widely scattered remote sensors such as low-cost data buoys which can be dropped at sea. The location of the buoys can be determined by HF radar, and data can be transmitted from the buoys at HF frequencies. HF radar is also of interest for other over-the-horizon applications such as air-traffic control over the oceans and remote monitoring of sea state [13].

The propagation of ionospherically reflected waves is a large and interesting subject which is treated by Budden [5], Davies [8], Kelso [17], and Ratcliffe [28]. In this text we limit ourselves to these brief remarks and move on to consider the subject of Faraday rotation, which affects both those waves which are reflected from the ionosphere and those which have a sufficiently high frequency to pass through the ionosphere.

4-3-4 Faraday Rotation

Propagation Parallel to Magnetic Field. In Sec. 4-3-1 it was demonstrated that the two characteristic waves that may propagate parallel to the applied magnetic field in a plasma are left and right circularly polarized waves. In this section parallel propagation of a linearly polarized wave will be considered. A linearly polarized wave is not a characteristic wave, and its polarization changes as it propagates. In the absence of losses, the polarization remains linear. The direction of the electric field intensity vector rotates, however, and the rotation, known as Faraday rotation, constitutes a change in polarization.

The rotation can be analyzed recognizing that a linearly polarized wave can be decomposed into left and right circularly polarized waves (lc and rc waves) of equal amplitude. Consider that at $z = 0$ the electric field intensity $\mathbf{E}$ of a linearly polarized wave propagating in the z direction is directed along the x axis as shown in Fig. 4-11. The direction of propagation is the z direction, and $\mathbf{B}_0$ lies along the z axis. The vectors $\mathbf{E}_l$ and $\mathbf{E}_r$, which are the $\mathbf{E}$ vectors of the component lc and rc waves, are shown at particular instants of time, taken as $t = 0$ and $t > 0$, and their directions of rotation are indicated by small auxiliary arrows. The relation between the values of $\mathbf{E}$, $\mathbf{E}_l$, and $\mathbf{E}_r$ at $z = 0$ is

$$\mathbf{E}(0, t) = \mathbf{E}_l + \mathbf{E}_r = |\mathbf{E}_l| e^{-j\omega t} + |\mathbf{E}_r| e^{j\omega t} \tag{4-102}$$

The quantities $e^{\pm j\omega t}$ represent the angular positions of $\mathbf{E}_l$ and $\mathbf{E}_r$. The mag-

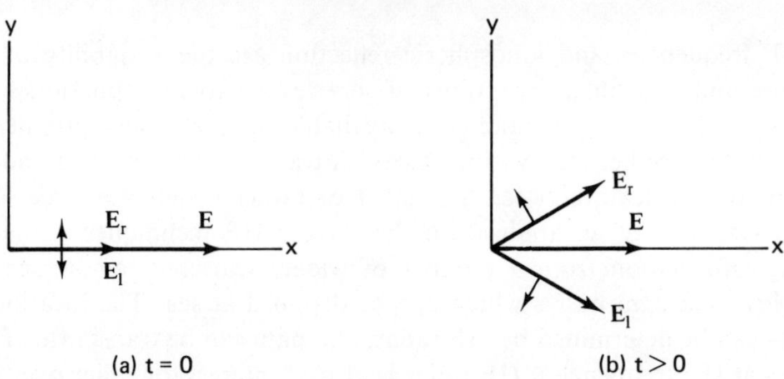

FIGURE 4-11. Vector diagram of possible instantaneous positions of **E**, $\mathbf{E}_l$, and $\mathbf{E}_r$.

nitude of **E** varies as cos ωt. At $z = z$ in a homogeneous region, **E** is described by the equation

$$\mathbf{E}(z, t) = |\mathbf{E}_l| e^{-j(\omega t - k_l z)} + |\mathbf{E}_r| e^{j(\omega t - k_r z)} \tag{4-103}$$

It is important to realize that the exponential factor $e^{j(\omega t - kz)}$ is used in different ways in discussing linearly polarized and circularly polarized waves. A linearly polarized wave having an x component of **E** and propagating in the z direction can be described by

$$\mathbf{E}(z) = \mathbf{a}_x E_m e^{-jkz}$$

Multiplying both sides of the equation by $e^{j\omega t}$, which factor was to be understood in the previous form, we obtain

$$\mathbf{E}(z)e^{j\omega t} = \mathbf{a}_x E_m e^{j(\omega t - kz)}$$

An equation for the instantaneous magnitude of **E** can be obtained by taking the real part of the latter equation. Thus

$$\mathbf{E}(z, t) = \mathbf{a}_x E_m \operatorname{Re}(e^{j(\omega t - kz)})$$

E_m itself can be complex but is taken as real here. It is important to note that information about the direction of the electric field intensity vector is given only by $\mathbf{a}_x$ in the above equations. The quantity $E_m e^{-jkz}$ is a phasor.

In the case of a right circularly polarized wave, however, the electric field intensity is described by

$$\mathbf{E}_r(z, t) = |\mathbf{E}_r| e^{j(\omega t - k_r z)}$$

where $\mathbf{E}_r$ is a vector having a magnitude indicated by $|\mathbf{E}_r|$ and a direction indicated by $e^{j(\omega t - k_r z)}$. When the $e^{j\omega t}$ factor is omitted, magnitude is indicated more simply by E_r, and the equation

$$\mathbf{E}_r(z) = E_r e^{-jk_r z}$$

is written, it should be understood to apply to a vector that is rotating uniformly with angular velocity ω in the rc direction but with a lag in the phase of rotation. For a left circularly polarized wave

$$\mathbf{E}_l(z, t) = |\mathbf{E}_l| e^{-j(\omega t - k_l z)}$$

and

$$\mathbf{E}_l(z) = E_l e^{jk_l z}$$

This expression refers to a wave rotating in the lc direction but lagging in the rc direction with increasing z, as depicted in Fig. 4-12.

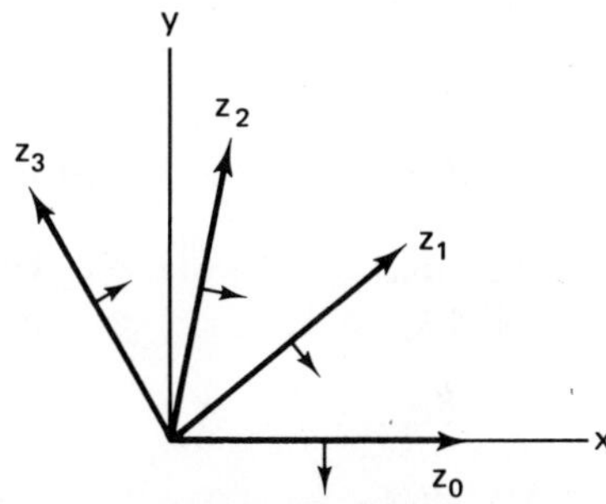

FIGURE 4-12. Instantaneous positions of electric field intensity vectors of left circularly polarized wave.

The propagation constants k_l and k_r, which are identical with β_l and β_r, should not be confused with K_l and K_r, the relative dielectric constants for the two wave types. To determine k_l, starting from K_l, for example, one can calculate $n_l = \sqrt{K_l}$ and $k_l = k_0 n_l$, where $k_0 = 2\pi/\lambda_0$ and $\lambda_0 = c/f$. Figure 4-12 can be regarded as a snapshot showing instantaneous positions of electric field intensity vectors for four different values of z, where $z_1 > z_0$, $z_2 > z_1$, etc. Thus the farther from the source or reference position in the $+z$ direction in which the wave is propagating, the greater the lag or retardation in the phase of rotation.

The direction of E at $z = z$ will now be compared with that at $z = 0$, at $t = 0$. At $t = 0$ and $z = 0$, let E, E_l, and E_r all lie along the x axis, as in Fig. 4-11(a). At $t = 0$ and $z = z$, Eq. (4-103) becomes

$$\mathbf{E}(z, 0) = E_l e^{jk_l z} + E_r e^{-jk_r z} \tag{4-104}$$

An an example of the application of this equation, let $k_l z = 80°$ and $k_r z = 40°$. The appropriate vector diagram is then shown in Fig. 4-13. The **E** vector is halfway between the $\mathbf{E}_l$ and $\mathbf{E}_r$ vectors as at $z = 0$. The $\mathbf{E}_l$ and $\mathbf{E}_r$ vectors rotate with the same angular velocity and in the same directions as at $z = 0$. However, the vectors $\mathbf{E}_l$ and $\mathbf{E}_r$ lag behind their positions at $z = 0$. k_l is larger than k_r, so the $\mathbf{E}_l$ vector lags more than the $\mathbf{E}_r$ vector. The lags

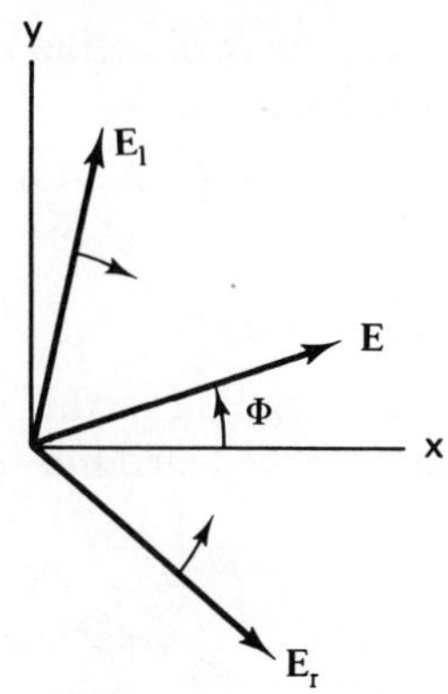

FIGURE 4-13. Faraday rotation through angle ϕ.

are opposite to the respective directions of rotation. Because the **E** vector is halfway between $\mathbf{E}_l$ and $\mathbf{E}_r$ and $k_l z > k_r z$,

$$\frac{k_l z - k_r z}{2} = \phi \tag{4-105}$$

where ϕ is the angle of rotation of E. Equation (4-105) is the basic equation that describes Faraday rotation in a homogeneous medium. The equation indicates that the rotation for the example depicted in Fig. 4-13 is

$$\frac{80^\circ - 40^\circ}{2} = 20^\circ$$

Faraday rotation is in the rc direction.

The above discussion applied to a homogeneous medium. For an actual atmosphere it is necessary to take into account the variation of the propagation constants and to integrate to obtain the total rotation. The calculation of the rotation for sufficiently high frequencies can be simplified by noting that

$$\begin{aligned}\frac{k_l - k_r}{2} &= k_0\left(\frac{n_l - n_r}{2}\right) = \frac{k_0}{2}\left[\left(1 - \frac{X}{1+Y}\right)^{1/2} - \left(1 - \frac{X}{1-Y}\right)^{1/2}\right] \\ &\approx \frac{k_0}{2}\left[1 - \frac{X}{2(1+Y)} - 1 + \frac{X}{2(1-Y)}\right]\end{aligned} \tag{4-106}$$

so that

$$\frac{k_l - k_r}{2} \approx \frac{k_0}{2} XY \tag{4-107}$$

The increment of rotation $d\phi$ in length dz can then be written as

$$d\phi \approx \frac{\pi}{\lambda_0}(XY)\, dz$$

where λ_0 is the free-space wavelength. The product XY is $(\omega_p^2/\omega^2)(\omega_B/\omega)$, and λ_0 can be expressed as $2\pi c/\omega$ so that

$$\left(\frac{\pi}{\lambda_0}\right)(XY) = \frac{\omega_p^2 \omega_B}{2c\omega^2} = \frac{NB_0 e^3}{2c\varepsilon_0 m^2 \omega^2}$$

The total rotation can then be given by [29]

$$\phi = \frac{e^3}{2c\varepsilon_0 m^2 \omega^2} \int NB\, dz \tag{4-108}$$

or

$$\phi = \frac{2.36 \times 10^4}{f^2} \int NB\, dz$$

with f in Hz and all quantites in MKS units. If propagation is not parallel to the magnetic field but the QL approximation applies, $B \cos\theta$ should be used in place of B, where θ is the angle between the direction of propagation and the direction of the magnetic field.

Faraday rotation can be used as a tool to study the ionosphere. Equation (4-108), for example, together with knowledge of the variation of B along a path, can be used to monitor variations in total ionospheric electron content along a path from a synchronous satellite to the earth. In communication applications, the fact that Faraday rotation occurs may need to be taken into account.

4-3-5 Ionospheric Scintillation

It was originally considered that the 4- and 6-GHz bands used for satellite communications would be totally unaffected by the ionosphere, but ionospherically induced scintillations have been observed in these bands. The largest effects occur in the equatorial and auroral regions, and lower frequencies such as those near 137 MHz are considerably more affected than microwave frequencies. At 137 MHz scintillations greater than 6 dB occur on zenith paths for less than 20% of the time near the geomagnetic equator, less than 2% of the time in the auroral zone, and less than 0.1% of the time at middle latitudes [7]. Equatorial scintillation is predominantly a nighttime phenomenon. Scintillation of signals from radio stars has been observed since about 1950, and thus scintillations at the lower frequencies should not be regarded as surprising, but the microwave scintillations were unexpected. Presentation of data on scintillation, as for propagation data in general, is facilitated by plots showing the fraction of time at which a certain level is, or is not, exceeded. Illustrative plots showing scintillation at two equatorial locations are given in Fig. 4-14 [35].

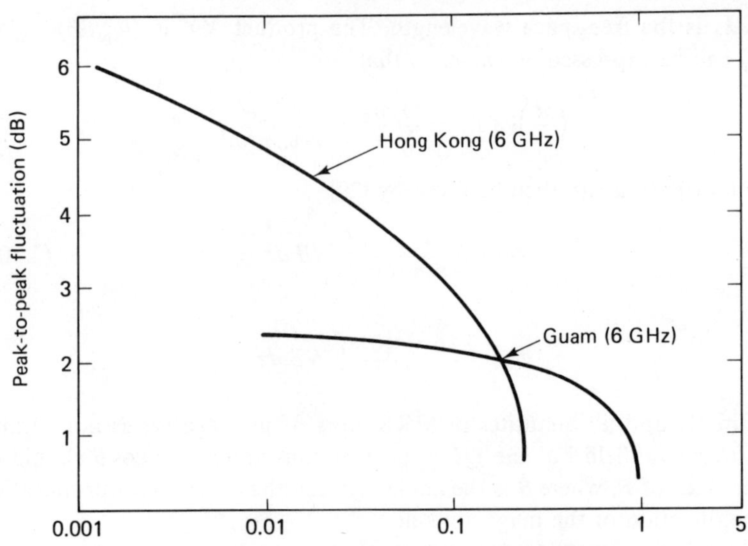

FIGURE 4-14. Cumulative amplitude distributions showing scintillation of satellite 6-GHz signals as received at Guam and Hong Kong. (From Taur, "Ionospheric Scintillation at 4 and 6 GHz," *COMSAT Technical Review*, 1973 [35].)

4-4 SCATTERING AND ABSORPTION BY PARTICLES AND GASES

4-4-1 Introduction

Scattering refers to a process of reradiation of incident electromagnetic energy. The reradiaton tends to have a rather broad directional pattern, so that some of the energy in an incident beam of radiation is scattered out of the beam. Scattering causes attenuation but does not involve the conversion of electromagnetic energy to another form, whereas absorption refers to the dissipation or conversion of electromagnetic energy into thermal energy. Scattering and absorption take place simultaneously in general, but one effect or the other may dominate in a particular case.

Scattering is comparatively simple to analyze when the scattering particles are essentially lossless and small compared to wavelength. The theory of this type of scattering is presented in Sec. 4-4-2. The more complicated subject of the scattering and absorption of electromagnetic waves by rain is considered briefly in Sec. 4-4-3.

Raindrops (0.05–0.65 cm in diameter) may be small compared to the wavelength of incident microwave radiation, and fog droplets (0.001–0.005 cm in diameter) are clearly small compared to microwave and millimeter-wave wavelengths. Water, however, is a lossy material, and there tends to be an important difference between the effects of water on microwaves and the scattering of light by the molecules of the air. The effect of small, lossy water droplets at microwave frequencies is primarily one of absorption, whereas scattering may be the predominant effect when optical waves are incident on air molecules. When the scattering particles are small compared to wavelength, the scattering process is known as Rayleigh scattering. When raindrops are not small compared to wavelength, both scattering and absorption may be important, and the more complicated general theory of Mie scattering must be used.

The incoherent scatter technique is considered in Sec. 4-4-5. The Raman and flourescence scattering phenomena, which involve a change of frequency of scattered radiation with respect to incident radiation, are mentioned only briefly in Sec. 4-4-6. Scattering of radiation by atmospheric turbulence is a large subject in itself and is considered in Sec. 4-5.

Scattering and absorption both cause attenuation, and their combined effect is known as extinction; that is,

$$\text{extinction} = \text{scattering} + \text{absorption}$$

This concept is applied to cross sections, attenuation coefficient, optical depth, etc. Thus

$$\alpha_{\text{ext}} = \alpha_{\text{sca}} + \alpha_{\text{abs}} \tag{4-109}$$

where α is an attenuation coefficient. The integral of an attenuation coefficient with respect to path length is known as optical depth and will be indicated here by the symbol τ. Thus

$$\tau_{\text{ext}} = \tau_{\text{sca}} + \tau_{\text{abs}} \tag{4-110}$$

A third parameter related to attenuation coefficient and optical depth is the transmission coefficient T. The relation between all three quantities is illustrated by the following expression, which gives the attenuated power density P in terms of the power density P_0 at a reference location:

$$P = P_0 e^{-\int \alpha_{\text{sca}} dl} = P_0 e^{-\tau_{\text{sca}}} = P_0 T_{\text{sca}} \tag{4-111}$$

for the case when scattering only is considered. Note that the attenuation coefficient discussed here is a power attenuation coefficient. In other cases an attenuation coefficient applying to field intensity is employed, and it must be stated or implied which type is being used in a particular situation. The coefficients given above may also be separated in other ways. Attenuation due to gases, for example, might be distinguished from attenuation due to particles.

4-4-2 Scattering and Absorption in Dry Air

Rayleigh scattering occurs when the size of the scattering particle is small compared to wavelength. When this condition applies, a major simplification results. Relative phase shifts throughout the particle can be neglected, and the effect of the scattering particle on electric field intensity can be determined from a solution of Laplace's equation for the scalar electric potential Φ as in a static problem, assuming a spherical shape for the particle. The application of Laplace's equation indicates that a small spherical particle acts like a small electric dipole having a dipole moment of

$$p_1 = 4\pi a^3 \left(\frac{n^2 - 1}{n^2 + 2}\right) \varepsilon_0 E_0 = 3V\left(\frac{n^2 - 1}{n^2 + 2}\right) \varepsilon_0 E_0 \tag{4-112}$$

where a is the radius of the particle, V is its volume, n is its index of refraction, and E_0 is the value of the sinusoidal incident electric field intensity which causes the particle to be polarized and to have a dipole moment. A person familiar with antenna theory will recognize a time-varying electric dipole as an elementary antenna that radiates an electromagnetic wave. The radiated electric field intensity at a large distance r from such an elementary antenna, assuming p_1 is oriented along the z axis and θ is measured from this axis, is given by

$$E = \frac{-p_1 \sin \theta k^2}{4\pi\varepsilon_0 r} e^{-jkr} \tag{4-113}$$

where k is the propagation constant. Substituting the expression for p_1 into Eq. (4-113) gives

$$|E_\theta| = \frac{3\pi V}{\lambda^2 r}\left(\frac{n^2 - 1}{n^2 + 2}\right) E_0 \sin \theta \tag{4-114}$$

To determine the attenuation of an incident wave in a region containing such scatterers, calculate first the total power scattered by one particle, the assumption being that this amount of power is scattered out of an incident beam. This total scattered power W can be obtained by integrating $E_\theta^2/2\eta$ over a spherical surface surrounding the scatterer, assuming E_θ to be a peak value. η is the characteristic impedance of the medium. Thus

$$W = \frac{9\pi^2 V^2 E_0^2}{2\eta\lambda^4 r^2}\left(\frac{n^2 - 1}{n^2 + 2}\right) \int_0^{4\pi} \sin^2 \theta \, d\Omega \tag{4-115}$$

where $d\Omega = r^2 \sin \theta \, d\theta \, d\phi$ is an element of solid angle. Carrying out the integration, multiplying by N_m (the number of particles per unit volume), and dividing by $E_\theta^2/2\eta$ to obtain a power attenuation coefficient α_{sca}, we obtain

$$\alpha_{sca} = \frac{24\pi^3 V^2 N_m}{\lambda^4}\left(\frac{n^2 - 1}{n^2 + 2}\right) \tag{4-116}$$

For air molecules for which $n \simeq 1$,

$$N_m\left(\frac{n^2-1}{n^2+2}\right)^2 V^2 = \frac{1}{N_m}\left[N_m\left(\frac{n^2-1}{n^2+2}\right)V\right]^2 \\ = \frac{1}{N_m}\frac{N_m(n-1)(n+1)V^2}{n^2+2} \simeq \frac{4}{9N_m}[N_m(n-1)V]^2 \tag{4-117}$$

To express the attenuation coefficient in terms of the average index of refraction for air, rather than in terms of the value for the individual molecules, let

$$N_m(n-1)V = (n-1)_{\text{air}} \tag{4-118}$$

The power attenuation coefficient then becomes

$$\alpha_{\text{sca}} = \frac{32\pi^3(n-1)^2_{\text{air}}}{3N_m\lambda^4} \tag{4-119}$$

A corresponding optical depth τ_{sca} for Rayleigh scattering for a vertical path through the entire atmosphere may be obtained by substituting for N_m the number 2.547×10^{25}, this being the number of molecules per unit volume of air at sea level, and then multiplying by 8.44×10^3 m, the thickness of the atmosphere if reduced to standard pressure and temperature. If λ is expressed in micrometers, as is convenient for optical wavelengths and $n-1$ is determined by use of Eq. (4-23), the result is

$$\tau_{\text{sca}} = \frac{8.46 \times 10^{-3}}{\lambda^4} \tag{4-120}$$

For a nonvertical path

$$P = P_0 e^{-\int \alpha_{\text{sca}}\, dh \sec \chi} = P_0 e^{-\tau_{\text{sca}} m} \tag{4-121}$$

where χ is the zenith angle (angle measured from the vertical) and m is effective path length relative to a vertical path. Rayleigh scattering is responsible for the blue color of the sky, as $\lambda_{\text{blue}} \simeq 0.425\ \mu$m and $\lambda_{\text{red}} \simeq 0.650\ \mu$m, for example, and blue light is scattered more than light near the other end of the spectrum.

The above calculation refers to scattering by air molecules, but the particles of natural and man-made origin mentioned in Sec. 3-3-3 also cause scattering. Depending on the size and properties of the particles, the attenuation they cause may or may not be described by Eq. (4-116). In addition the normal atmospheric gases, including water vapor, cause absorption which varies with frequency. The total values of attenuation coefficient α and optical depth τ can in general be separated into components such that, for example,

$$\tau = \tau_{\text{abs}} + \tau_{\text{sca}} + \tau_d$$

where τ_{abs} accounts for absorption by gases, τ_{sca} accounts for scattering by gases as determined above, and τ_d is a turbidity coefficient which accounts for scattering and absorption by aerosols. The parameters τ_{abs} and τ_{sca} (or

α_{abs} and α_{sca}) are known and tabulated in tables so that measurements can allow determination of τ_d. τ is the extinction (total) optical depth. Tables giving propagation parameters for optical and infrared frequencies are given in Valley [38], and Table 4-2 consists of brief excerpts.

Table 4-2 Parameters for a Clear Standard Atmosphere.*

Alt. (km)	*Rayleigh Atten. Coeff. (per km)*	*Rayleigh Optical Thick. (0–h)*	*Aerosol Atten. Coeff. (per km)*	*Ext. Coeff. (per km)*	*Ext. Optical Thick. (0–h)*
		Parameters at 0.40 μm			
0	4.303 −2	0.0000	2.00 −1	2.43 −1	0.000
1	3.905 −2	0.0410	8.80 −2	1.27 −1	0.185
2	3.536 −2	0.0782	3.80 −2	7.34 −2	0.285
3	3.194 −2	0.1119	1.60 −2	4.79 −2	0.346
4	2.878 −2	0.1422	7.20 −3	3.60 −2	0.388
5	2.587 −2	0.1696	3.20 −3	2.91 −2	0.420
6	2.319 −2	0.1941	1.10 −3	2.43 −2	0.447
7	2.072 −2	0.2160	4.00 −4	2.11 −2	0.470
8	1.847 −2	0.2356	1.40 −4	1.86 −2	0.490
9	1.641 −2	0.2531	5.00 −5	1.65 −2	0.507
10	1.452 −2	0.2685	2.60 −5	1.46 −2	0.523
12	1.096 −2	0.2941	2.20 −5	1.10 −2	0.548
14	8.003 −3	0.3129	2.60 −5	8.03 −3	0.567
16	5.847 −3	0.3267	6.80 −5	5.92 −3	0.581
18	4.273 −3	0.3368	8.00 −5	4.35 −3	0.591
20	3.122 −3	0.3441	8.60 −5	3.21 −3	0.599
25	1.408 −3	0.3549	3.60 −5	1.44 −3	0.610
30	6.465 −4	0.3598	1.90 −5	6.66 −4	0.615
		Parameters at 0.80 μm			
0	2.545 −3	0.0000	1.27 −1	1.30 −1	0.000
1	2.309 −3	0.0024	5.59 −2	5.82 −2	0.094
2	2.091 −3	0.0046	2.41 −2	2.63 −2	0.136
3	1.889 −3	0.0066	1.02 −2	1.21 −2	0.155
4	1.702 −3	0.0084	4.57 −3	6.30 −3	0.164
5	1.530 −3	0.0100	2.03 −3	3.58 −3	0.169
6	1.371 −3	0.0115	6.98 −4	2.09 −3	0.172
7	1.226 −3	0.0128	2.54 −4	1.50 −3	0.174
8	1.092 −3	0.0139	8.89 −5	1.20 −3	0.175
9	9.702 −4	0.0150	3.17 −5	1.03 −3	0.177
10	8.590 −4	0.0159	1.65 −5	9.10 −4	0.177
12	6.480 −4	0.0174	1.40 −5	7.24 −4	0.179
14	4.733 −4	0.0185	1.65 −5	5.86 −4	0.180
16	3.458 −4	0.0193	4.32 −5	4.92 −4	0.182
18	2.527 −4	0.0199	5.08 −5	4.26 −4	0.182
20	1.847 −4	0.0204	5.46 −5	4.03 −4	0.183
25	8.325 −5	0.0210	2.29 −5	2.86 −4	0.185
30	3.824 −5	0.0213	1.21 −5	1.41 −4	0.186

Table 4-2 (Continued)

Alt. (km)	*Rayleigh Atten. Coeff. (per km)*	*Rayleigh Optical Thick. (0–h)*	*Aerosol Atten. Coeff. (per km)*	*Ext. Coeff. (per km)*	*Ext. Optical Thick. (0–h)*
		Parameters at 1.26 μm			
0	4.076 −4	0.0000	1.08 −1	1.08 −1	0.000
1	3.699 −4	0.0004	4.75 −2	4.79 −2	0.078
2	3.349 −4	0.0007	2.05 −2	2.09 −2	0.113
3	3.025 −4	0.0011	8.64 −3	8.94 −3	0.127
4	2.726 −4	0.0013	3.89 −3	4.16 −3	0.134
5	2.450 −4	0.0016	1.73 −3	1.97 −3	0.137
6	2.196 −4	0.0018	5.94 −4	8.14 −3	0.138
7	1.963 −4	0.0020	2.16 −4	4.12 −4	0.139
8	1.749 −4	0.0022	7.56 −5	2.51 −4	0.139
9	1.554 −4	0.0024	2.70 −5	1.82 −4	0.140
10	1.376 −4	0.0025	1.40 −5	1.52 −4	0.140
12	1.038 −4	0.0028	1.19 −5	1.16 −4	0.140
14	7.582 −5	0.0030	1.40 −5	8.99 −5	0.140
16	5.539 −5	0.0031	3.67 −5	9.21 −5	0.140
18	4.048 −5	0.0032	4.32 −5	8.37 −5	0.141
20	2.958 −5	0.0033	4.64 −5	7.60 −5	0.141
25	1.334 −5	0.0034	1.94 −5	3.28 −5	0.141
30	6.125 −6	0.0034	1.03 −5	1.64 −5	0.141

*Numbers following the minus sign are the power of 10 by which the entry should be multiplied. For Rayleigh optical thickness, the 0–∞ values for 0.40, 0.80, and 1.26 μm are 0.2238, 0.0215, and 0.0034, respectively. For the extinction optical thickness, the 0–∞ values are 0.619, 0.187, and 0.141, respectively. h–∞ values can be obtained by subtracting 0–h values from the 0–∞ values.
From Valley [38].

4-4-3 Effects of Rain on Microwaves and Millimeter Waves: Mie Scattering

Rain, snow, and hail may cause strong attenuation of optical and millimeter waves, and rain can be troublesome at frequencies at least as low as the *L* band (frequencies near 1400 MHz, wavelengths near 23 cm), where backscatter echoes from rain can interfere with the detection of othei radar targets. From the viewpoint of a meteorologist, however, radar echoes from rain may provide useful information. The U.S. Weather Service has a network of *S*-band radars and also uses some of the FAA *L*-band radars, for example, to obtain weather data, and radar meteorology is a well-developed and interesting subject. Nevertheless, for someone using radar for aircraft surveillance or a microwave link for communications, rain may be a nuisance. The effects of rain tend to become increasingly intense at higher frequencies, up to about 150 GHz at least, and millimeter-wavelength transmissions tend to be seriously limited by rain.

As the demand for communication capacity grows, increasing attention is being given to the use of higher frequencies, including those corresponding to millimeter wavelengths, for earth-satellite communication paths. Thus the effects of rain on microwave and millimeter-wave transmissions is a subject of much interest at this time.

The power attenuation coefficient for rain can be expressed as

$$\alpha_p = \int_a n(a, x, t) C_{\text{ext}}\left(n_c, \frac{a}{\lambda}\right) da \tag{4-122}$$

where a is drop radius and integration is indicated over the entire range of drop sizes. Actually the calculation is carried out as a summation and $n(a, x, t)\, da$ gives the number of drops within an increment da of a as a function of distance along the path x and time t. C_{ext} is an extinction cross section and accounts for both scattering and absorption; C_{ext} is a function of the complex index of refraction n_c of the drop, and this in turn is a function of frequency. C_{ext} is also a function of the ratio of drop radius to wavelength. A serious practical difficulty is that of estimating $n(a, x, t)\, da$, or even $n(a)\, da$, corresponding to a given rainfall rate. In practice an empirical distribution, the Laws and Parsons distribution, which is a function of rainfall rate, has often been used and has given good results. There are relatively more large drops in heavy rain than in light rain, and attenuation is not a linear function of rainfall rate.

The Laws and Parsons data, shown in Table 4-3, does not give $n(a)\, da$ directly for the application of Eq. (4-122) at a particular location and time, but gives $m(a)\, da$, the fraction of the total volume of water striking the ground due to drops of a given size. To calculate $n(a)\, da$, one must use $m(a)\, da$ and also information about the terminal velocity $v(a)$, as shown in Fig. 4-15 [4]. Larger drops reach higher limiting or terminal velocities. To calculate $n(a)\, da$ in units of $1/\text{m}^3$ use

$$n(a)\, da = \frac{m(a)\, da\, p}{v(a) a^3 \times 15.1} \tag{4-123}$$

where p is precipitation rate in mm/h, a is in cm, $v(a)$ is in m/s, and $m(a)\, da$ is nondimensional. The value of da used in Table 4-3 is 0.025 cm. Laser techniques are now capable of obtaining drop size distributions, but the Laws and Parsons distribution, obtained by the tedious process of letting raindrops fall on pans of flour and measuring the sizes of the resulting agglomerations, is still used. Of course, even if one does have information giving particle size distributions corresponding to rainfall rates, a practical difficulty is the lack of knowledge about the statistics of rainfall (see Sec. 6-4-4).

The determination of $C_{\text{ext}}(n_c, a/\lambda)$ in the general case is based on the Mie scattering theory as discussed by Goldstein [in 19] and Kerker [18]. C_{ext} is expressed in terms of coefficients which involve Bessel functions of complex arguments. An important feature to notice is the dependence on the a/λ ratio. It is impractical to treat the quite complex Mie scattering theory here. Furthermore, the Mie theory assumes raindrops to be spherical in

Table 4-3 Percent of Volume Reaching Ground Contributed by Drops of Various Sizes.*

Drop Radius Limits, a (mm)	*p (mm/h) (% of Total Volume)*							
	0.25	1.25	2.5	12.5	25	50	100	150
0–0.125	1.0	0.5	0.3	0.1				
0.125–0.375	27.0	10.4	7.0	2.5	1.7	1.2	1.0	1.0
0.375–0.625	50.1	37.1	27.8	11.5	7.6	5.4	4.6	4.1
0.625–0.875	18.2	31.3	32.8	24.5	18.4	12.5	8.8	7.6
0.875–1.125	3.0	13.5	19.0	25.4	23.9	19.9	13.9	11.7
1.125–1.375	0.7	4.9	7.9	17.3	19.9	20.9	17.1	13.9
1.375–1.625		1.5	3.3	10.1	12.8	15.6	18.4	17.7
1.625–1.875		0.6	1.1	4.3	8.2	10.9	15.0	16.0
1.875–2.125		0.2	0.6	2.3	3.5	6.7	9.0	11.9
2.125–2.375			0.2	1.2	2.1	3.3	6.8	7.7
2.375–2.625				0.6	1.1	1.8	3.0	3.6
2.625–2.875				0.2	0.5	1.1	1.7	2.2
2.875–3.125					0.3	0.5	1.0	1.2
3.125–3.375						0.2	0.7	1.0

*Drop radius interval, $da = 0.25$ mm $= 0.025$ cm. Multiply percentage values by 0.01 to obtain $m(a)\,da$.
After J. O. Laws and D. A. Parsons, "The Relation of Drop Size to Intensity," *Transactions of the American Geophysical Union*, pp. 452–460, 1943.

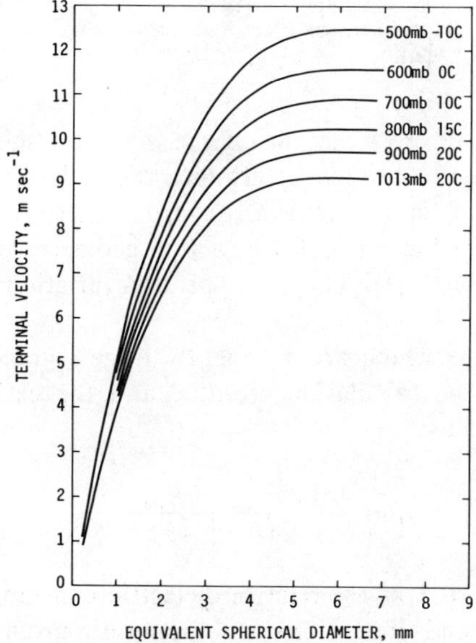

FIGURE 4-15. Terminal velocity of raindrops at six pressure levels in a summer atmosphere as a function of the equivalent spherical diameter. (From Beard, "Terminal Velocity and Shape of Cloud and Precipitation Drops Aloft," *Journal of the Atmospheric Sciences*, May 1976 [4].) The pressures of 1013, 900, 800, 700, 600, and 500 mb correspond roughly to altitudes of 0, 975, 1950, 3000, 4200, and 5600 m of the *U.S. Standard Atmosphere, 1976.*

shape, and an additional complication is that raindrops are actually not truly spherical, especially in the case of large drops. Calculations of attenuation by Zufferey [40] are reproduced in Fig. 4-16.

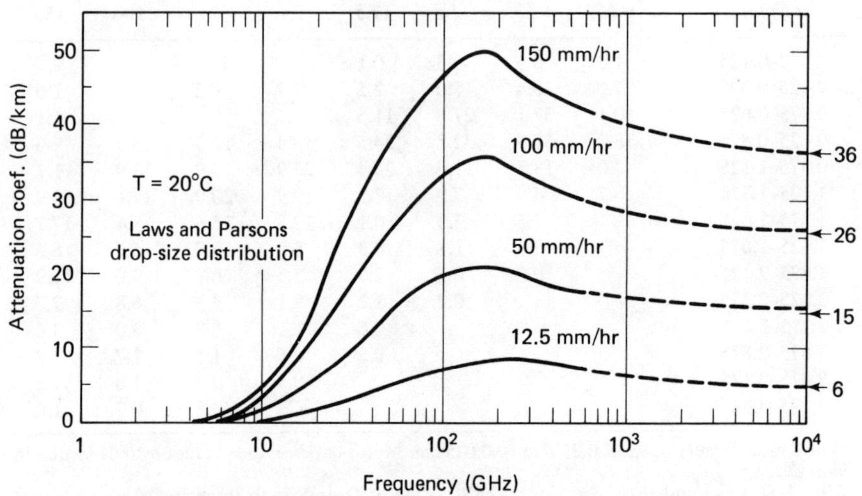

FIGURE 4-16. Average attenuation coefficient as a function of frequency, for selected rainfall rates. (From Zuffery, "A Study of Rain Effects on Electromagnetic Waves in the 1–600 GHz Range," M.S. thesis, Department of Electrical Engineering, University of Colorado, 1972 [40].)

A dependence of cross section on the a/λ ratio shows also in σ, the backscatter cross section, which determines the radar echo intensity of a raindrop. [The C_{ext} of Eq. (4-122) is a forward-scatter cross section.] Figure 4-17 shows results for the ratio of σ to actual geometrical cross section πa^2 as determined by Aden [1]. The form of the variation is a function of frequency.

For wavelengths which are sufficiently large compared to drop size, a simplification in the calculations results, and the extinction coefficient C_{ext} can be expressed as

$$C_{ext} = \frac{4\pi^2 a^3}{\lambda} \frac{6K_{im}}{(K_r + 2)^2 + K_{im}^2} \tag{4-124}$$

where K_r is the real part of the relative dielectric constant of water, K_{im} is the imaginary part, and dimensions are commonly given in cm. For $\lambda =$ 10 cm, K_r is about 78.45 and K_{im} is about 11.19 at 20°C.

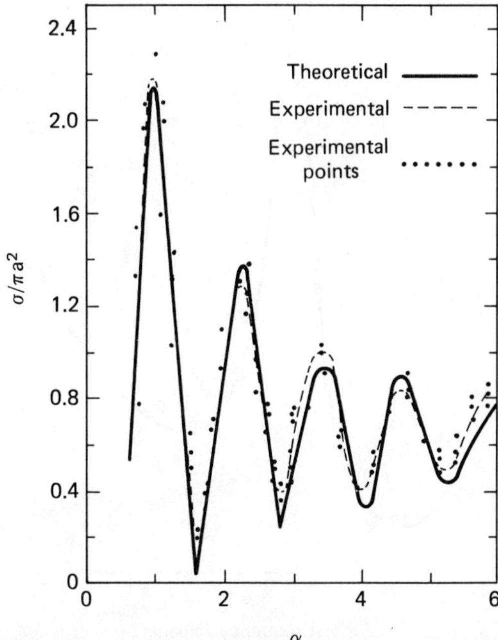

FIGURE 4-17. Backscatter from water spheres. (After Aden, "Electromagnetic Scattering from Spheres with Sizes Comparable to the Wavelength," *Journal of Applied Physics*, May 1951 [1].)

4-4-4 Absorption by Gases

Millimeter-wave transmission in the troposphere is subject to attenuation by water vapor and oxygen, as illustrated in Fig. 4-18. Normally frequencies removed from those where maximum attenuation occurs would be chosen for practical applications. During World War II, however, a *K*-band radar system was developed without realizing that it would operate near the peak of the water vapor absorption line (about 22 GHz). The result was that range was considerably reduced from what was expected.

4-4-5 Incoherent Scatter

In 1958, Gordon predicted that the state of the radar art was such that incoherent backscatter from free electrons of the ionosphere should be detectable, and Bowles observed such backscatter at Long Branch, Illinois in the same year [9]. The basis of backscatter by an electron is that the electron achieves an ordered motion due to the electric field intensity of an incident

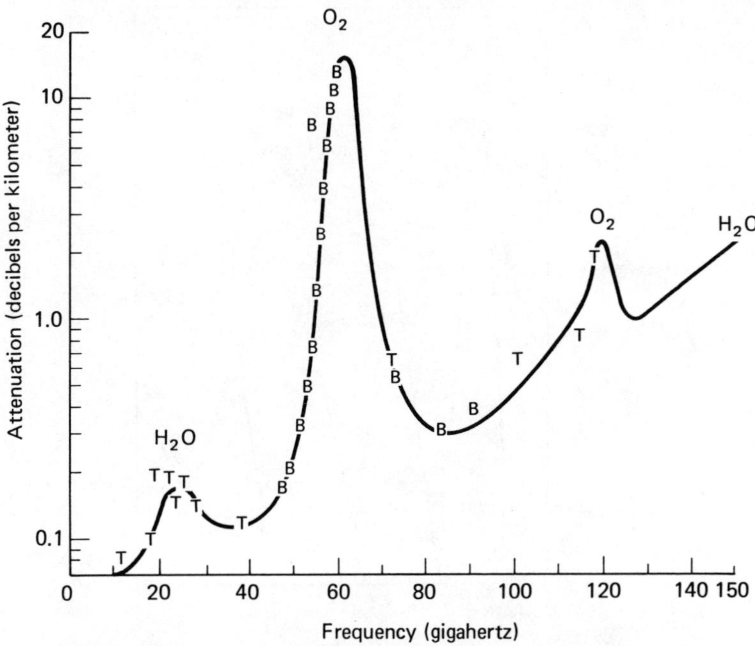

FIGURE 4-18. Attenuation at millimeter wavelengths for a sea-level atmosphere with a water vapor content of 7.5 g/m³. (After D.C. Hogg, "Millimeter-Wave Communication Through the Atmosphere," *Science*, Jan. 5, 1968.) The solid curve is the theoretical and the T's and B's are values measured by the University of Texas and Bell Labs.

wave. An electron exhibiting such motion constitutes an electrical current and acts as a source of electromagnetic waves. That is, each electron becomes a tiny antenna that radiates electromagnetic waves. It was originally thought that the radiated powers would add directly and that the radar cross section per unit volume would be $\sigma_0 N$, where σ_0 is the cross section of an electron and N is the electron density. It turns out, however, that to a first approximation the cross section per unit volume η is given by

$$\eta = \frac{\sigma_0 N}{1 + T_e/T_i} \tag{4-125}$$

where T_e is electron temperature and T_i is ion temperature. The incoherent scatter technique is a powerful tool for studying the ionosphere [9,31]. Among its advantages is the fact that echoes can be received from throughout the entire ionosphere, including the region above the peak of the F layer. As the effects of ions introduce a degree of coherence into the backscatter from free electrons, designation as incoherent is not strictly correct, and the term Thomson scatter is preferred by some. Because familiarity with radar techniques is needed to understand incoherent scatter observations clearly and because incoherent scatter provides a means of remote sensing of the ionosphere, further discussion of the topic is included in Sec. 6-4-6.

4-4-6 Other Scattering Mechanisms

RAMAN SCATTERING. Raman scattering is the weakest of the mechanisms considered here but is of interest because the Raman scatter signal occurs at frequencies of $\nu \pm \nu_1$, $\nu \pm \nu_2$, etc., where ν is the incident frequency and ν_1, ν_2, etc., are characteristic of the scattering molecule. Thus observation of Raman backscatter provides a means of identifying the scattering molecules. The lower-frequency Raman lines are more intense than the higher-frequency lines and are therefore the lines of most interest. The Raman mechanism is illustrated in Fig. 4-19.

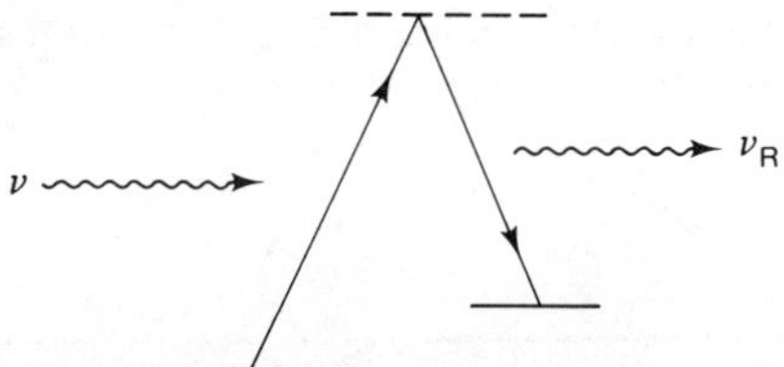

FIGURE 4-19. Mechanism of Raman scatter. $\nu_R = \nu \pm \nu_1$, etc.

FLOURESCENCE AND RESONANCE RAMAN. These mechanisms can also be illustrated by Fig. 4-19, except that in the latter cases the incident frequency is matched to a molecular resonance. The flourescence mechanism can involve relatively large scattering cross sections. Tunable lasers are required to take full advantage of the flourescence mechanism, in contrast to the case for Raman scatter, which can occur over a range of optical and infrared frequencies.

4-5 SCATTERING BY ATMOSPHERIC TURBULENCE

4-5-1 Basic Concepts

Scattering of electromagnetic waves by turbulence is a subject of interest because it is the basis of troposcatter communication, because powerful radars can detect turbulence, and because of the importance and possibly harmful effects of turbulence in aircraft operations. Acoustic sounders are adapted to detecting and monitoring turbulence at short ranges but are not treated in this section.

In considering the scattering of electromagnetic waves by turbulence, part of the problem is one of electromagnetic theory and part is one of turbulence theory. The subject is complex, and only a brief semiquantitative discussion is given here. The application of electromagnetic theory to the case of homogeous, isotropic turbulence results in an expression for the radio cross section per unit volume (m^2/m^3) as follows:

$$\eta = 8\pi^2 k^4 \sin^2 \psi \Phi_n(|\mathbf{k} - \mathbf{k}_0|) \tag{4-126}$$

where $\Phi_n(|\mathbf{k} - \mathbf{k}_0|)$ can also be expressed as $\Phi_n(2k \sin \theta/2)$ or $\Phi_n(2\pi/\lambda')$. The geometry considered is depicted in Fig. 4-20. A turbulent region of the atmosphere is illuminated by an incident wave, having a vector propagation constant $\mathbf{k}_0$. This vector indicates the direction of wave propagation, and its magnitude is the ordinary propagation constant $2\pi/\lambda$. Radiation scattered at an angle θ is indicated by a similar vector propagation constant $\mathbf{k}$. $\mathbf{k}_0$ and $\mathbf{k}$ are equal in magnitude, and only their directions are different. ψ is the angle between the direction of the electrical field intensity of the incident wave and the direction of $\mathbf{k}$.

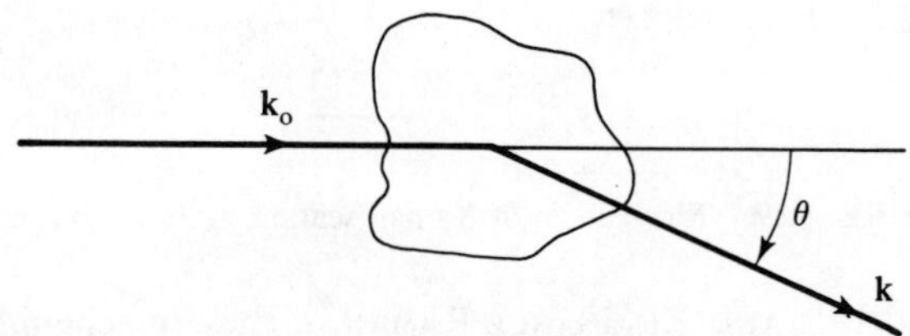

FIGURE 4-20. Scattering geometry.

The expression gives the intensity of the scattered radiation in the direction θ for a given incident power density. $\Phi_n(2\pi/\lambda')$ is the spatial spectrum of index of refraction variations and gives the intensity of the variations that have a periodicity of λ', where $\lambda' = \lambda/[2 \sin(\theta/2)]$, λ being the electromagentic wavelength. For the backscatter case, $\theta = 180°$ and $\psi = 90°$ and

$$\eta = 8\pi^2 k^4 \Phi_n\left(\frac{4\pi}{\lambda}\right) \tag{4-127}$$

If $4\pi/\lambda$ is represented by k_r, a spatial wave number, then $k = k_r/2$ and

$$\eta = \frac{\pi^2}{2} k_r^4 \Phi_n(k_r) \tag{4-128}$$

for the backscatter case.

It can be seen that the echo that a monostatic radar receives from a turbulent region, on the basis of the theory presented, depends only on those index of refraction variations which have a periodicity of one-half the radar wavelength. To determine Φ_n over a range of spatial wave numbers or wavelengths by radar techniques, one would need radars whose frequencies cover a corresponding range. In attempting to calculate η by use of Eqs. (4-125)–(4-127), it is evident that the equations give η in terms of Φ_n but tell nothing about the magnitude of Φ_n. Neither do the equations provide any information about the nature or sources of turbulence.

To consider the subject of turbulence itself, consider a layer of fluid that is confined between two parallel plates. If the lower plate is held stationary but the upper plate is moved with a velocity V, the velocity within the the fluid will vary as suggested by the arrows in Fig. 4-21. The fluid immediately adjacent to the moving plate will move with essentially the same velocity as the plate. The fluid immediately adjacent to the lower plate will have essentially zero velocity. A force F will be required to move the upper plate, and the force will be given by

$$F = \frac{\mu A V}{z} \tag{4-129}$$

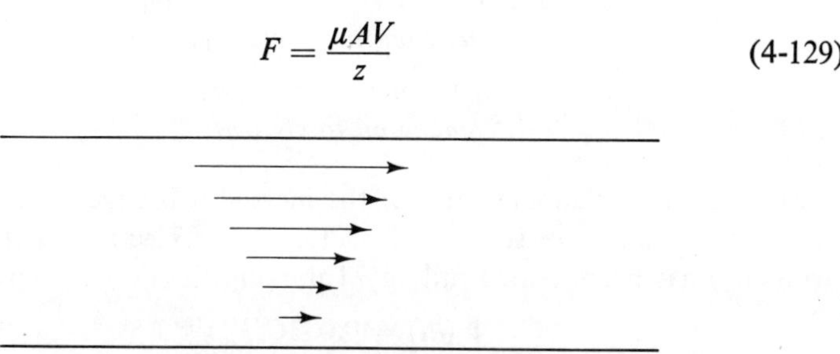

FIGURE 4-21. Flow between fixed and moving plates.

where z is the spacing of the plates, A is the surface area of the upper plate, and μ is called the dynamic coefficient of viscosity. If the force per unit area or stress τ is considered and a linear variation of V with z is assumed in the above case,

$$\tau = \mu \frac{\partial V}{\partial z} \tag{4-130}$$

where $\partial V/\partial z$ represents the degree of shear of the fluid, shear referring to a difference in velocity between adjacent volumes of fluid. What has been considered so far is a smooth, laminar motion, but if the velocity V is increased above a certain value, chaotic, turbulent motion, involving swirling eddies, takes place. The turbulent state occurs when a certain value of the Reynold's number $VL\rho/\mu$ is reached, where ρ is the density of the fluid and L is a characteristic length [corresponding to z in Eq. (4-129)]. μ/ρ is the kinematic coefficient of viscosity ν, and the Reynold's number R can also be given by

$$R = \frac{VL}{\nu} \tag{4-131}$$

In the atmosphere turbulence develops in a roughly similar way. When wind shear (variation of horizontal wind with height), atmospheric waves, and convection resulting from heating occur, turbulence may develop. Turbulence commonly occurs at the boundaries between atmospheric layers.

The theory of homogeneous, isotropic turbulence considers that turbulence is introduced in the form of large turbulent eddies of size L_0 and that energy is transferred from larger to smaller eddies throughout the inertial subrange corresponding to eddies of size l, where

$$L_0 \geq l \geq l_0$$

For eddies smaller than l_0, viscous effects dominate, and turbulent energy is dissipated. The above concept is described by the following verse, credited to L. F. Richardson:

Big whirls have little whirls that
feed on their velocity,
And little whirls have lesser whirls
and so on to viscosity.

The concept and characteristics of the inertial subrange were elucidated by Kolmogorov, and the subject of scattering of electromagnetic waves by turbulence has been treated fully by Tatarski [33,34], who showed that

$$\Phi_n(k_r) = 0.033 C_n^2 k_r^{-11/3} \tag{4-132}$$

C_n^2 is a measure of the spatial variation of index of refraction and is related to C_T^2, which is a measure of the spatial variation of temperature. When this expression is introduced into that for η for backscatter, recalling that for this case $k_r = 4\pi/\lambda$, were λ is the radar wavelength, the result is

$$\eta = 0.38 C_n^2 \lambda^{-1/3} \tag{4-133}$$

When the factors contributing to C_n^2 are analyzed, it develops that η may also be given by

$$\eta = 0.38 a^2 \varepsilon^{2/3} \beta^{-2} \left(\frac{d\bar{n}}{dz}\right)^2 \lambda^{-1/3} \tag{4-134}$$

a^2 is a constant having a value of about 2.8, β is wind shear and can be expressed as du/dz (where u is horizontal wind), and $d\bar{n}/dz$ is the vertical gradient of potential refractive index (refractive index referred to a reference altitude). ε is the rate of dissipation of turbulent energy and is a very important quantity because it is actually a measure of the intensity of turbulence. Equation (4-134) suggests that if β and $d\bar{n}/dz$ are known or if it is known that they vary together then ε could be determined from a measurement of η. At low altitudes $d\bar{n}/dz$ may be strongly affected by humidity variations which may not be correlated with β. At high altitudes, however, $d\bar{n}/dz$ may be determined largely by temperature and $d\bar{n}/dz$ and β may tend to vary together under some conditions so that a measurement of η alone might provide an indication of turbulence intensity [24]. It now appears also, however, that an important factor affecting backscatter from the clear air is the fraction of the volume that actually is turbulent. This aspect is mentioned further in Sec. 6-4-6.

Further insight into the significance of C_n^2 is provided by the fact that

$$D_n(r) = C_n^2 r^{2/3} \tag{4-135}$$

where $D_n(r)$, the structure function for index of refraction variations, is given by

$$D(r) = \overline{[f(R + r) - f(r)]^2} \tag{4-136}$$

where the overbar indicates average value. At a distance of 1 m,

$$D_n(1) = C_n^2 = \overline{(n_1 - n_2)^2}$$

Thus if the index of refraction variations at two points spaced 1 m apart were known, they would provide the value of C_n^2 at that location.

4-5-2 Troposcatter Systems

The forward scatter of electromagnetic radiation by atmospheric turbulence is believed to be the basis of operation of troposcatter systems, which provide communication beyond the horizon at frequencies that are too high to be scattered significantly by the ionosphere. Controversy has existed concerning the relative importance of scattering from turbulence and reflection from layers, but the trend has developed to emphasize scattering from turbulence. The latter does not necessarily occur uniformly throughout the atmosphere, however, but may exhibit a layered structure itself.

As troposcatter systems do not utilize line-of-sight paths, they experience a loss in addition to the "free-space" (L_{FS}) loss introduced in Sec. 2-4. The additional loss is a function of the angular distance θ and is determined by the path geometry (Fig. 4-22), as the transmitted beam is directed to only clear the horizon. For a smooth spherical earth, θ would have the value in radians of d/r, where d is the length of the path and r is the effective earth radius or kr_0 (where r_0 is the actual earth radius).

When the terrain is not smooth, angles of θ_{HT} and θ_{HR} must be added, where θ_{HT} is the angle of the horizon above the horizontal from the transmitter location and θ_{HR} is the corresponding angle for the receiver location. When the latter angles are corrected for earth curvature, the total angle is given by

$$\theta = \frac{d}{r} + \left(\frac{\Delta h_T}{d_{HT}} - \frac{d_{HT}}{2r}\right) + \left(\frac{\Delta h_R}{d_{HR}} - \frac{d_{HR}}{2r}\right) \tag{4-137}$$

where Δh_T is the height of the horizon above (or below) the transmitting location and d_{HT} is the distance to the horizon from the transmitter. Δh_R and d_{HR} have corresponding meanings referred to the receiver location. Various means for estimating the loss factor for troposcatter systems are utilized and are

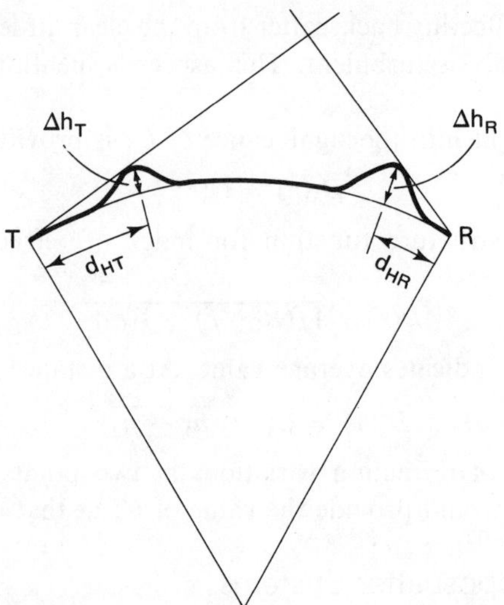

FIGURE 4-22. Geometry of troposcatter path, corresponding to Eq. (4-137); *d* is the distance from *T* to *R*.

discussed by Panter [25] and Freeman [10]. We show here only an empirical curve developed by ITT for 900 MHz. The actual loss will vary with climate, especially with the water vapor content of the atmosphere, and will tend to be greater than that of Fig. 4-23 for very dry or arctic areas and less for very humid areas, Fig. 4-23 referring to temperate latitudes. For frequencies

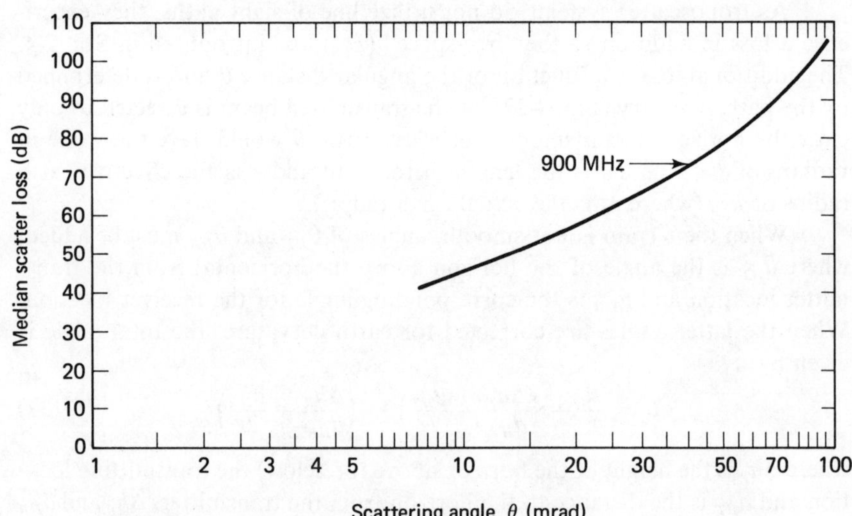

FIGURE 4-23. ITT additional loss factor for troposcatter path. (Courtesy of International Telephone and Telegraph Corporation.)

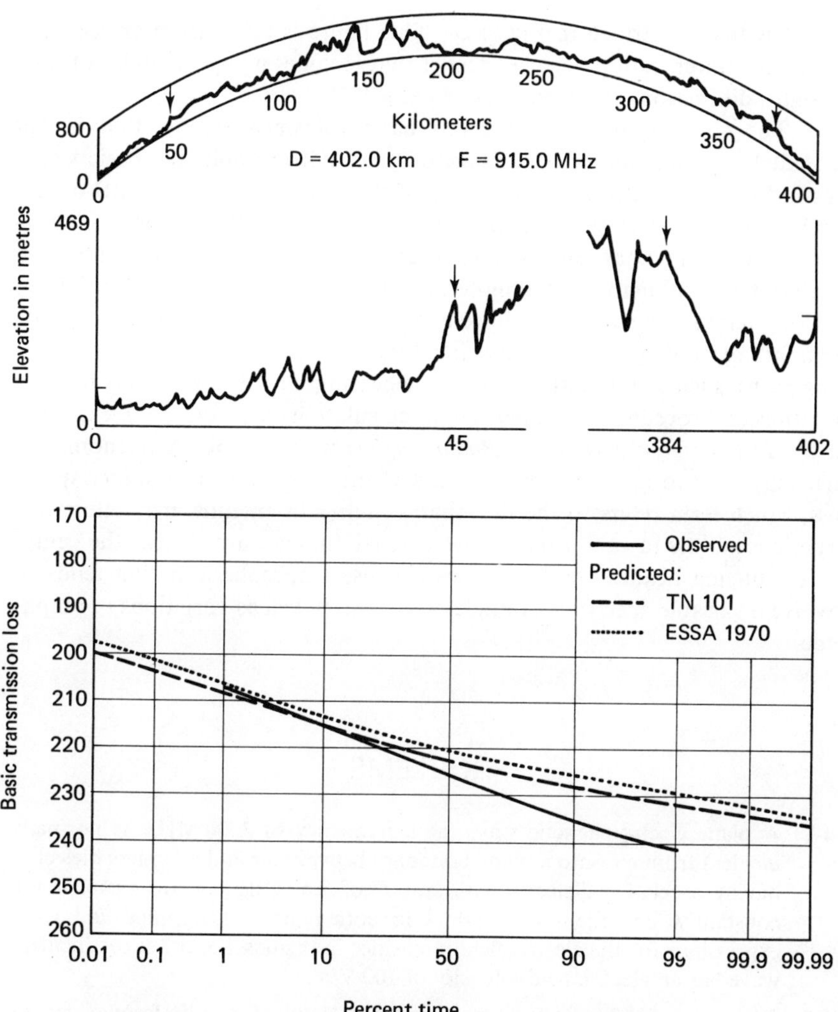

FIGURE 4-24. Observed and predicted performance of 402-km troposcatter path at 915 MHz from Lexington, Massachusetts, to Syracuse, New York. (From Longley et al., "Measured and Predicted Long-Term Distributions of Tropospheric Transmission Loss," 1971 [22].)

other than 900 MHz, a correction factor of 10 log $(f_2/900)$ dB is utilized. Loss is observed to increase with frequency, so that if $f_2 = 1800$ MHz, for example, an additional loss of 3 dB above that for 900 MHz can be expected.

Actual signal levels on radio paths, whether line-of-sight or scatter paths, vary statistically, and a useful procedure for presenting path performance is that illustrated in Fig. 4-24. This figure is taken from an Institute for Telecommunication Sciences report that includes many examples of line-of-sight, diffraction, and troposcatter paths [22].

The curve shows the observed basic transmission loss that is not exceeded as a function of percentage of time. For example, the basic transmission loss is 240 dB or less for 99% of the time and about 208 dB or less for 1% of the time. Basic transmission loss is defined as the loss between loss-free isotropic antennas. Such curves often refer to hourly median values and represent a combination of long-term and short-term fading or variation, the long-term fading tending to have a log-normal distribution and the short-term fading tending to have a Rayleigh distribution. For troposcatter or other long paths, such as HF paths, great circle bearings and distances may need to be utilized. Procedures for great circle calculations are given, for example, in the *ITT Reference Data for Radio Engineers* handbook. A phenomenon that may need to be taken into account is the antenna-to-medium coupling loss, which term refers to the fact that a high-gain antenna may effectively experience a loss in gain if the received signal does not arrive near the center of its antenna beam. The loss arises because tropospheric fading tends to involve a variable spectrum of angle of arrival as well as variations in amplitude.

PROBLEMS

4-1 A plane electromagnetic wave, at a frequency of 3000 MHz, is normally incident from air onto a plane boundary between air and a lossless dielectric having a relative dielectric constant K of 2.5. Calculate the propagation constant k and the wavelength λ in both regions. Determine and write expressions for the electric field intensities in regions 1 and 2 if the incident wave has an electric field intensity of 100 V/m.

4-2 Determine the index of refraction in large N units for radio frequencies for the *U.S. Standard Atmosphere* for altitudes of 0, 2000, and 4000 m for zero water vapor pressure and for a relative humidity of 80%. (Relative humidity $= e/e_s$, where e_s is the saturation water vapor pressure. e_s is tabulated as 17.06 mb for 15°C, 7.05 mb for 2°C, and 2.40 mb for −11°C.) Calculate also the approximate index of refraction in large N units for the same altitudes for optical frequencies.

4-3 The temperature increases by 10°C in a thin temperature inversion layer. Calculate the change in index of refraction in large N units for radio and acoustic waves as one proceeds from the bottom to the top of the layer, assuming a temperature of 290 K below the layer. Take the water vapor pressure e to be 10 mb below the layer and 0 at the top of the layer. Neglect the change in pressure, and use sea-level pressure.

4-4 Calculate the relative dielectric constant K, the index of refraction n, the wave length λ, and the propagation constant k for an electromagnetic wave at a frequency of 20 MHz in a region where the electron density is $1 \times 10^6/\text{cm}^3$, neglecting the magnetic field.

4-5 Calculate the gyrofrequency in Hz for electrons, atomic hydrogen ions, and molecular oxygen ions in a magnetic field of 0.5 G.

4-6 Calculate the relative dielectric constants K, indices of refraction n, propagation constants k, and wavelengths λ for the left and right circularly polarized waves propagating parallel to the magnetic field, assuming $f = 40$ MHz, $f_p = 10$ MHz, and $f_B = 1.5$ MHz.

4-7 The angle α of Fig. 4-3 is measured to be $+100\ \mu$rad for an optical path that is 16 km in length at Boulder, Colorado. Determine the average value of dT/dh along the path. Repeat if α is $-100\ \mu$rad.

4-8 For $N_0 = 350$, $H = 7$ km, and an assumed exponential atmosphere for which $N = N_0 e^{-h/H}$, calculate the β's, τ's, and δ for a ray which is launched at $\beta_0 = 0°$ and which reaches the height of 3 km. Also calculate the corresponding β's for a uniform atmosphere (constant N). Use three layers, each 1 km in thickness, between levels 0 and 1, 1 and 2, and 2 and 3.

4-9 A microwave path 64 km in length has transmitting and receiving sites on hills at elevations of 900 m. A peak at an elevation of 760 m lies at a point 0.4 of the distance from the transmitter to the receiver. Make flat earth plots for k values of $\frac{4}{3}$ and $\frac{2}{3}$ to determine path clearance. (kr_0 is an effective earth radius.)

4-10 Find the approximate maximum usable frequency for reflection from the ionosphere at an angle of 60° from the vertical if the maximum electron density at the peak of the F layer is 8×10^5 electrons/cm³. Neglect the earth's magnetic field and curvature.

4-11 Find the distance for a linearly polarized electromagnetic wave to rotate through an angle of 2π rad if the wave frequency is 50 MHz, $f_B = 1.4$ MHz, the electron density is $1 \times 10^6/\text{cm}^3$, and the wave propagates parallel to the magnetic field.

4-12 Derive Eq. (4-113) by using the Hertz vector

$$\boldsymbol{\pi} = \int_V \frac{\mathbf{p}e^{-jkr}\,dV}{\varepsilon 4\pi r}$$

and

$$\mathbf{E} = \nabla \times \nabla \times \boldsymbol{\pi}$$

[$\mathbf{p}$ in the expression for the Hertz potential is electric dipole moment per unit volume, whereas the p_1 of Eq. (4-113) represents $\int \mathbf{p}\,dV$.]

4-13 The far electric field intensity of an electric dipole is also commonly expressed as

$$E_\theta = \frac{Ih\mu_0 j\omega}{4\pi r} e^{-jkr}$$

where the dipole is represented by a current I flowing along a short length h. Show that this expression is equivalent to Eq. (4-113). (This form of the equation for E_θ is conveniently derived from

$$\mathbf{A} = \int \frac{\mu_0 \mathbf{J} e^{-jkr}}{4\pi r}$$

and

$$E_\theta = -j\omega A \sin\theta$$

where **J** is electric current density and

$$\mathbf{J} = \frac{\partial \mathbf{p}}{\partial t}.\Big)$$

4-14 Find the transmission coefficients T for $\lambda = 0.4$ and $\lambda = 0.8\ \mu$m for paths through the entire earth's atmosphere and for zenith angles of 0 and 40° if the only loss mechanism considered is Rayleigh scattering. (Use the expression derived for the purpose.)

4-15 a. Determine a transmission coefficient T for a wavelength λ of 0.4 μm for a horizontal 5-km path at sea level, taking into account Rayleigh and aerosol scattering. In addition some attenuation may result from water vapor. For example, if the relative humidity is 30% at a temperature of 21°C, charts show that this humidity corresponds to 20 mm of precipitable water vapor along the path and that $T_{H_2O} = 0.723$. Include this factor with those already obtained to give an overall transmission coefficient.

b. Determine transmission coefficients T for a vertical path through the entire atmosphere for a wavelength λ of 0.8 μm for Rayleigh scattering alone and for all loss mechanisms other than water vapor. Use information from Table 4-2 for this problem.

4-16 Calculate the attenuation coefficient for a 50-mm/h rainfall at a frequency of 3000 MHz at a temperature of 20°C. Assume that the Laws and Parsons distribution applies and that the wavelength is sufficiently long for the simplified expression for C_{ext} to apply. [To insert C_{ext} into Eq. (4-122) for calculation of α_p/m with $n(a)da$ from Eq. (4-123), C_{ext} must be expressed in m^2.]

REFERENCES

[1] Aden, A. L., "Electromagnetic scattering from spheres with sizes comparable to the wavelength," *Journal of Applied Physics*, vol. 22, pp. 601–605, May 1951.

[2] Allis, W. P., S. J. Buchsbaum, and A. Bers, *Waves in Anisotropic Plasmas*. Cambridge, MA: M.I.T. Press, 1963.

[3] Bean, B. R., and E. J. Dutton, *Radio Meteorology*. Washington, D.C.: Supt. of Documents, U.S. Government Printing Office, 1966.

[4] Beard, K. V., "Terminal velocity and shape of cloud and precipitation drops aloft," *Journal of Atmospheric Sciences*, vol. 33, pp. 851–864, May 1976.

[5] Budden, K. G., *Radio Waves in the Ionosphere*. Cambridge: Cambridge University Press, 1961.

[6] CAMPEN, C. F., and A. E. COLE, "Tropospheric variations of refractive index at microwave frequencies," *Air Force Surveys in Geophysics, No. 79*. Bedford, MA: Geophysics Research Directorate, Air Force Cambridge Research Laboratories, 1955.

[7] CRANE, R. K., "Ionospheric scintillation," *Proc. IEEE*, vol. 65, pp. 180–199, Feb. 1977.

[8] DAVIES, K., *Ionospheric Radio Propagation*. Washington, D.C.: Supt. of Documents, U.S. Government Printing Office, 1965.

[9] EVANS, J. V., "Theory and practice of ionospheric study by Thomson scatter radar." *Proc. IEEE*, vol. 57, pp. 496–530, April 1969.

[10] FREEMAN, R. L., *Telecommunication Transmission Handbook*. New York: Wiley, 1975.

[11] GTE Lenkurt, *Engineering Considerations for Microwave Communication Systems*. San Carlos, CA: GTE Lenkurt, Inc., 1972.

[12] HAYDON, G. W., M. LEFTIN, and R. ROSICH, "Predicting the performance of high frequency sky-wave telecommuncation systems," *OT Report 76-102*. Boulder, CO: Institute for Telecommunication Sciences, Sept. 1976.

[13] HEADRICK, J. M., and M. I. SKOLNIK, "Over-the-horizon radar in the HF band," *Proc. IEEE*, vol. 62, pp. 664–673, June 1974.

[14] JESKE, H. E. G. (ed.), *Atmospheric Effects on Radar Target Identification and Imaging*. Dordrecht, Holland: D. Reidel Publishing Company, 1976.

[15] JOHNK, C. T. A., *Engineering Electromagnetic Fields and Waves*. New York: Wiley, 1975.

[16] JORDAN, E. C., and K. G. BALMAIN, *Electromagnetic Waves and Radiating Systems*, 2nd ed. Englewood Cliffs, NJ: Prentice-Hall, 1968.

[17] KELSO, J. M., *Radio Ray Propagation in the Ionosphere*. New York: McGraw-Hill, 1964.

[18] KERKER, M., *The Scattering of Light and Other Electromagnetic Radiation*. New York: Academic Press, 1969.

[19] KERR, D. E. (ed.), *Propagation of Short Radio Waves*. New York: McGraw-Hill, 1951.

[20] LAWRENCE, R. S., "Remote sensing by line-of-sight propagation," in *Remote Sensing of the Troposphere* (V. E. Derr, ed.), pp. 25-1 to 25-15. Washington, D.C.: Supt. of Documents, U.S. Government Printing Office, 1972.

[21] LITTLE, C. G., "Acoustic methods for the remote probing of the lower atmosphere," *Proc. IEEE*, vol. 57, pp. 571–578, April 1969.

[22] LONGLEY, A. G., R. K. REASONER, and V. L. FULLER, "Measured and predicted long-term distributions of tropospheric transmission loss," *OT/TRER 16*, ITS, Boulder, CO. Washington, D.C.: Supt. of Documents, U.S. Government Printing Office, July 1971.

[23] MORELAND, W. B., *Estimating Meteorological Effects on Radar Propagation*, Vols. I and II, *Technical Report 183*. Scott AFB, IL: Air Weather Service, U.S. Air Force, Jan. 1965.

[24] OTTERSTEN, H., "Radar backscattering from the turbulent clear atmosphere," *Radio Science*, vol. 4, pp. 1251–1255, Dec. 1969.

[25] PANTER, P. F., *Communication Systems Design*. New York: McGraw-Hill, 1972.

[26] RAMO, S., J. R. WHINNERY, and T. VAN DUZER, *Fields and Waves in Communication Electronics*. New York: Wiley, 1965.

[27] RAO, N. N., *Elements of Engineering Electromagnetics*. Englewood Cliffs, NJ: Prentice-Hall, 1977.

[28] RATCLIFFE, J. A., *An Introduction to the Ionosphere and Magnetosphere*. Cambridge: Cambridge University Press, 1972.

[29] RISHBETH, H., and O. K. GARRIOTT, *Introduction to Ionospheric Physics*. New York: Academic Press, 1969.

[30] SMITH, E. K., and S. WEINTRAUB, "The constants in the equation for atmospheric refractive index at radio frequencies," *Proc. IRE*, vol. 41, pp. 1035–1037, Aug. 1953.

[31] *Special Issue of Radio Science, Conference on Incoherent Scatter, Tromso, Norway*, vol. 9, Feb. 1974.

[32] STRATTON, J., *Electromagnetic Theory*. New York: McGraw-Hill, 1941.

[33] TATARSKI, V. E., *Wave Propagation in a Turbulent Medium* (translated from Russian by R. A. Silverman). New York: McGraw-Hill, 1961.

[34] TATARSKI, V. E., *The Effects of the Turbulent Atmosphere on Wave Propagation* (translated from Russian). Springfield, VA: National Technical Information Service, U.S. Department of Commerce, 1971.

[35] TAUR, R. R. ,"Ionospheric scintillation at 4 and 6 GHz," *COMSAT Technical Review*, vol. 3, pp. 145–163, 1973.

[36] THOMPSON, M. C., JR., "Measurements of wave-front distortion," in *Atmospheric Effects on Radar Target Identification and Imaging* (H. E. G. Jeske, ed.), pp. 289–299. Dordrecht, Holland: D. Reidel Publishing Company, 1976.

[37] THOMPSON, M. C., JR., and H. B. JANES, "Radio path length stability of ground-to-ground microwave links," *NBS Technical Note No. 219*. Washington, D.C., Supt. of Documents, U.S. Government Printing Office, 1964.

[38] VALLEY, S. L. (ed.), *Handbook of Geophysics and Space Environments*. New York: McGraw-Hill, 1965.

[39] WEISBROD, S., and L. J. ANDERSON, "Simple methods for computing tropospheric and ionospheric refractive effects on radio waves," *Proc. IRE*, vol. 47, pp. 1770–1777, Oct. 1959.

[40] ZUFFEREY, C. H., "A Study of Rain Effects on Electromagnetic Waves in the 1–600 GHz Range," M. S. thesis. Boulder, CO: Department of Electrical Engineering, University of Colorado, 1972.

5-1 BLACKBODY RADIATION LAWS

The concepts of thermal radiation are of importance in treating radiation processes in the atmosphere, remote sensing, radio astronomy, and thermal radiation from the sun and earth. All objects are continuously emitting and absorbing radiation. In the past, if one read references in these various fields he would find different units and terminologies used to refer to the radiation quantities. The recent tendency is to employ SI units (the International System of Units), which are metric units. Even so, there is still variation in terminology and usage. The units of this text are generally SI-MKS (meter-kilogram-second) units, and the terminology of the present section tends to follow that of Kraus [24]. A variation from some practices is that W is used as the quantity symbol and W is the unit symbol for power in watts, while P is the quantity symbol for the power density of an electromagnetic wave (W/m^2).

Consider the radiation incident on a surface from the direction of the sky. The power per unit area incident on a surface per cycle per second per steradian, B (W/m^2/Hz/rad^2), has commonly been called *brightness* by radio astronomers, and the integral of brightness with respect to frequency has been called total brightness B'. It is necessary to integrate with respect to solid angle in addition to obtain total power density incident on a surface. Some parties do not recommend the term brightness because it may suggest a visual effect, and the term *irradiance* is used by some to indicate the total power density incident on a surface. Spectral radiance N_λ, as used by Hudson [21], for example, is similar to brightness except that it has units of

BLACKBODY RADIATION LAWS AND SOLAR RADIATION

5

W/cm²/μm/rad², where μm stands for a micrometer, which is a unit of wavelength. If emission from a surface is considered, either brightness or radiance I, having the same units as brightness, may be used. Another system uses the term radiant exitance M for emission from a surface in watts per unit area, while infrared engineers may use radiant emittance for the same quantity. Note that the spectral quantities (such as B and N_λ above) can be expressed in terms of either frequency or wavelength. It can be argued that there is virtue in using the per-unit frequency system as energy is proportional to frequency rather than wavelength.

Planck's radiation law, which applies to radiation from a blackbody, assumes slightly different forms depending on what units it is expressed in. In terms of brightness it has the form

$$B = \frac{2h\nu^3}{c^2}\frac{1}{e^{h\nu/kT}-1} \quad \text{W/m}^2\text{/Hz/rad}^2 \tag{5-1}$$

where h = Planck's constant (6.63×10^{-34} Js),
ν = frequency in Hz or cps,
c = velocity of light (3×10^8 m/s),
k = Boltzmann's constant (1.38×10^{-23} J/K),
T = temperature in kelvins (K).

If brightness is measured in terms of W/m²/λ/rad², the expression is

$$B_\lambda = \frac{2hc^2}{\lambda^5}\frac{1}{e^{hc/kT\lambda}-1} \tag{5-2}$$

which follows from the fact that $\lambda f = c$ and

$$d\nu = \frac{-c}{\lambda^2}d\lambda$$

In terms of exitance

$$M_\lambda = \frac{2\pi c^2 h}{\lambda^5}\frac{1}{e^{hc/kT\lambda}-1} \quad \text{W/m}^2/\lambda \tag{5-3}$$

which differs by a factor of π from B_λ. Planck's law, which he presented in December 1900, applies to blackbody radiation or to radiation from an ideal thermal radiator or perfectly black object for which the radiation depends only on the temperature. The radiation in a cavity, with only a small opening for observing the radiation, should correspond closely to that from a blackbody. Historically, quantum theory had its beginning in the attempt to account for the equilibrium distribution of electromagnetic radiation in such a cavity. (See, for example, Chap. 1 of *Quantum Theory* by Bohm [5].)

The Rayleigh-Jeans radiation law stating that

$$B = \frac{2kT}{\lambda^2} \quad \text{W/m}^2\text{/Hz/rad}^2 \tag{5-4}$$

had been developed earlier in 1900 on the basis of classical electrodynamics.

It gives the correct value for large λ, corresponding to the radio-frequency portion of the electromagnetic spectrum, but it gives an obviously impossible result as λ approaches zero. The actual observed variation of B with frequency is as shown by the solid curve in Fig. 5-1. The derivation of the Rayleigh-Jeans law is given in Bohm [5]. In searching for a modification of the Rayleigh-Jeans law which would give the correct answer, Planck assumed that the energies of the oscillators in a cavity could not have arbitrary values but were restricted to multiples of $h\nu$, where h is now known as Planck's constant. That is, he assumed that the energies of the oscillators could be expressed by

$$E_n = nh\nu \tag{5-5}$$

He also assumed that Maxwell-Boltzmann statistics applied so that the probability of an oscillator of frequency ν having an energy of $nh\nu$ would be proportional to $e^{-E_n/kT}$. Using these assumptions, Planck developed his

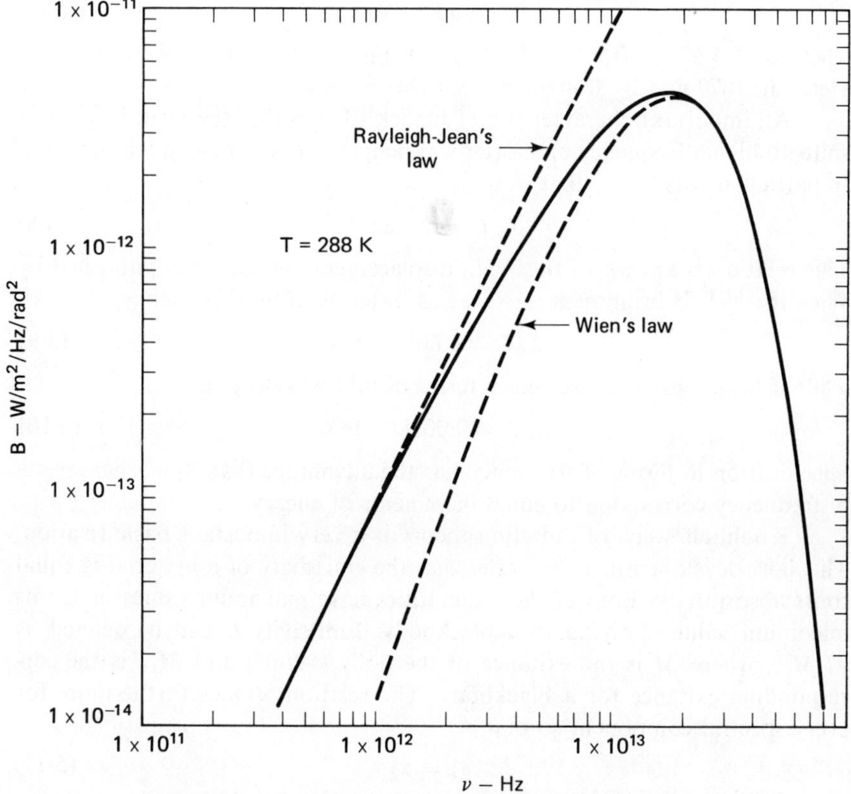

FIGURE 5-1. Brightness as a function of frequency for $T =$ 288 K.

radiation law. For sufficiently low frequencies or for $h\nu \ll kT$ for which

$$e^{h\nu/kT} - 1 \simeq \frac{h\nu}{kT}$$

Planck's law gives the same results as the Rayleigh-Jeans law, as can be verified by substitution.

Another approximation to Planck's law but one that applies when $h\nu \gg kT$ so that

$$e^{h\nu/kT} - 1 \simeq e^{h\nu/kT}$$

is

$$B = \frac{2h\nu^3}{c^2} e^{-h\nu/kT} \quad \text{W/m}^2\text{/Hz/rad}^2 \tag{5-6}$$

This is known as Wien's radiation law and was formulated in 1896.

If Planck's radiation law is integrated over all frequencies to obtain a total brightness or exitance, the result is the Stefan-Boltzmann law. In terms of exitance it is expressed as

$$M_{bb} = \sigma T^4 \quad \text{W/m}^2 \tag{5-7}$$

where $\sigma = 5.67 \times 10^{-8}$ W/m²/K⁴. Contributions to this law were made by Stefan in 1879 and by Boltzmann in 1894.

An important characteristic of Planck's law is that the peak brightness shifts to higher frequency or shorter wavelength with increase in temperature. In particular it is found that

$$\lambda_m T = \text{constant} \tag{5-8}$$

This relation is known as the Wien displacement law and was published by Wien in 1894. If brightness is expressed in terms of unit frequency,

$$\lambda_m T = 0.0051 \quad \text{mK} \tag{5-9}$$

while if brightness is expressed in terms of unit wavelength,

$$\lambda_m T = 0.00288 \quad \text{mK} \tag{5-10}$$

Specification in terms of frequency has the advantage that equal increments of frequency correspond to equal increments of energy.

Kirchhoff's law of radiation theory is a very important basic relation. This law, developed in 1860, states that the emissivity of a material is equal to its absorptivity. Both of these quantities have maximum values of 1, this maximum value applying to a blackbody. Emissivity ϵ' can be defined as M/M_{bb}, where M is the exitance of the body (W/m²) and M_{bb} is the corresponding exitance for a blackbody. The relation applies furthermore for every spectral component so that

$$\epsilon'_\lambda = \alpha'_\lambda \tag{5-11}$$

where ϵ'_λ stands for emissivity at a wavelength of λ and α'_λ stands for absorptivity at a wavelength of λ. It is mentioned later (Sec. 5-5-2), however, that frequency-selective coatings have been developed for which the emissivity and absorptivity values are different when the emission and absorption take place in different frequency or wavelength bands.

An application to the earth helps to illustrate the interpretation of Kirchhoff's law. The earth having a temperature of only about 287 K emits very little in the neighborhood of $\lambda = 0.48\ \mu$m where the radiation from the sun is most intense. However, the earth emits almost as much per unit area as a blackbody of the same temperature would emit at the same wavelength. Thus its emissivity is high and Kirchhoff's law says that its absorptivity is high also. The earth absorbs efficiently at this wavelength even though it emits very little at the same wavelength.

5-2 ABSORPTION, EMISSION, AND ANTENNA TEMPERATURE

Radiation is attenuated in passing through a purely absorbing homogeneous region in accordance with the relation

$$P = P_0 e^{-\alpha x} \tag{5-12}$$

P has units of W/m^2, α is an attenuation coefficient, and the relation refers to propagation in the x direction. A relation of this type was used in Sec. 3-2 in calculating ionization of the atmosphere. Alternatively, a volume absorption coefficient κ can be used such that $\alpha = \kappa\rho$, where κ has units of m^2/kg and ρ is density in kg/m^3. Also, the quantity αx can be replaced by the optical depth τ. Attenuation is also commonly measured in dB, where $(P_0/P)_{\text{dB}} = 10 \log P_0/P$, and attenuation in dB thus equals 4.3 αx. Dividing P by the solid angle subtended by its source gives an expression like that of Eq. (5-12) but one that involves brightness, namely

$$B = B_0 e^{-\alpha x} \tag{5-13}$$

If the region emits as well as absorbs radiation, however, the resulting equation for P or B has the form of an equation of transfer, as treated by Chandrasekhar [10]. An applicable form of the equation of transfer is

$$\frac{dB}{dr} = -\kappa\rho B + \frac{j\rho}{4\pi} \tag{5-14}$$

and its solution is

$$B = B_s e^{-\tau} + \frac{j}{4\pi\kappa}(1 - e^{-\tau}) \tag{5-15}$$

or

$$B = B_s e^{-\tau} + B_i(1 - e^{-\tau}) \tag{5-16}$$

where j is an emission coefficient having units of W/kg/Hz. The equation gives the observed brightness when a source of brightness B_s is observed through an absorbing region having an optical depth of τ and an intrinsic brightness B_i of $j/4\pi\kappa$. It is also possible to use the equation with $B_s = 0$. As brightness is proportional to temperature for long wavelengths for which the Rayleigh-Jeans law applies, a corresponding equation can be written for temperature in that case as follows:

$$T_b = T_s e^{-\tau} + T_i(1 - e^{-\tau}) \tag{5-17}$$

In most situations which engineers encounter, Eq. (5-12) or (5-13) is applicable because the signal power is sufficiently large and the emission coefficient j is sufficiently small. However, other situations arise when Eq. (5-16) or (5-17) is needed. The subject of radiative transfer, as utilized in considering the solar atmosphere and steller atmospheres in general, for example, is a complex subject, and only one simple case is mentioned here.

At this point, consider the use of an antenna to observe brightness or temperature, say of a source region in the sky. To this end, note that the infinitesimal power dW, incident on a flat surface dA, in a bandwidth $d\nu$, from an element of solid angle $d\Omega$, located at an angle of θ from the zenith is given by

$$dW = B\cos\theta \, d\Omega \, dA \, d\nu \tag{5-18}$$

where $dA \cos\theta$ is the area perpendicular to the direction from the source of the radiation. $d\Omega$ can be expressed in terms of the angles θ and ϕ of a spherical coordinate system by

$$d\Omega = \sin\theta \, d\theta \, d\phi$$

If the incident radiation is constant over an area A, then the total power received by the area is given by

$$W = A\int_\nu \int_\Omega B\cos\theta \, d\Omega \, d\nu \tag{5-19}$$

Consider now that the incident radiation is received by an antenna. A similar equation applies in this case, with the area A representing the effective area of the antenna. The equation must also take into account the antenna pattern $P(\theta, \phi)$. By introducing this quantity, the expression for W, the power received by the antenna, becomes

$$W = A\int_\nu \int_\Omega B(\theta, \phi)P(\theta, \phi) \, d\Omega \, d\nu \tag{5-20}$$

However, if B is constant over the antenna pattern, it may be sufficient to replace integration with respect to Ω by multiplication by Ω_A, an effective solid angle of the antenna defined by

$$\Omega_A \approx \tfrac{4}{3}\theta_{HP}\phi_{HP} \tag{5-21}$$

where θ_{HP} and ϕ_{HP} are half-power beamwidths in the θ and ϕ directions. Note that if the beamwidths are given in degrees, conversion from degrees squared to steradians or square radians can be accomplished by dividing by $(57.3)^2 = 3282$. If in addition P and B are essentially constant over a bandwidth $\Delta\nu$, W can be given by

$$W = AB\Omega_A\,\Delta\nu \tag{5-22}$$

Applying this expression to radio-frequency radiation for which the Rayleigh-Jeans law applies, one can replace B by $2kT/\lambda^2$, resulting in

$$W = A\frac{2kT}{\lambda^2}\Omega_A\,\Delta\nu$$

But $A\Omega_A = \lambda^2$ if A is the A'_{eff} for calculating directivity of Sec. 2-4, as can be seen by comparing the two expressions for directivity

$$D = \frac{4\pi A'_{\text{eff}}}{\lambda^2} \tag{5-23}$$

and

$$D = \frac{4\pi}{\Omega_A} \tag{5-24}$$

Thus $W = 2kT\,\Delta\nu$, or if a unit bandwidth is considered and a factor of 0.5 is introduced, as is appropriate for the case of a linearly polarized antenna receiving randomly polarized radiation,

$$w = kT \tag{5-25}$$

Thus the power per unit bandwidth received by the antenna is proportional to the temperature of the source from which the antenna is receiving thermal radiation. Therefore an antenna and associated receiving equipment can be used to record the effective temperature of objects within its beamwidth. When used in this way a microwave receiving system can be called a microwave radiometer.

Suppose, however, that the source region does not fill the antenna beam. In this case the antenna signal power per Hz will not be given by Eq. (5-25), with T representing the region temperature, but T will be the average temperature the antenna sees. The averaging can be carried out in terms of solid angles or in terms of areas. For example, if a region at a temperature of T_1 fills solid angle Ω_1 and a region at a temperature of T_2 fills solid angle Ω_2,

$$T = \frac{T_1\Omega_1 + T_2\Omega_2}{\Omega_A} \tag{5-26}$$

assuming that Ω_1 and Ω_2 fall within Ω_A.

In using a radiometer to record the temperature of a source or, more generally, an effective temperature T_A, it is commonly said that the antenna temperature is T_A. This does not mean that the antenna temperature as measured with a thermometer is T_A but that the antenna is receiving radiation equivalent to that from a source at a temperature of T_A.

5-3 THERMAL RADIATION FROM THE SUN AND EARTH

The sun radiates approximately as a blackbody having a temperature of about 5800 K if the Stefan-Boltzmann law is assumed to apply. The wavelength of maximum emission of about 0.48 μm, however, corresponds to a temperature of about 6000 K if the Wien displacement law is applied. The earth radiates approximately as a blackbody at a considerably lower temperature (see Fig. 5-4) with a wavelength of maximum emission of about 10 μm. The situation for the earth is complicated by atmospheric absorption and emission.

The power density of solar radiation or irradiance at normal incidence outside the atmosphere, referred to the mean solar distance (earth center to sun center) of 1.496×10^{11} m, is called the solar constant. Because of the eccentricity of the earth's orbit, the actual incident solar radiation varies by about $\pm 3.4\%$ from the mean, being greatest in January when the sun and earth are closest. The point of closest approach (perihelion) is at a distance of $1\ 471 \times 10^{11}$ m from the sun, and the farthest point from the sun (aphelion) is at a distance of 1.521×10^{11} m. Some of the incident solar radiation is absorbed and scattered in the earth's atmosphere before reaching the surface of the earth, and the spectrum at ground level may be as shown by the solid curve of Fig. 5-2, where the dotted curve represents the incident spectrum. The irregular variations in the solid curve are due to frequency-selective absorption by water vapor, carbon dioxide, oxygen, and ozone.

The determination of the solar constant was the object of research by C. G. Abbot and collaborators at the Astrophysical Observatory of the Smithsonian Institution for a period of about a half century starting in 1902. Their classical method involved assuming that the spectral power density at ground level P_λ is related to $P_0(\lambda)$, the incident power density, by

$$P_\lambda = P_0(\lambda)e^{-\tau_\lambda \sec \chi} \tag{5-27}$$

where τ_λ is the optical depth corresponding to wavelength λ and χ is the zenith angle of the sun. The expression can also be written as

$$P_\lambda = P_0(\lambda)e^{-\tau_\lambda m} = P_0(\lambda)a_\lambda^m \tag{5-28}$$

where m, called the air mass, is the ratio of path length at zenith angle χ to that at a zenith angle of 0°. In practice P_λ was measured at 40 different wavelengths from mountaintop locations during the course of a day during which time the zenith angle and therefore the value of m were changing. Taking logarithms of both sides of Eq. (5-28) results in

$$\log P_\lambda = m \log a_\lambda + \log P_0(\lambda) \tag{5-29}$$

For m between about 1.2 and 3, the curve of m versus P_λ is a straight line

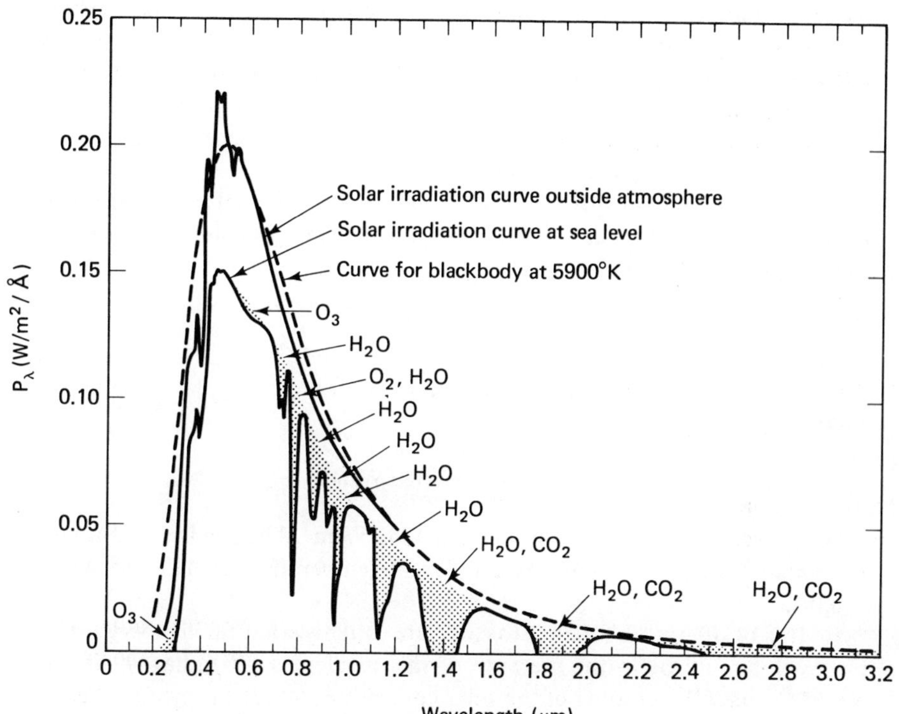

FIGURE 5-2. Spectral distribution curves related to the sun; shaded areas indicate absorption, at sea level, due to the atmospheric constituents shown. [From Valley (ed.), Air Force Cambridge Research Laboratories, *Handbook of Geophysics and Space Environment*, published by McGraw-Hill Book Company, 1965 [51].]

which may be extrapolated to $m = 0$ to obtain $P_0(\lambda)$. P_λ varies considerably because of absorption by atmospheric gases, but the plot of $P_0(\lambda)$ is a reasonably smooth curve. The solar constant is determined by taking the area under the $P_0(\lambda)$ curve.

The technique of extrapolating to zero air mass has limitations, and the use of rockets in the 1950's and jet (CV-990) and rocket (X-15) aircraft, high-altitude balloons, and spacecraft in the 1960's has provided increased accuracy. A summary of the various measurements has been given by Drummond and Thekaekara [13]. The value for the solar constant used in *Inadvertent Climate Modification* [45] was 1360 W/m^2, and the NASA/ASTM value is 1353 W/m^2 (1.94 cal/cm^2/min or 1.94 langleys/min or 429.2 Btu/hr/ft^2) [46].

A number of instruments are used for solar energy measurements, but we mention here only the Eppley pyroheliometer, which is widely used in the United States. It has an inner blackened ring which is heated by the incident solar radiation and a white or silver outer ring which reflects the incident radiation and remains at ambient temperature. A thermocouple of copper-constantan junctions is utilized, with the warm junction connected to the black ring and the cold junction to the outer ring. The rings are enclosed in a hemisphere of glass. The electrical voltage from the thermocouple can be displayed by a recording potentiometer.

A calculation roughly confirming the value of the solar constant is as follows. By using the Stefan-Boltzmann law and multiplying by the surface area of the sun, the total power emitted by the sun is

$$W = 4\pi r^2 \sigma T^4 = 3.91 \times 10^{26}\ \text{W}$$

where $r = 6.96 \times 10^8$ m, $\sigma = 5.67 \times 10^{-8}$ W/m^2/K^4, and T is taken as 5800 K. Dividing by 4π times the square of the mean solar distance of 1.496×10^{11} m gives

$$P = \frac{W}{4\pi d^2} = 1389\ \text{W/m}^2$$

which is reasonably close to the value of 1353 W/m^2. The sun is not a perfect blackbody and acts as if it has a different temperature at different wavelengths.

It is of interest to attempt to calculate roughly what the temperature of the earth should be in the presence of this incident solar radiation. If one were to equate the product of the solar constant P_s and the projected area of the earth A'_e with the product of the entire surface area of the earth A_e and $M = \sigma T_e^4$, assuming thermal equilibrium and considering that the earth absorbs like a disk and radiates like a sphere, a value of 278 K is obtained if P_s is 1353 W/m^2 and the earth radius is taken as 6370 km. A more meaningful calculation, however, takes into account the fact that a fraction A of the incident solar energy is reflected, rather than absorbed as by a blackbody at the earth's temperature. Taking this global albedo factor A into account, the earth's effective temperature T_e is given by

$$T_e = \left[\frac{(1 - A)P_s}{4\sigma}\right]^{1/4} \tag{5-30}$$

The factor 4 takes into account the ratio of A_e to A'_e. The result obtained in this way for T_e if $A = 0.3$ is about 253 K or -20°C. The observed mean temperature of the earth, however, is about 14°C (or 34°C warmer than the calculated value). The discrepancy is accounted for by the greenhouse effect. The earth's atmosphere allows solar radiation to penetrate quite readily, but longer wavelength radiation emitted by the earth cannot escape through the atmosphere as efficiently. To achieve a sufficient outgoing flux of radiation to reach equilibrium, the earth's average surface temperature rises to the higher value of about 287 K.

The heat balance of the earth is a subject of importance to meteorology and to environmental studies and considerations of solar energy. Figure 5-3,

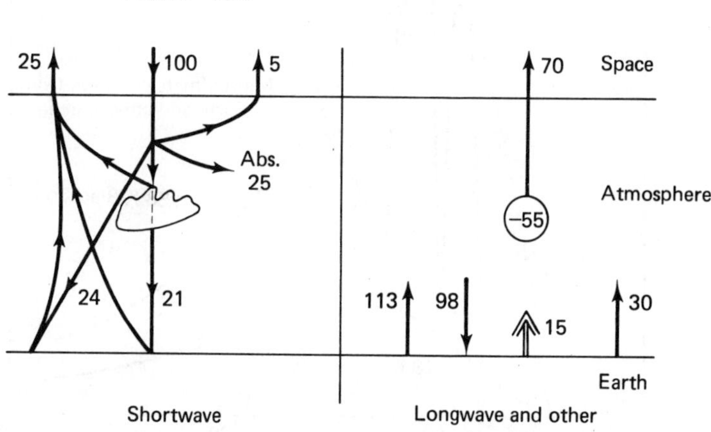

FIGURE 5-3. Heat balance of earth.

slightly modified from that of London and Sasamori [28], illustrates the major features involved. The precise values in such a description are subject to change. The earth's overall albedo (the fraction of the incident radiation reflected) is shown as 0.30, compared with a previously used value of 0.33. The principal reason for this difference is that the latter value involves the belief that a larger fraction of the incoming solar energy is absorbed by dust.

Clouds cover about 50% of the earth and are a major factor in reflecting incident solar radiation. The albedo of clouds varies from about 0.2 for thin cirrus clouds to 0.7 for cumulus clouds. Considering snow-free land surfaces, deserts have the highest albedo of about 0.28, and other values vary from about 0.13 to 0.24. The albedo of the oceans varies with angle of incidence and surface conditions but is low in general (0.07–0.2 at 60°N, for example). Atmospheric absorption of solar radiation is caused principally by water vapor, but it is also due to O_2 in the ionosphere (above 110 km), to O_3 or ozone in the stratosphere, and to CO_2, dust, haze, and clouds. Some solar radiation is scattered by the atmosphere back to space (5%), and some is scattered to the ground. The net upward infrared radiation from the earth is about 15%. The earth's surface radiates a much larger amount, but much of this is absorbed by the water vapor and carbon dioxide of the atmosphere and reradiated, a large fraction reradiated back to the earth. The earth's atmosphere is relatively transparent, however, to some radiation in the 8- to 13-μm range. Figure 5-4 from Johnson [22] shows that at the relatively transparent wavelengths the envelope of the curve of the earth's radiation approximates a blackbody curve for about 288 K, while for the more nearly opaque wavelengths the envelope approximates a blackbody curve for about 218 K. Most of the radiation to space from the earth actually comes from the atmosphere, especially from water vapor, CO_2, and clouds and dust, and not directly from the earth's surface. The 30% figure on the right in Fig. 5-3 refers to the

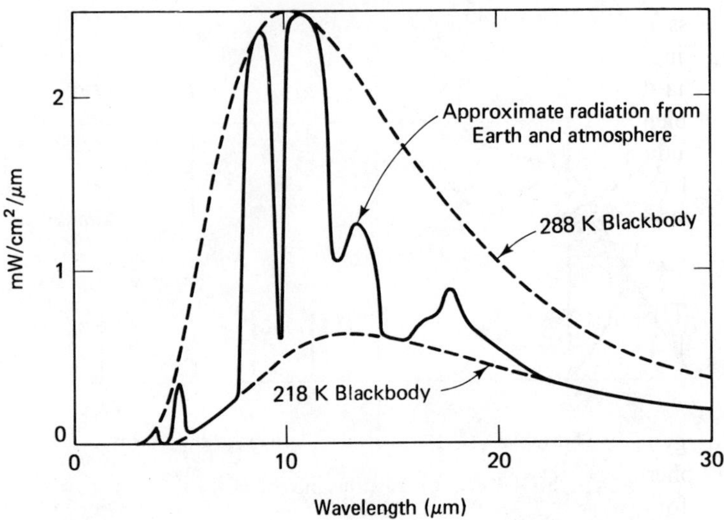

FIGURE 5-4. Typical spectral-radiance curve for thermal radiation leaving the earth. The 288-K blackbody curve approximates the radiation from the earth's surface, and the 218-K blackbody curve approximates the radiation from the atmosphere in those spectral regions where the atmosphere is opaque. (Reprinted from *Satellite Environment Handbook*, 2nd ed., edited by Francis S. Johnson, with permission of the publishers, Stanford University Press. Copyright © 1965 by the Board of Trustees of the Leland Stanford Junior University.)

transfer of energy from the earth's surface to the atmosphere by nonradiative processes—by convection and by the upward transport of water vapor which subsequently condenses and loses its latent heat.

Possible effects on the climate from increases in CO_2, particulate content, and other pollutants due to man's activity have been discussed by SMIC [45]. One aspect of interest involves the arctic ice pack. At latitudes above 60°, the albedo of stable snow cover is about 0.8, while that of the ocean is from 0.09 to 0.23. Pack ice covers about 90% of the Arctic Ocean. The ice reflects solar radiation that in its absence would be absorbed by the ocean. It has been suggested that a slight change in the temperature of the Arctic would cause some melting of the ice, which would cause additional heat to be absorbed, which would cause further melting. That is, positive feedback can be expected. The SMIC report cautions against acting on suggestions of spreading soot or black dust on the ice (to better open the Arctic Ocean for shipping in the summer) or of damming the Bering Strait without making every effort to understand fully the implications of such actions.

The SMIC report makes a number of recommendations, including monitoring of the earth-atmosphere albedo, the global distribution of cloudiness, and the optical properties and trends of atmospheric particles and determining the absolute value of the solar constant to better than $\pm 0.5\%$. Determination of the effects of particles and other pollutants is a very complex problem and requires detailed theoretical, experimental, and observational studies. Among the recommendations on this aspect is one for increased research on the refractive index (including absorption) of atmospheric particles.

5-4 UTILIZATION OF SOLAR ENERGY IN THE BIOSPHERE

The energy received at the earth's surface from the sun and its utilization in the biosphere will now be considered. The NASA/ASTM value quoted in Sec. 5-3 for the solar constant, the solar power received outside the earth's atmosphere at the mean solar distance, is 1353 W/m^2. However, the solar power received at the earth's surface varies from 0 to 1000 W/m^2 or slightly higher. On a worldwide basis, the earth intercepts about 1.51×10^{18} kWh/year of solar energy. About 0.4 of this amount, or 6.2×10^{17} kWh/year, reaches the earth's surface, and the amount reaching the land areas only is about 2.16×10^{17} kWh/yr [7]. If only fertile land areas, perhaps 20% of the earth's surface, are considered, the corresponding figure is 4.3×10^{16} kWh/year.

Incident solar energy is utilized in the process of photosynthesis in plants (see Sec. 3-4-1) to synthesize organic carbon compounds, and these are man's major sources of food and energy. Assuming that plants absorb 0.3 of the incident solar radiation over the fertile land areas of the earth and that plants convert 0.01 of the absorbed light on the average, Rabinowitch and Govindjee [36] calculate that 3×10^{10} tons of synthesized organic carbon would be produced per year, which is in reasonable agreement with the 2×10^{10} tons which have been estimated on the basis of the productivity of the various types of vegetation. A larger figure of 1.64×10^{11} metric tons is quoted by Woodwell [53] (a metric ton is 1.1 U.S. ton). The production efficiency of the oceans is not well known but is considerably lower than that for land. The maximum efficiency for photosynthesis, which is achieved only in reduced light by *Chlorella* algae, is considered by Rabinowitch and Govindjee [36] to be 0.34. Some field crops, such as sugar cane, and natural ecosystems, such as varieties of tropical forests, may have efficiencies as high as three times the average figure of 0.01 quoted above.

The producers, green plants which carry out photosynthesis, provide food for the herbivores (plant eaters or primary consumers), which in turn provide food for predators or secondary consumers, which in turn may provide food for other predators, or tertiary consumers. This sequence is called a food chain and is characterized by a decrease in numbers but an increase in individual size, proceeding from producers to tertiary consumers. The situation may actually be considerably more complicated than that of one simple food chain, and a number of food chains may be interrelated in a food web. Along with the decrease in numbers in a food chain the total biomass decreases, and the corresponding energy, namely, that stored in the biomass and potentially available to other organisms, decreases. The numbers, biomass, and energy are sometimes represented graphically by pyramids of numbers, biomass, and energy. Figure 5-5 shows a commonly utilized example of a pyramid of energy, namely that for an aquatic community at Silver Springs, Florida as determined by Odum and Odum [34].

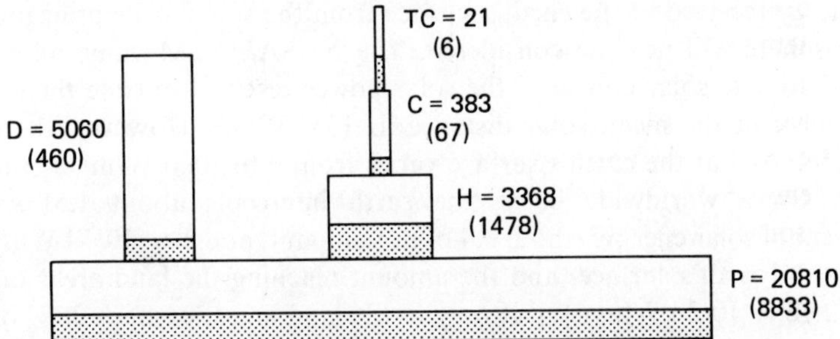

FIGURE 5-5. Energy pyramid for Silver Springs, Florida on an annual basis. The portion of the total energy flow which is actually fixed as organic biomass, and which is potentially available as food for other populations in the next trophic level, is indicated by figures in parentheses and by the shaded portion of each tier. Trophic levels are as follows: P = producers, H = herbivores, C = carnivores, TC = top carnivores, D = decomposers. Note that about half of the total assimilation ends up in the bodies of organisms (the rest being lost as heat or export), except for the decomposers level where percent loss in respiration is much greater. (From Odum and Odum, *Fundamentals of Ecology*, 2nd ed., W. B. Saunders Company, 1959 [34].) 10,000 kcal = 11.63 kWh.

In some ecosystems the import of dead organic material or detritus from nearby is an important source of energy. For example, a small spring may receive the major portion of its energy from detritus falling on it from the nearby forest. (A generalized energy flow diagram taking import and export into account is shown by Kormondy [23], p. 30.) The complicated food web in a Long Island estuary was studied by Woodwell [52]. Nutrients (minerals) are not diminished in the same way as energy, biomass, etc., and may be concentrated at the higher levels of a food chain, as may toxic materials such as DDT and mercury.

Man obtains the minimum of approximately 2.4 kWh of energy that he needs per day on the average from the plants and animals of the biosphere. Much of his food comes from ecosystems that are extensively managed to serve his purposes, and the food chains involved may be short simple ones. These agricultural ecosystems are not inherently stable as are natural ecosystems but are maintained by additional inputs of energy in the form of fertilizer, pesticides, and herbicides and cultivation.

All of our fossil fuels were produced originally by the process of photosynthesis, and wood was at one time an important energy source. It is worthwhile considering if photosynthesis could be utilized more efficiently to contribute to the solution of our energy and food problems [9, 33]. Possibilities include tree farming, cultivation of grasses, algae culture, and use of organic wastes. With respect to the last, the Montfort Co. of Greeley, Colorado, operators of the world's largest feedlots, is undertaking the production of methane from animal wastes. Calvin [9] mentions that sugar cane has the highest yield of yearly averaged photosynthesis and discusses the production of alcohol from sugar. He also mentions that the *Hevea* rubber plant is an attractive source of hydrocarbon. Another interesting possibility is the photochemical production of hydrogen by a modified process of photosynthesis in some algae [4, 33] or by similar but artificial means. Soy beans have received attention as a valuable source of vegetable protein, which requires less energy than meat protein.

5-5 SOLAR COLLECTORS

5-5-1 Introduction

The various applications of solar energy are described in the classic and very interesting book by Daniels [12]. Included are chapters on cooking, heating water, agricultural and industrial drying, heating buildings, distillation of water, solar furnaces, cooling and refrigeration, heat engines, thermoelectric conversion, photovoltaic conversion, and photochemical conversion as well as chapters on solar radiation, collectors, storage, and selective surfaces. Several more recent volumes by Behrman [3], Brinkworth [6], Duffie and Beckman [14], Meinel and Meinel [31], McVeigh [30], and others also provide valuable coverage of the subject of solar energy.

The treatment here is limited. After discussing the use of solar radiation in the biosphere in the previous section, attention is now directed to collectors of solar energy. These are needed for many of the applications of solar energy, including the heating and cooling of buildings and the thermal conversion of solar energy to electrical energy. Heating of buildings and conversion from solar to electrical energy are treated briefly in the following two sections.

Collectors of solar energy may be of the flat-plate or concentrating types, and the medium for transferring heat from a collector may be water or air or some other fluid.

5-5-2 Flat-Plate Collectors

The features of a flat-plate collector utilizing water as a heat-transfer medium are illustrated in Fig. 5-6. A blackened metal surface in such a collector absorbs incident solar radiation, and similarly blackened metal tubing bonded to the surface contains water. Structures having tubing built in

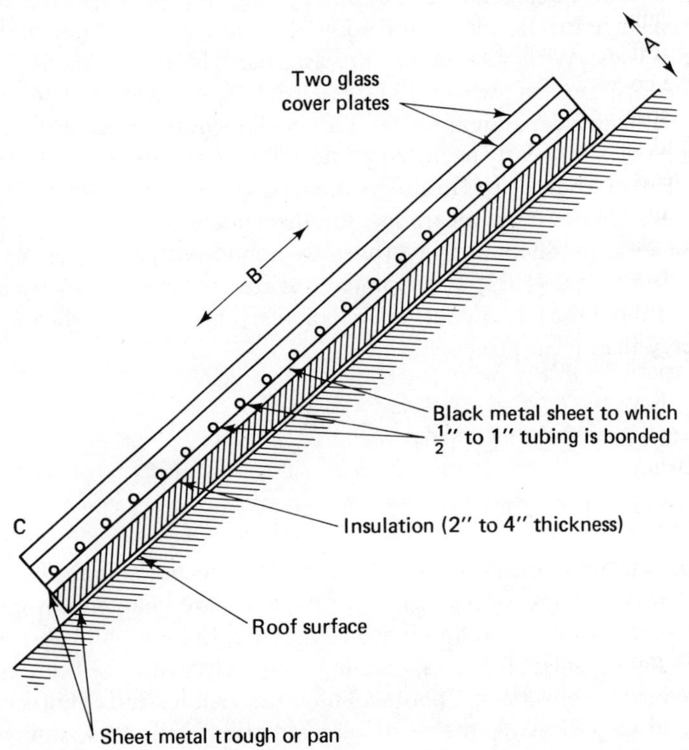

Notes: Ends of tubes manifolded together, one to three glass covers depending on conditions

Dimensions: Thickness (A direction) 3 inches to 6 inches; length (B direction) 4 feet to 20 feet; width (C direction) 10 feet to 50 feet; slope dependent on location and on winter-summer load comparison

FIGURE 5-6. Form of flat-plate solar collector. (From NSF/NASA Solar Energy Panel, 1972 [33].)

as an integral part, such as Roll-Bond, are now available also. Aluminum is commonly used as the metal in solar collectors, but it is subject to corrosion, and it appears to be necessary to add corrosion inhibitors to the water when aluminum is employed. Copper is now being promoted as a more suitable material for solar collectors than aluminum; one major supplier has announced that it has ceased to use aluminum in its solar collectors. The water in the collector tubes is heated, and the flow of the heated water conducts heat from the collector to where it is stored. Where there is danger of water freezing, a mixture of water and ethylene glycol (antifreeze) can be used. Use of air as a heat-transfer medium avoids the danger of freezing, the problem of corrosion, and the hazard of leaks. Water has the advantage of being an efficient heat-storage medium, however, as well as a heat-transfer medium. Efficiencies of the air and water systems are comparable.

The covering glass or plastic plates pass the incident solar radiation on to the absorbing surface, but the longer-wavelength (infrared) radiation from the heated surface cannot pass through the plates efficiently, in the case of glass at least. One to three covering plates may be desirable depending on the severity of the climate, two being the most generally suitable number. Ordinary glass is subject to fracture from hail, wind, and vandalism but tends to be more nearly opaque to long-wave radiation than plastic. The use of one layer of ordinary glass and an outer layer of plastic or more expensive tempered glass, which is less subject to fracture, has advantages.

Variations in collector design that have been the subject of recent research have included the use of fiberglass air filter material between the absorbing surface and cover and the use of a transparent solid (methyl methacrylate) in the same way. Both materials inhibit convection and are said to contribute to the achievement of high temperatures. The fiberglass is a feature of a commercially available collector. A collector designed to have structural strength and thus to double as a collector and roof offers cost savings. One version of the collector using air filter material, mentioned above, also involves a fiberglass panel which can serve as roofing material. Use of a partial vacuum between absorber and cover and of a black liquid (such as a solution of India ink) as an absorber has also been investigated.

As in the case of the earth and its atmosphere, the temperature of the absorbing surface of a collector rises until an equilibrium is reached between the incoming and outgoing quantities of heat. The relation between these quantities is

$$PT\alpha' A = P_{\text{in}}A = P_{\text{net}}A + (P_r + P_c + P_k)A \tag{5-31}$$

where P is incident solar power density at the outer surface of the covering plate of glass or plastic, T is the transmission coefficient of the covering plates (to be distinguished from temperature T by the context in which it is used), α' is the absorptivity of the blackened absorbing surface, and A is the area of the collector. P_{in} is the density of incident power at the absorbing surface, and P_{net} is the useful power output density. The other P's are similarly defined, P_r referring to heat power lost by radiation, P_c to loss by convection and conduction to the air, and P_k to loss by conduction through the insulation and collector materials. Note that although the conduction loss actually occurs largely in the side and back walls, an equivalent P_k (per unit area of the cover plate) is utilized in Eq. (5-31). Temperatures in the 38–93°C (100–200°F) range can be obtained by conventional collectors, and temperatures up to about 150°C (300°F) might be achieved by using sophisticated techniques. As discussed later, however, the collector efficiencies tend to be low at the higher temperatures. Note that the area A is the same for reception of incident solar energy and for radiation at infrared wavelengths. Insulation is used on the back of the collector and the exposed sides to minimize the loss P_k. The radiation power density P'_r per unit area emitted from a surface is given by

$$P'_r = \epsilon'\sigma T^4 \quad \text{W/m}^2 \tag{5-32}$$

corresponding to Eq. (5-7) but including the emissivity factor ϵ', which is less than unity. The conduction losses in the walls have the form of

$$P'_k = -k\frac{dT}{dl} \quad \text{W/m}^2 \tag{5-33}$$

where k is thermal conductivity, and the convection power density loss from a surface, P'_c, is given by

$$P'_c = h(T_s - T) \quad \text{W/m}^2 \tag{5-34}$$

where h is a convection coefficient, T_s is the surface temperature, and T is the temperature of the surrounding air. Primes are used in the above expression to indicate that although these terms involve the same phenomena as Eq. (5-31) they are general expressions not specifically applied to a solar collector. The combined radiation and convection losses P_t from the top of an actual collector may be given by

$$P_t = h_{ac}(T_a - T_c) + \frac{\sigma(T_a^4 - T_c^4)}{(1/\epsilon'_a) + (1/\epsilon'_c) - 1} \quad \text{W/m}^2 \tag{5-35}$$

where h_{ac} is the heat transfer coefficient between the parallel surfaces, the subscript a refers to the absorbing surface, and the subscript c refers to the glass or plastic cover. Further information concerning the coefficient h_{ac} and the practical calculation of collector loss and efficiency, with examples of calculations for particular conditions, is given by Duffie and Beckman [14].

The transmission coefficient T is determined by the amount of reflection caused by the covering plates and by the amount of absorption within the plates. The index of refraction for glass at optical frequencies falls between 1.5 and 1.9. T is normally not greater than 0.9 for two covering plates. The reflection loss can be reduced by using a coating having a thickness of $\lambda/4$, where λ is a wavelength near the middle of the visible frequency range. In research on collectors at the Marshall Space Flight Center [32], a plastic, Tedlar, has been used with an embedded wire grid. The wire blocks about 7% of the incoming solar radiation but adds structural strength.

Blackened surfaces may have an absorptivity α', higher than 0.9. High absorptivity is an advantage with respect to the reception of solar radiation, but the high emissivity that normally accompanies high absorptivity is a disadvantage as it allows a high value of P_r. However, the outgoing radiation is at a lower frequency than the incident solar radiation, which fact allows the possibility of selective surfaces which absorb efficiently in the wavelength range of solar radiation but emit inefficiently in the infrared range. Work on selective coatings has been conducted at several organizations including Minneapolis-Honeywell, the University of Arizona, and NASA. MSFC of NASA reports a coating, on an aluminum substrate, which involves a thin layer of zinc, an electrodeposited layer of bright nickel, and a final coating of electrodeposited black nickel of a thickness of about 0.15 nm. Values of absorptance α' of about 0.92 for solar radiation and of emissivity ϵ' of 0.06 for infrared radiation are quoted. It has been reported that black chrome, a widely available commercial decorative finish, has desirable selective properties.

Examples of the magnitudes involved in Eq. (5-31) are provided by data from the MSFC. One example involves an air temperature of 6°C (43°F), an absorbing surface temperature of 74°C (165°F), incoming solar radiation of 946 W/m² (300 Btu/h/ft²), and a flat-plate collector using Roll-Bond with selective coating and Tedlar. Values corresponding to Eq. (5-31) for the absorbing surface (but with the area A left out so that the values are on a per-unit area basis) are illustrated below, with the terms of Eq. (5-31) repeated for reference:

$$PT\alpha' = P_{\text{in}} = P_{\text{net}} + (P_r + P_c + P_k)$$
$$725 \qquad = 425 + (35 + 211 + 54)$$

Thus of the incident 946 W/m², 725 are actually absorbed by the absorbing surface, 425 can be extracted as useful heat, 35 are lost by the process of radiation, 211 are lost by convection, and 54 are lost by conduction. If conditions remain the same otherwise but the absorbing surface temperature is 130°C (280°F), the corresponding numerical equation is

$$725 = 0 + (101 + 532 + 104)$$

No useful heat can now be obtained, or extracted, from the collector, and the radiation, convection, and conduction losses have increased to the point where their sum equals the input power. Thus the temperature of 130°C is the maximum temperature obtainable by the absorbing surface, and this temperature can be maintained only if no useful heat P_{net} is taken from the collector. Heat can be taken from the collector having the absorbing surface temperature of 130°C, of course, but the absorbing surface temperature will then drop and losses will decrease correspondingly.

Efficiency is sometimes stated for a flat-plate collector as the ratio of P_{net} to P for a given temperature difference (collector-surface temperature minus ambient temperature), and this type of efficiency can be plotted as a function of temperature difference. The greater the temperature difference, the lower the efficiency as the above numerical examples indicate. Efficiency defined in this way has limited significance, but it does indicate that the higher the temperature of the heat provided, with a given collector, the lower the efficiency. What may be more important, however, is the cost of the heat provided, and an inefficient collector might appear to be favorable from an economic viewpoint.

5-5-3 Concentrating Collectors

Concentrating collectors can be used to obtain higher temperatures than those available from flat-plate collectors (up to 500°C with a rather crude collector and up to 3500°C or more with a large precision collector). Concentrating collectors can be used for cooking, solar furnaces, heat engines, and thermal conversion of solar energy to electrical energy, and such collectors may prove to be desirable for heating of buildings as well. Certain shapes that are useful for microwave antennas are also useful for solar collectors. Consider further the paraboloidal antenna or collector mentioned in Chap. 2 with respect to microwave communication systems. The parabola in Fig. 5-7 is described by the equation $y^2 = 4fx$, where f is the distance from the origin to the focal point or focus F. By definition every point on the parabola, such as P, is equidistant from the focus and the directrix, the dotted line on the left in Fig. 5-7. A circular paraboloidal antenna or collector can be formed by rotating the parabola in Fig. 5-7 about the x axis. Such a structure has the virtue that all of the plane-wave radiation that is incident on the collector with rays parallel to the axis (the x axis in Fig. 5-7) is concentrated upon reflection at the focus F where the object to be heated is placed. A cylindrical parabolic collector is a cylindrical structure having the parabolic cross section of Fig. 5-7. In this case the incident energy is concentrated on a focal line rather than at a focal point.

A limitation of focusing collectors is that they can utilize only direct sunlight and not diffuse or scattered radiation. A paraboloidal reflector, furthermore, must track the sun and keep the aperture of the collector at right angles to the direction of the sun. The necessity for tracking creates a problem which is particularly serious in the case of a large collector. A similar problem is also encountered sometimes at microwave and lower frequencies. A satisfactory solution in both cases may be to use a collector or reflector of circular cross section. The 305-m reflector operating at a frequency of about 430

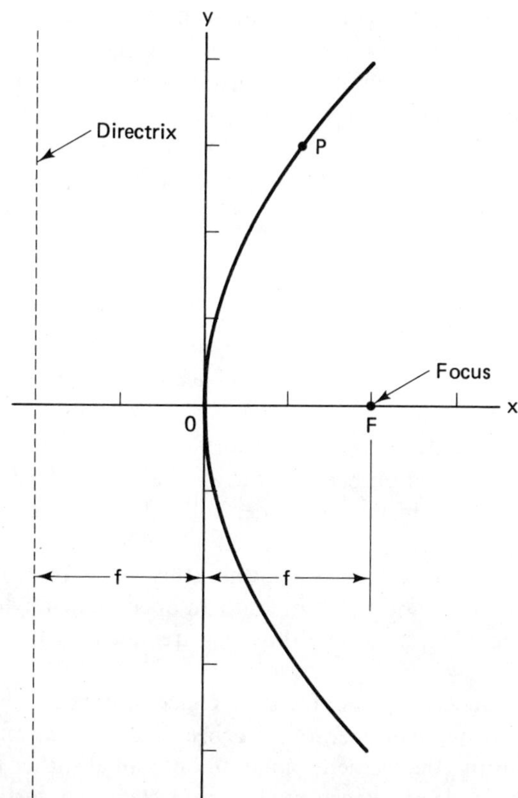

FIGURE 5-7. The form of a parabola for $f = 2$. Paraboloidal antennas have parabolic cross sections.

MHz at the Arecibo Radio Observatory in Puerto Rico, for example, is a spherical reflector. Such reflectors do not concentrate energy at a focal point but, because of spherical aberration, over a radial focal line parallel to the incident rays. Spherical collectors can be utilized if a focal structure of the appropriate length is placed in the proper radial position and made rotatable. The advantage of a spherical collector is that energy can be received over a rather wide range of angles by rotation of the focal structure, which is much smaller and more easily rotatable than the reflector surface itself. In radio applications, the Arecibo beam can be shifted 20° from the zenith by movement of the antenna line feed [17], and in another case operation has been accomplished over a 110° angular range [29].

Although flat-plate collectors have been the type commonly used on buildings, the question has been raised as to whether their cost can or will be reduced sufficiently in volume production to make them attractive economically, and it has been suggested that concentrating collectors may be more efficient. A 10.5-m spherical collector, for example, has been used for a home near Boulder, Colorado. Other collectors having the superficial appearance of flat-plate collectors but actually consisting of a combination of reflecting (or concentrating) surfaces and collecting elements have been designed. The Owens-Illinois system consists of a series of glass tubes overlaying a shaped reflecting surface as in Fig. 5-8. Each tube actually is three concentric tubes, an outer covering tube, an absorbing tube, and an innermost feeder tube.

FIGURE 5-8. Collecting tubes and shaped reflector.

The region between the cover and absorbing tubes is held at a vacuum. Fluid flows in one direction between the absorbing and feeder tubes and in the opposite direction inside the feeder tube. The arrangement is said to allow receiving both direct and diffuse sunlight efficiently over a rather large range of angles. Another type of concentrating collector is one utilizing a Fresnel lens to concentrate radiation onto an opening in a cylindrical cavity. Meinel and Meinel [31] provide a thorough treatment of the optics of solar collection.

An important class of nonfocusing concentrators has been developed rather recently. Its mode of operation has been described as one of funneling, rather than focusing, the incident radiation onto an absorber having an area smaller than the collector aperture (Fig. 5-9). The two surfaces of such a collector, in the two-dimensional form, are parabolic in shape, and the collector has been called a compound parabolic concentrator [37]. The collector was developed by Dr. R. Winston of the Enrico Fermi Institute, located at the University of Chicago, and it is commonly called a Winston concentrator. The collector concentration C, described by

$$C = \frac{A}{A_{\text{abs}}} = \frac{\text{aperture area}}{\text{absorber area}} \tag{5-36a}$$

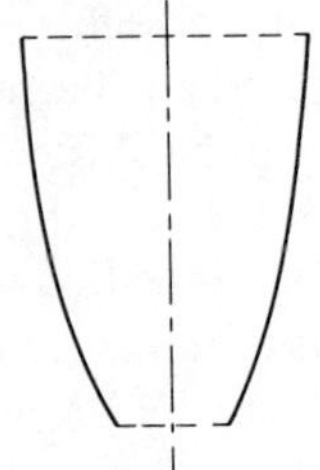

FIGURE 5-9. Compound parabolic concentrator.

is related to the maximum acceptance half-angle θ_{max} by

$$C = \frac{1}{\sin \theta_{max}} \tag{5-36b}$$

for the ideal, two-dimensional concentrator. For $C = 2$, $\theta_{max} = 30°$, while for $C = 10$, $\theta_{max} = 5.74°$. In the case of a conventional parabolic concentrator, the corresponding angle, neglecting mirror inaccuracies, is about 4.7 mrad or 0.25°, the half-angle of the sun. The Winston collector can achieve concentration of the sun's rays without the necessity for tracking for a reasonable period of time and can thus be described as a fixed-mirror collector. The concentration is achieved, however, at the expense of a relatively large aperture area [37]. Other fixed-mirror collectors achieving rather large solar acceptance angles without tracking are described by Meinel and Meinel [31]. Of this class of collector, the performance of the Winston collector is optimum, in terms of continuous hours of operation for a given concentration, for example, but as the Winston collector requires larger aperture and mirror areas than the others, it may not be optimum from the economic viewpoint.

5-5-4 Availability of Solar Energy

Incident solar power density or irradiance at the earth's surface varies with latitude, season, time of day, and weather. It is affected strongly by clouds and precipitation, and it is also attenuated by the factors mentioned in Sec. 4-5-2, namely Rayleigh scattering, scattering and absorption by aerosols, and absorption by gases. In addition, as solar radiation passes through the higher atmosphere, it is attenuated by ozone. The amount of atmospheric attenuation is a function of path length, as indicated by the factor m, the air mass of Eq. (5-28). Figure 5-10 shows irradiance plotted versus air mass for the case of water vapor equivalent to 20 mm of water and ozone equivalent to 3.4 mm under normal temperature and pressure and for various turbidity factors τ_d, where τ_d is expressed as β/λ^{α}, with λ the wavelength in micrometers. Figure 5-10 applies to a surface that is perpendicular to the sun's rays at all times.

The power intercepted by a collector is also a function of the slope of the collector surface. The irradiance P_s on an arbitrary surface is related to that on a surface normal to the sun's rays, P_n, by

$$P_s = P_n \cos C \tag{5-37}$$

where C is the angle between the normal to the surface and the sun's rays. If the surface is horizontal, the angle C is equal to the zenith angle of the sun, χ (the angle of the sun measured from the perpendicular). The zenith angle at a particular latitude, date, and time can be determined from

$$\cos \chi = \sin \delta_s \sin \theta' + \cos \delta_s \cos \theta' \cos h \tag{5-38}$$

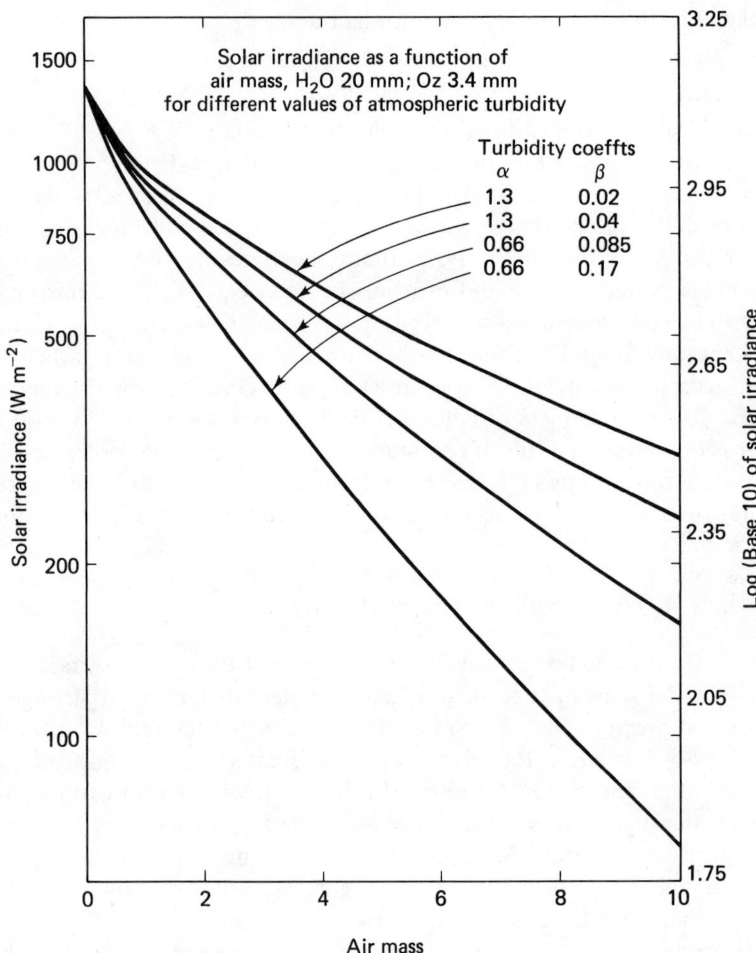

FIGURE 5-10. Semilog plots of irradiance versus air mass. Curves show the variation of total energy received directly from the sun by unit area exposed normally to the sun's rays for different values of solar zenith angle and atmospheric turbidity. (From Thekaekara, *The Energy Crisis and Energy from the Sun*, 1974 [46].)

where δ_s is the solar declination, θ' is the latitude, and h is the hour angle. δ_s varies between 0 and $\pm 23.5°$ and is the angle of the sun at local noon at the equator (0° at the vernal and autumnal equinoxes of March 21 and Sept. 21). Values of δ_s can be determined from the *American Ephemeris and Nautical Almanac* (the ephemeris published by the Naval Observatory and issued annually) from the column entitled "Apparent Declination" in the section "Sun." The hour angle h is 0° at local noon and is the angle the earth would have to be rotated through for the sun to be at its highest angle. As 1 h corresponds to rotation of the earth through 15°, the hour angle in degrees can be calculated by taking 0.25 times the difference in time in minutes from noon in Apparent Solar Time (AST). AST can be calculated by taking the local standard time and subtracting (adding) four times the amount in degrees by which the longitude is greater (less) than that of the proper standard time meridian and also applying an "equation of time" correction. The meridians for Eastern, Central, Mountain, and Pacific Standard Times are, for example, respectively, 75°, 90°, 105°, and 120°. The equation of time correction, accounting for the difference between time as indicated by a sundial and by a clock, can be obtained from the ephemeris by taking 12: 00 minus the time in the column entitled "Ephemeris Transit" in the section "Sun." (The time given under "Ephemeris Transit" is Ephemeris Time, which could be corrected to give Universal Time, but for the purposes of this chapter the correction is unimportant. Also, for present purposes, values of declination and the equation of time for any year can be used for any other year.)

Example: For Chicago for Oct. 1, find the hour angle corresponding to 3 p.m., CDT. (The longitude is 87.6° and the equation of time for Oct. 1 is 10 min.)

$$\text{AST} = 2\text{ p.m.} + 4(2.4) + 10 = 2{:}19.6$$
$$h = 0.25(139.6) = 34.9° \text{ west}$$

It is also of interest to know the azimuth of the sun ϕ' measured from the noon value, and this is given by

$$\sin \phi' = \cos \delta_s \frac{\sin h}{\sin \chi} \tag{5-39}$$

The maximum amount of solar energy can be collected by keeping the collector surface perpendicular to the sun's rays at all times, but it is usually impractical to do so for applications for which flat-plate collectors are used. Instead a favorably inclined fixed surface is utilized. For a collector for which the normal to the surface is tilted to the south by the angle B, in the northern hemisphere, the expression for $\cos C$ is like that of Eq. (5-38) for $\cos \chi$ but with θ' replaced by $\theta' - B$ or

$$\cos C = \sin \delta_s \sin (\theta' - B) + \cos \delta_s \cos (\theta' - B) \cos h \tag{5-40}$$

For an arbitrary tilt of the collector surface the corresponding expression for $\cos C$ is

$$\begin{aligned}\cos C = {} & \sin \delta_s[(\sin \theta' \cos B) + (\cos \theta' \sin B) \cos A] \\ & + \cos \delta_s[(\cos \theta' \cos B) \cos h \\ & + (\sin \theta' \sin B)(-\cos A \cos h) \\ & - (\cos B \sin \theta') \sin h]\end{aligned} \tag{5-41}$$

where the inclined surface is defined by first rotating a normal to a horizontal surface to the north through an angle B and then rotating the normal about the vertical in a clockwise direction by an angle A. If a fixed surface is to be used, the surface would usually be tilted to the south and Eq. (5-40) would apply, but if morning sunlight were partially blocked, for example, a tilt favoring the reception of afternoon sunlight might be used in which case Eq. (5-41) would apply. Also it may not be desirable architecturally for all homes in a community to have only south-facing slopes, and a combination of slopes facing south, east, and west may be used. For operation throughout the year, for both winter heating and summer cooling, a south-facing slope about equal to the latitude is said to be the most favorable [6], whereas for heating alone a value equal to the latitude plus 15° is said to be optimum. For winter heating at latitudes of about 40° and higher the optimum angle is rather large, and the installation of vertical collectors in the walls may be a reasonable measure, especially if snow loading is a problem. A solar house constructed in Tucson, Arizona has a collector tilt of 27° to favor summer cooling, and the Colorado State University house at Fort Collins (near 40.6°), which house is both heated and cooled by solar energy, uses a slope of 45°.

Values of solar altitude angle ($90° - \chi$), azimuth ϕ', and irradiance (Btu/h/ft^2) for south-facing surfaces of various slopes, as well as the direct normal irradiance, are given for 40° north latitude for the hours of the day for the 21st day of each month in the 1974 applications handbook of ASHRAE [2]. Daylong values of solar energy are also given for selected latitudes between 24 and 64°. Data on solar radiation at numerous locations in the United States are given in the *Climatic Atlas of the United States* [50].

The most favorable areas of the world for the application of solar energy tend to be between about 15 and 40° latitude. These areas are generally drier and sunnier than the equatorial belt, which is the next most favorable area. Even the polar regions can make use of solar energy in the summer. Meinel and Meinel [31] quote figures of 970 W/m^2 for the direct (D) solar flux under desert sea-level conditions, with a figure of 1050 W/m^2 for the direct plus scattered ($D + S$) radiation. Corresponding figures for standard sea level are 930 (D) and 1030 ($D + S$) W/m^2, while typical urban figures are given as 610 (D) and 810 ($D + S$) W/m^2. One way of indicating the solar radiation, or insolation, that is available is in terms of an annual average power density (W/m^2). On this basis, the Sahara Desert has the highest value of about 280 W/m^2, and the southern United States and northern Mexico are in another high-intensity region of about 260 W/m^2. Some areas in the northwestern and northeastern parts of the United States have values less than 150 W/m^2. When considering a solar energy installation in the United States, specific data concerning insolation in the area under consideration should be obtained, if possible, from the *Climatic Atlas of the United States* or the National Weather Service. More detailed and thorough coverage by recording stations will probably be needed as interest in solar energy increases.

An illustration of the energy received by a solar collector is given by the figure of 700 W/m^2, which is what Daniels [12] uses as a representative average for an entire day for a favorably inclined surface. For a useful day length of 500 min (8 h and 20 min), the 700-W figure corresponds to a total energy input of 5.8 kWh/m^2. Trybout and Löf [49] use 4.4 kWh/m^2/day as a representative average for the entire year. Daniels quotes figures as high as 8.1 kWh/m^2/day for the total incident radiation in the southwestern United States on June 21 and the same figure for the Antarctic in its summer. If the 700-W/m^2 figure is applicable, an area of 1.44 m^2 is required for an input of 1 kW of solar power. The efficiency of the collector must be taken into account also, however, in determining how much output power can be obtained. As indicated previously, efficiency varies inversely with the difference between collector and ambient temperatures.

5-6 HEATING OF BUILDINGS

When considering heating of buildings a distinction can be made between passive and active systems. Although some argument can develop over precise definitions, passive systems [1] rely mainly on natural processes of convection and radiation and tend not to need electrical motors for circulating fluids. Passive systems are thus generally cheaper and simpler than active systems, but certain types of passive systems are most obviously applicable only to rather small buildings having simple floor plans with a minimum of partitions. One type of passive system or subsystem involves a thermosyphon, illustrated in Fig. 5-11. The fluid can be either water or air, and the application can be to heating a building or to heating water. In any case the heated fluid is lighter than the cold fluid and thus rises to the top of the storage vessel or tank, displacing cool fluid from the bottom of the storage tank in the process. The displaced cool fluid circulates to the collector. The storage tank must be above the collector for the system to operate. The application

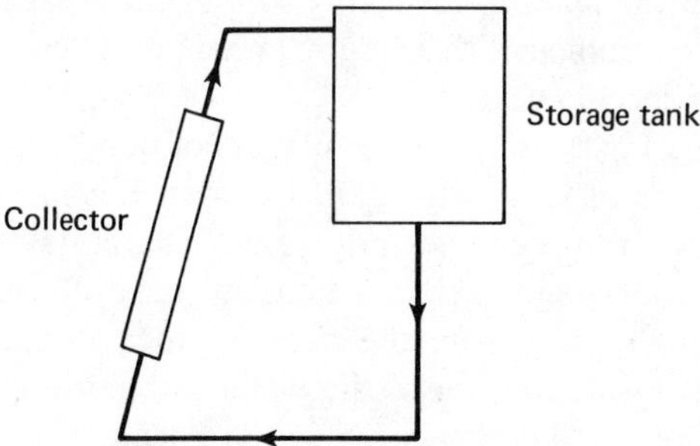

FIGURE 5-11. Thermosyphon system.

to the heating of a building is facilitated if the building is on a slope and the collector can be placed below the level of the base of the building. Once heat is accumulated in the storage vessel it can be distributed throughout the building by either passive or active distribution systems.

A second type of passive system involves using a vertical structure or assembly as a south-facing inner wall. The inner wall may be a rather massive concrete wall, as shown in Fig. 5-12. The wall absorbs solar radiation in the

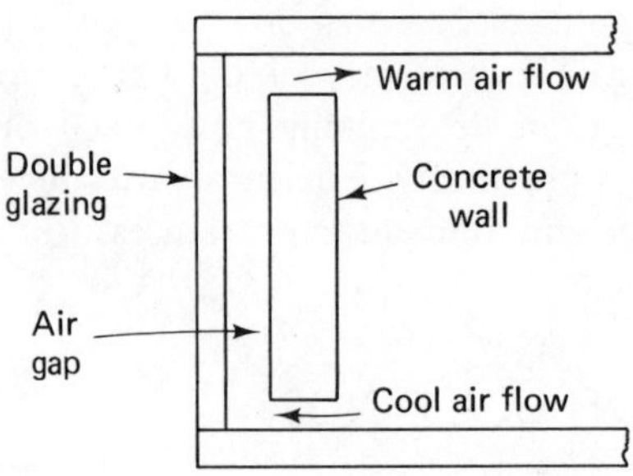

FIGURE 5-12. Passive system employing concrete inner wall.

daytime and radiates heat into the room at night. Having gaps in the wall at the top and bottom also provides for a convective flow of air. The double window inhibits the outward flow of heat and can be augmented by a heavy insulating drape or movable insulation in some other form at night. Steve Baer has used the concept discussed but with an assembly of drums filled with water for storage, in place of a concrete wall. In the installation at his home an outer wall or layer of insulating material is folded down to the ground in the daytime to allow the sun's rays to enter but is raised at night to provide insulation. Another means for providing insulation for use with either an active or passive system is the use of bead walls, consisting of two sheets of glass with beads blown into the space between the glass at night. The use of skylids, which can open automatically in the daytime and close at night, is a means of augmenting the capability for collecting the sun's energy. To store sufficient energy, the structure or material on which the solar radiation impinges, whether the main south-facing wall or the areas illuminated by the skylids, must have a sufficient thermal mass, which is determined by the product of mass and specific heat.

A third type of passive system involves roof ponds, as incorporated into the Skytherm system of H. R. Hay [20]. The system is most obviously applicable to regions where water does not freeze. The building has a flat metal ceiling, and the roof pond, perhaps actually bags of water, rests on the ceiling. Heat flows readily from the pond to the interior of the building. Movable (sliding) insulation above the water is removed in the daytime in the winter, and the sun then heats the water. At night the insulation covers the water, and radiation from the ceiling keeps the interior of the building warm. Reversing the process (covering the pond with insulation in the day and uncovering at night) provides air conditioning in the summer.

All systems for using solar energy for heating buildings involve the functions of collecting solar energy, storing energy, and transferring energy from one location to another [25]. The passive systems using vertical wall structures and roof ponds utilize the same mass as the main collector and reservoir and natural processes of radiation and convection to distribute the heat. Conventional active systems, however, normally involve separate collectors and storage reservoirs and use electrical motors to accomplish heat transfer from collector to storage reservoir and storage reservoir to the areas to be heated. Solar collectors were discussed in the previous section. The heat-storage medium for conventional installations may be water or rocks or pebbles which store heat as sensible heat, involving the product of specific heat and temperature rise. Alternatively heat can be stored as latent heat, involving a change of phase of a material at constant temperature. Water has the highest specific heat of any material and can store 1.16 kWh/m^3/°C. $Na_2SO_4 \cdot 10H_2O$, which changes to $Na_2SO_4 + 10H_2O$ at 32.3°C or 90°F, can store about 4.25 times as much heat upon making this transition as water can for a 20°C (36°F) temperature rise [12]. Rocks or pebbles are commonly used for storage when air is the medium used to transfer heat from the collector to the storage volume.

An illustration of heating needs has been provided by Trybout and Löf [49], who use a figure of 4.4 kWh/degree day for a 1000-ft^2 house. This value is equivalent to 0.62 Btu/h/ft^2 for a 1°F temperature difference and is close to the value of 0.5 Btu/h/ft^2, a figure used for a house meeting FHA requirements. A very well-designed and insulated house might need only one-half this much heat [26]. (The number of degree days is 65°F minus the average temperature in °F. This average is halfway between the maximum and minimum temperatures.) Trybout and Löf use 4000 degree days as a representative value for a calculation showing that the incident solar energy on the roof of a house may be nearly 10 times the amount needed for heating. The number of degree days can be obtained from the U.S. National Weather Service records and varies from 0 in the tropics to between 1000 and 2000 in southern areas of the United States to over 9000 in northern areas having severe winter weather. At Denver, Colorado, the average number of degree days from 1933–1934 to 1972–1973 was 5889. The *Climatic Atlas of the United States* is a good source of data on temperatures for all months and in many U.S. locations.

In utilizing solar heating, as is the case for other types of heating as well, sufficient insulation, double windows, and other energy conservation measures should be utilized. The heat transferred by conduction, P'_k, through a material having a heat conductivity of k is given by

$$P'_k = -kA\frac{dT}{dl} \tag{5-33}$$

equivalent to

$$k = \left| \frac{P'_k}{A(dT/dl)} \right| \tag{5-42}$$

where A is the plane area considered, T is the temperature, and l is distance. Related quantities are the thermal resistance

$$R = \frac{l}{k} \tag{5-43}$$

which applies to a uniform layer of unit cross section and thickness l, and

$$U = \frac{1}{R}$$

the thermal conductance. As for electrical resistance, thermal resistances in series add and $U_{total} = 1/R_{total}$. k is often quoted in units of Btu/h/ft^2 for a temperature difference of 1°F and a 1-in. thickness. In these units, good insulating materials have k values of 0.3 or less. The minimum acceptable R values for well-insulated structures are sometimes taken as 30 for ceilings and 20 for walls and floors, in units compatible with those for k, given above. If R values are known for the ceiling, walls, floor, windows, and doors of a house, the heating requirements can be calculated for given outdoor and indoor temperatures. An additional allowance may be needed for the heat lost because of the opening of doors, etc., but lights, human activity, etc., add some heat to compensate for the latter.

Although it is normally desirable and necessary to have windows allowing solar radiation to enter a building in the daytime, it is also essential to minimize heat losses at night. It is thus worth emphasizing that in addition to insulating a building well in general some movable insulation may be desirable for covering or insulating the window areas at night. In addition to the bead walls and folding or sliding window covers already mentioned, the application of panels of styrofoam to the interior of windows may be a practical measure in some cases.

Solar hot water heaters have been used in Israel, Japan, Florida, California, Arizona, etc., and are practical at this time. As an illustration of solar hot-water-heater performance, a collector area of 8.2 m^2 would be needed in a region having a solar radiation input of 4.4 kWh/m^2/day if a collector operating at an efficiency of 33% were used to supply a temperature rise of 38°C for 189 l (50 gal) of water. A lower temperature rise or a higher collector efficiency or a higher value of irradiance would, of course, allow use of a smaller collector. The arrangement for supplying hot water can be built as an integral feature of the system for heating the building, or it can be a completely separate system, as implied above.

The costs of solar collectors could be better justified in many areas if they could be used to provide both heating and cooling. Conventional air conditioning equipment presently causes a severe strain on electric power facilities in large areas of the country, and replacement by solar units could be highly desirable if economically feasible. The absorption cooling system is rather well adapted to the use of solar energy. As an introduction to the subject of cooling and refrigeration, however, consider first the vapor compression refrigeration cycle which is commonly used in home refrigerators (Fig. 5-13). Refrigeration is accomplished by the evaporation of a fluid such as Freon-12, the latent heat of vaporization being supplied to the evaporator from the volume to be refrigerated. A compressor, operated by an electric motor, compresses the vapor which is collected in a condenser. The fluid in the condenser then passes through an expansion valve back to the evaporator. The energy necessary to maintain the process is supplied by the compressor.

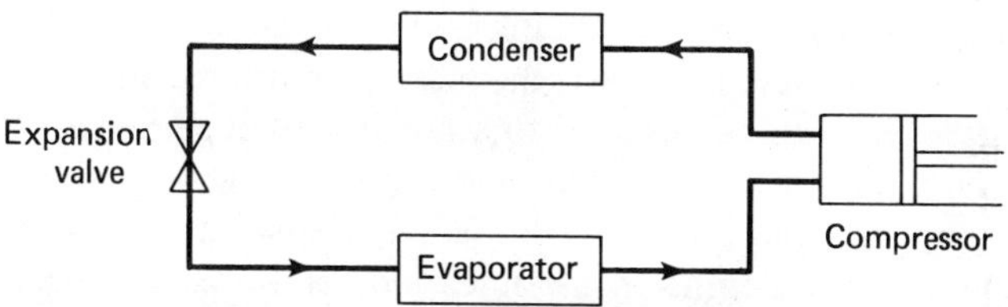

FIGURE 5-13. Vapor compression cycle.

In the absorption system, the evaporator and condenser remain much as they are and perform the same functions as in the vapor compression system, but the compressor is replaced by an absorber and generator and pump, as shown in Fig. 5-14. Ammonia and lithium bromide-water absorption systems have been utilized for operation with solar energy heat sources; attention will be concentrated here on the latter system. In the lithium bromide system, the fluid in the condenser-evaporator portion of the system is water, and water vapor from the evaporator is absorbed by lithium bromide in the absorber with the result that a lithium bromide-water liquid mixture, with a high water content, is found there. A pump supplies this solution at a high pressure to the generator, and heat from the solar collectors heats the

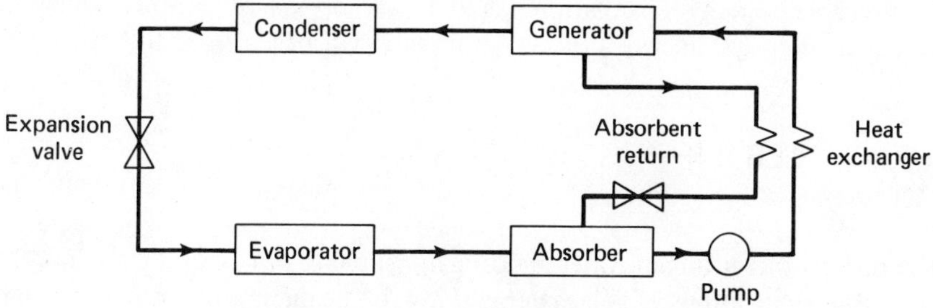

FIGURE 5-14. Absorption cooling system.

solution in the generator. The result is that water vapor is driven off from the generator and supplied to the condenser, from where it passes through an expansion valve back to the evaporator. The lithium bromide stays in the absorber-generator part of the system, and a lithium bromide solution, containing only a relatively small amount of water, is returned to the absorber through a second expansion valve. The pump requires only a small amount of energy, as compared to the compressor of a vapor compression system, and the heat input from the solar collectors is the principal source of energy. Although the system can operate with the heat supplied by solar flat-plate collectors, a higher temperature is needed than for heating applications. Consequently flat-plate collectors utilized for cooling tend to operate at low efficiencies. The wide application of solar energy for cooling purposes is more distant than the wide application for heating. The text by Threlkeld [47] is a good source of information on heating and cooling systems.

An interesting development is the interest shown recently in small solar greenhouses which can be attached to a home. Although home greenhouses may have been considered as a luxury in the past, solar greenhouses can be used for growing vegetables and flowers indoors, can contribute to the heating and air conditioning of a home, and are being developed for low-income housing in New Mexico.

The Skytherm roof pond system for natural heating and air conditioning was mentioned previously, and in dry climates with a large diurnal temperature range, such as that of Colorado, daytime cooling can be accomplished with air systems employing rock storage. The rocks can be cooled by the early morning air and can then cool the house later in the day. The same storage medium can be used for summer cooling and winter heating. A solar-assisted heat pump system has been proposed for retrofitting of solar energy to all-electric residence units [48].

As an example of what might be accomplished with solar heating and cooling (NSF/NASA panel, 1972), a 1500-ft^2 house in a temperate, sunny central U.S. location could be provided with about three fourths of its heating and cooling needs with a 600- to 800-ft^2 (56- to 74-m^2) collector and 2000 gal (7.57 m^3 or 7570 l) of hot-water storage. An auxiliary heat source is needed in any solar energy installation, as overcast or severe weather which a practical solar system will not accommodate will occur from time to time.

5-7 WATER LIFTING

The pumping or lifting of water has long appeared to be one of the potentially favorable applications of solar energy [12, 41]. The pumping of water by solar means has been commonly accomplished by solar heat engines which pump water directly or which drive electrical generators that develop electricity for pumping water. The Stirling hot air engine, invented in 1816, was used for pumping water in 1818 and adapted to solar operation by Erricsson in 1870.

Two of the best known applications of solar energy to the pumping of water were carried out in southern California and in Egypt. The Solar Motor Company of Boston was responsible for several solar pump installations in southern California and Arizona in the first years of the twentieth century. A machine using a reflector 10.2 m in diameter for concentrating solar energy was installed at the Edwin Cawston Ostrich Farm in South Pasadena, California in 1901 and attracted much attention. The Solar Motor Company enterprise, however, was rather short-lived, as it could not compete with other techniques that became available for pumping water at the time. The second solar pump singled out for mention here was installed by Shuman and Boys of the United States in Egypt in 1912. Having commenced work on a solar engine that used flat-plate collectors in 1906, Shuman, with the collaboration of Boys, built what was the world's largest solar pumping plant in Egypt. This solar engine, which utilized parabolic-cylinder collectors having an area of about 1200 m^2, is said to have developed 37–45 kW continuously for 5 h according to one report, but only 19 hp (14 kW) according to another. The project was abandoned in 1915 because of World War I and competition from cheaper sources of energy.

More recently, in 1974, Girardier of France was the leader of a project for the installation of solar pumps in drought-stricken areas of the Sahel in Africa. By June 1975, 30 solar engines had been delivered to the Sahel. The same company also installed pumps in Mexico at Caborca in Sonora and Ceballos in Durango.

A simple system for lifting comparatively small amounts of water is illustrated in Fig. 5-15; a fluorocarbon compound is placed in a polymeric vapor bag in a pressure generator which is also a simple solar collector. Upon being heated by incident solar radiation the compound vaporizes at a chosen low temperature, e.g., 23.8°C (74.8°F) for trichlorofluoromethane (Freon-11) at atmospheric pressure. Upon change of state from liquid to gaseous form, this particular fluid expands, at constant pressure, by the factor of 250. When the expansion is used to lift water, however, the expansion is limited to less than 250, the pressure in the system increases above the ambient value, and the excess pressure generated is used to lift the water, which may be used for household needs, irrigation of crops, etc. A number of variations of the basic system are possible, one form being useful for powering and controlling the flow of water to and from liquid-heating solar collectors. Water is supplied to the collectors when sufficient solar radiation is incident and removed from the collectors when radiation is insufficient or lacking. Thus freeze protection is provided by this type of system.

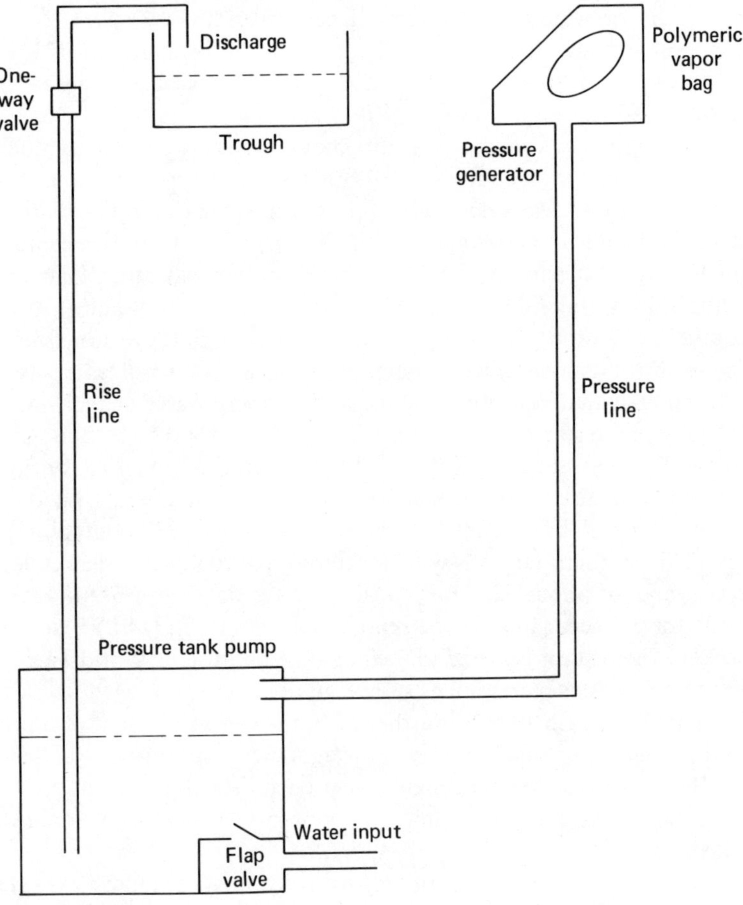

FIGURE 5-15. Simple solar pump (patent by W. A. Harper).

5-8 CONVERSION TO ELECTRICAL ENERGY

Conversion of solar energy to electrical energy can be accomplished by thermal, photovoltaic, and thermoelectric and thermionic means or by using the wind and ocean thermal differences which are due to solar energy originally.

5-8-1 Photovoltaic Conversion

The space program fostered extensive development of silicon photovoltaic-cell panels which achieved outputs as high as 16.5 kW in the case of Skylab by 1974. The cells, originally developed in the early 1950's at the Bell Telephone Laboratories, consist of silicon wafers about 10–12 mils thick.

The bulk of a cell is a *p*-type semiconductor obtained by doping with boron, and a 1-mil layer of *n*-type silicon is formed on one side of the wafer by diffusion of a material such as arsenic, antimony, or phosphorus. A cell develops an open-circuit voltage of about 0.6 V and operates with voltages between 0.4 and 0.5 V. A current density of about 30 mA/cm^2 is provided with incident radiation of 1000 W/m^2. The efficiency of the cells is about 11% in space. The spectrum of sunlight at the earth's surface is different from that outside the atmosphere, and terrestrial efficiencies of the cells have been higher by a few percent. To make large-scale terrestrial applications economically feasible, costs of silicon cells needed to be reduced by factors of 20 to 40 from 1977 values. By that time, however, use of the cells was attractive for special applications such as isolated radio repeaters, navigational aids, lights for offshore oil rigs, corrosion protection for pipelines, etc. (See Fig. 5-16.) The efficiencies (5 or 6%) of conventional cadmium sulfide-copper sulfide (CdS-Cu_2S) cells are lower than those of silicon cells, but cadmium sulfide cells are cheaper than silicon cells to manufacture. Cadmium sulfide cells were used in the experimental house (Solar One) at the University of Delaware where they are employed in such a way as to function as both a source of electrical energy and as heat collectors [27]. The use of cadmium sulfide in thin films has appeared to be an attractive possibility. An alloy of amorphous silicon and hydrogen has also appeared to be suitable for use in thin films.

Promising areas of solar-cell development include thin-film cells, concentrating systems, and hybrid or total-energy systems. Costs of concentrating collectors, per unit area, are less than costs of solar cells, and cells designed

FIGURE 5-16. A solar-cell installation of approximately 900-W capacity supplies power for a remote microwave installation in the state of Alaska. (Courtesy of Spectrolab.)

for high power densities do not necessarily cost more than cells designed for normal levels of solar radiation. Thus systems employing concentrating collectors have shown considerable promise [40, 41]. Although gallium arsenide cells are more expensive than silicon cells, their high efficiency (19% or higher) and ability to withstand high temperatures may make them attractive for use in concentrating systems. When solar energy is concentrated by large factors, perhaps 1000 or more, in concentrating systems, water cooling of the cells may be necessary. This consideration suggests the use of solar cells in systems that supply heat as well as electricity, as already mentioned for the case of cadmium sulfide cells. A hybrid system approach, involving the use of solar energy for heating, lighting, and electricity, has been described by Duguay [15]. He pointed out that, as the efficiency of conversion of solar energy to electricity is between 10 and 20%, most of the remaining solar energy can be utilized as heat in hybrid systems. With respect to lighting, he asserted that the gradual exclusion of natural light from the work environment is a trend that should be reversed. In systems for projecting light from the sun into the interiors of buildings, utilizing mirrors, lenses, and diffusers, the solar spectrum could be separated into visible portions for use for lighting and infrared portions for use with photovoltaic cells for generating electricity.

Costs of silicon cells in 1977 were about $15/W, in full sunlight. A reduction to $1/W as early as 1980, by using concentrating collectors, was anticipated, and a reduction to $0.50/W by 1986, by using flat-plate arrays, was a goal [40]. Automation and streamlining of production plus new technological developments, such as those mentioned, may well allow achieving these reductions in cost.

The use of a large solar array on a synchronous satellite, with a 3-GHz microwave link for transmitting power from the satellite to the earth, has been proposed by Glaser [8, 18]. Unlike a ground installation, a synchronous satellite receives solar energy for 24 h a day on most days of the year. Also the satellite receivers would not experience overcast or atmospheric attenuation and would be oriented perpendicular to the sun's rays at all times. Thus up to 15 times as much solar energy could be received on a given surface as on the earth. A microwave transmitting antenna about 1 km in diameter would be used. The microwave loss on the approximately 35,900-km path from satellite to earth would be about 6%, and the overall transmission efficiency from dc in space to dc on earth would be in the 55–75% range. Power outputs on the earth of 4000–15,000 MW have been considered.

5-8-2 Thermal Conversion

Two types of systems have been proposed for the large-scale thermal conversion of solar energy to electrical energy. One is the central-receiver system in which energy from a number of plane, or nearly plane, mirrors would be concentrated onto a central receiver, or absorber, located in a tower on the order of 100 m or more in height [39]. The second type would involve distributed receivers; a number of collectors or receivers would be arranged in what has been referred to as a solar farm. The heat from the various collectors would be gathered by pipes and transported to a central location. In both cases the heat obtained from the incident solar radiation would be used to drive steam turbogenerators.

The central-receiver concept is similar to that employed in the large solar furnace at Odeillo, France and seems to be the type of large-scale system viewed most favorably in recent years. The solar furnace in France uses 63 mirrors, or heliostats, 6 m high and 7.5 m wide, mounted on a hillside. These direct energy to a paraboloidal surface, 40 m high and 53 m wide, on the side of a building, and the paraboloidal surface concentrates the energy onto a small area about 0.6 m in diameter. About 750 kW of thermal power at 2200°C are provided.

In September 1977, the Energy Research and Development Administration (ERDA) awarded a contract for a pilot 10-kW plant using a 86.3-m tower and 1500 sun-tracking heliostats to be constructed near Barstow, California. ERDA's plan calls for a 50-MW commercial plant to follow the pilot plant. The pilot plant will be a joint enterprise of Stearns-Rogers and McDonnel Douglas Astronautics Co. Martin Marietta has also formulated designs for solar thermal plants, including a 100-MW plant utilizing a 122-m tower at the southern border of each of 8 collector fields, each field having 1840 mirrors, for a total of 14,720 altogether. The plant would be located on a terraced hillside.

The Meinels, in proposing a solar-farm type of installation and in numerous presentations throughout the country, did much to arouse interest in large-scale thermal conversion [31]. They planned cylindrical concentrators, having an absorbing heat pipe at the focal line, and gave considerable attention to the use of selective coatings for the heat pipe.

In considering the design of thermal conversion plants, a major factor to consider is the relative advantages of installing sufficient solar capacity to store energy for periods when solar energy is not available on the one hand or of having nonsolar backup systems to assume the load when solar power is not available. Storage of energy as heat in the ground and in the form of hydrogen comprises two possibilities. The integration of solar and hydroelectric systems appears to be a favorable prospect. Solar energy would be available and utilized during the daytime, and water stored in reservoirs would be saved for periods when solar energy was not available. The two systems would complement each other rather than one being a backup for the other.

For small-scale conversion of solar energy to electrical or other forms, heat engines are a possibility. The noiseless and efficient Stirling hot-air engine is worthy of consideration for solar applications [12]. A $\frac{1}{3}$-hp solar hot-air engine, a feature of a University of Florida solar house, is used to charge a solar-electric automobile battery [16]. Heat engines must compete with photovoltaic cells, however, as a means of generating electricity on a small scale.

The maximum theoretical efficiency η of any process for converting heat energy into work is given by

$$\eta = \frac{T_{\text{source}} - T_{\text{sink}}}{T_{\text{source}}} \tag{5-44}$$

where T_{source} is the temperature of the heat supplied and T_{sink} is the temperature at which heat is rejected, normally the ambient temperature. It is difficult to provide large quantities of heat from solar sources at a sufficiently high temperature T_{source} to obtain a very high efficiency. The overall efficiency of a thermal process for converting solar energy into electricity involves the product of the efficiency of the solar collectors, the efficiency of conversion of heat energy to mechanical energy, and the efficiency of conversion of mechanical energy into electrical energy.

5-9 SOLAR ENERGY AS A RESOURCE

Solar energy has a long history, but until recent years efforts to utilize it directly in place of fossil fuels on a significant scale were not successful [31]. Interest in solar energy was spurred by the oil crisis of 1973, however, and increased considerably by about the mid-1970's. Some have come to regard solar energy not only as the ultimate resource but as one that can make major contributions at an early date. Finally, after interest was generated in certain public sectors and scientific and technical circles, the U.S. federal government started to show serious interest. By 1977 many new buildings of various types—homes, schools, office buildings, etc.—were being equipped with solar heating systems, and any account of numbers of buildings was in need of frequent revision [43]. Although space and water heating are the most obvious applications of solar energy and conversion to electricity has begun to show promise, the indirect forms of solar energy, including wind energy, energy derived from ocean-temperature differences, and energy derived from products of photosynthesis (Sec. 5-4), need to be taken into account as well in assessing solar energy as a resource. Water power is a significant indirect form of solar energy that accounts for 26% of the world's electrical generating capacity, and opportunities for further development exist, especially in developing countries [26]. Some brief remarks will now be made on wind energy and ocean-temperature differences in order to complete more nearly the picture painted so far.

The system of winds described in Sec. 3-2-2 provides opportunities for the generation of energy in the region of prevailing westerly winds. Especially favorable areas in or near the United States are said to be offshore from New England and the Texas Gulf Coast, the Great Plains and Rocky Mountains, and the Aleutian chain of Alaska [33]. A gap in the Rocky Mountains at Medicine Bow, Wyoming, having an average annual wind speed of 10 m/s (21 mi/h), has attracted attention as a favorable location [42]. The use of wind was formerly widespread for pumping water and generating electric power on farms in the United States and in the Netherlands, Denmark, and elsewhere, but the development of rural electrification largely put an end to this use in the United States, and the use of wind power declined in the other locations also. A well-known wind power installation was constructed on Grandpa's Knob, Vermont, however, in the 1940's. A 1.25-MW wind machine with a blade diameter of 53.3 m operated for about 18 months before bearing problems and a later blade failure put it out of operation [35].

The energy density per unit volume in a stream of air of mass density ρ moving with a velocity v is given by

$$E = \frac{\rho v^2}{2} \quad \text{J/m}^3 \tag{5-45}$$

and the power density in the air stream flowing past a given point W_w is given by the product of the energy density and velocity or by

$$W_w = \frac{\rho v^3}{2} \quad \text{W/m}^3 \tag{5-46}$$

The velocity of the air stream cannot be reduced to zero, however, in the process of extracting energy, and the maximum efficiency of a wind machine turns out to be 59% [26]. Efficiencies actually achieved in practice may be near 35%. A power density W_w of 500 W/m² corresponds to the 10-m/s wind mentioned for Medicine Bow, Wyoming. An annual maximum yield of 402×10^9 kWh has been estimated for the Aleutian chain [33]. The wind energy of this remote location might be used to provide hydrogen which could be transported elsewhere. The total maximum potential of two areas offshore from New England was estimated as 477×10^9 kWh [33]. Some of the references concerning wind power are those by Palmer Putnam, concerning the Grandpa's Knob installation [35], Golding [19], Simmons [44], and Clews [11], the last a practical treatise on small wind machines.

In tropical and subtropical waters, warm water overlies colder water at depths of from 1000 m near 30°N to a little more than 100 m near the equator. The Gulf Stream, an area of interest to the United States, has surface temperatures of 21.9°C with temperatures at depth as much as 22°C lower. Although this temperature difference is small and efficiencies of only about 3% or less can be expected [Eq. (5-44)], a large volume of water is available.

In the case of solar energy, including wind energy [42], as in the field of energy in general, there are proponents of large centralized facilities and proponents of smaller, more numerous, and widely dispersed systems. A

100-kW wind machine has been built by NASA near Cleveland, Ohio, and designs of machines with powers up to 2.5 MW have been considered. The Department of Energy Rocky Flats laboratory near Denver, Colorado, however, has tested small machines having 1-, 8-, and 40-kW ratings. An article in *Science* discussed criticisms of the U.S. effort in solar energy to the effect that the effort seemed to give excessive emphasis to large facilities, in the nuclear pattern [38]. While some large installations (such as power towers for thermal conversion, photovoltaic cells in space, etc.) may be appropriate, solar energy is ideally suited for small, widely dispersed systems also. The supply of solar energy, including both direct and indirect manifestations is vast; the problem of utilizing solar energy is primarily one of economics. It seems clear that solar energy will make a significant contribution to meeting energy needs, but it remains to be seen how large the contribution will be and how soon it will be realized.

PROBLEMS

5-1 Given that the sky brightness B is uniform over the entire sky and over a bandwidth of 2 MHz, find w, the power per unit frequency, received on a square meter surface, and find W, the total power received on the surface, over the 2-MHz bandwidth. (Find w by actually integrating B over the entire sky.) Take B as 10^{-21} W/m²/Hz/rad².

5-2 If an antenna has horizontal and vertical half-power beamwidths of 1 and 10° and an area of 360 m² and receives energy from a sky with a uniform brightness of 10^{-22} W/m²/Hz/rad², find the total power received per unit bandwidth, assuming the incident radiation to be randomly polarized and the antenna to be linearly polarized.

5-3 A radiometer receives a signal from a source through an intervening emitting and absorbing region having a uniform intrinsic brightness temperature T_i of 125 K. The optical depth of the intervening region is 1.2, and it subtends a solid angle of 4 deg². The effective brightness temperature of the source T_s is 250 K, and it subtends a solid angle of 1 deg². If the receiving system operates at a frequency of 1200 MHz and its antenna has an effective area for directivity of 20 m², find the effective antenna temperature T_A. Assume the effective source temperature outside the emitting and absorbing region, but within the antenna beamwidth, to be zero.

5-4 Calculate the brightness B using Planck's law and also using the Rayleigh-Jeans law for a temperature of 12,600 K and for a wavelength of 1.25 cm. Also calculate the corresponding value of B_λ.

5-5 Determine the power reflection and transmission coefficients for perfectly plane, lossless glass layers for glass having an index of refraction of 1.5 and for thicknesses of $\lambda/8$, $\lambda/4$, and $\lambda/2$. About what values of the coefficients would you expect for commercial glass sheets at optical frequencies, on the basis of the values calculated?

5-6 a. Calculate the equilibrium temperature of a flat plate having an absorptivity α' of 0.92 and an emissivity ϵ' of 0.1 for normally incident solar radiation of 1353 W/m^2, assuming the surface to be in space (a vacuum) and thus neglecting any losses other than radiation losses. Consider that the plate absorbs solar radiation on one side but radiates from both sides.

b. Repeat the calculation for the flat plate for $\alpha' = \epsilon' = 0.92$.

c. Carry out the same calculation as in part a ($\alpha' = 0.92$, $\epsilon' = 0.1$) for the case of a spherical body, assuming that the entire sphere reaches the same equilibrium temperature and that the sphere absorbs like a disk but radiates like a sphere.

d. Repeat the calculation for the sphere for $\alpha' = \epsilon' = 0.92$.

5-7 Prove that all rays emanating from F and reflected from the parabolic surface of Fig. 5-7 are parallel. (*Hint:* Form the equation of a line tangent to the parabolic surface at P and determine the point Q, where this tangent strikes the x axis. Then consider the nature of the triangle PQF.)

5-8 Derive Eq. (5-38) for cos χ. (Express a unit vector $\mathbf{a}_r$ that is perpendicular to the earth's surface and a vector $\mathbf{P}$ representing the sun's rays in rectangular coordinates but in terms of appropriate angles. Let the coordinate system be centered at the earth's center with the z axis passing through the poles, and let $\mathbf{P}$ lie in the x-z plane. The earth thus rotates with respect to the coordinate system, and the angle ϕ of $\mathbf{a}_r$ thus varies with the time of day. Take the dot product of $\mathbf{a}_r$ and $\mathbf{P}$ and interpret, taking positive h to refer to afternoon hours.)

5-9 a. Calculate χ, the zenith angle of the sun, for sites at lat. 40°N, long. 105°W (near Boulder, Colorado) and lat. 65°N, long. 148°W (near Fairbanks, Alaska) for the longest and shortest days of the year for $h = 0°$.

b. For 10:00 a.m. local standard daylight savings time, calculate χ for Fairbanks for the longest day. (The equation of time for the longest day has the value of -1.4 min.)

5-10 Calculate the azimuth of the sun at sunrise for Boulder and Fairbanks for the longest and shortest days. (See Prob. 5-9.)

5-11 a. If the curve of Fig. 5-9 for the clearest atmosphere applies, find the solar irradiance on a normal surface for $h = 60°$ for a latitude of 40°N for the longest day of the year and for the equinoxes.

b. If conditions are the same as in part a otherwise, find the solar irradiance on a surface tilted to the south by 45°.

5-12 a. Calculate the heating load in kW for a 1200-ft^2 (111.5-m^2) house [ceiling and floor both 1200 ft^2 (111.5 m^2)], walls 952 ft^2 (88.4 m^2), windows 168 ft^2 (15.6 m^2), and doors 42 ft^2 (3.9 m^2), for a temperature difference of 25°C (45°F), assuming R factors of 30 for the ceiling, 20 for the walls, 17.5 for the floors, 1.85 for the double windows, and 3.6 for the doors. Add 1.5 kW for infiltration losses. These values might apply to a well-insulated building. (1 Btu = 0.000293 kWh.)

b. Repeat for an R of 2 for the ceiling, 4.5 for the walls, 4.2 for the floor, 1 for single windows, and 2.4 for doors, using the same infiltration loss. These values might apply to a building without insulation.

c. Calculate the kWh required to heat the house for a 25°C degree day (45°F degree day) and for a year if the total Celsius degree days for the year were 3333 (6000 Fahrenheit degree days) for insulated and uninsulated conditions, neglecting infiltration losses.

REFERENCES

[1] ANDERSON, B., *Solar Energy: Fundamentals in Building Design.* New York: McGraw-Hill, 1977.

[2] ASHRAE, "Solar energy utilization for heating and cooling," in *ASHRAE Handbook & Product Directory 1974*, Chap. 59. New York: American Society of Heating, Refrigerating, and Air-Conditioning Engineers, Inc., 1974.

[3] BEHRMAN, D., *Solar Energy, The Awakening Science.* Boston: Little, Brown, 1976.

[4] BENEMAN, J. R., and N. M. WEARE, "Hydrogen evolution by nitrogen-fixing *Anabaena cylindrica* cultures," *Science*, vol. 184, pp. 174–175, April 12, 1974.

[5] BOHM, D., *Quantum Theory.* Englewood Cliffs, NJ: Prentice-Hall, 1951.

[6] BRINKWORTH, B. J., *Solar Energy for Man.* New York: Wiley, 1972.

[7] BROOKS, F. A., "Solar, wind, and water-power resources," in *Natural Resources* (M. R. Huberty and W. L. Flock, eds.), pp. 444–447. New York: McGraw-Hill, 1959.

[8] BROWN, W. C., "Adapting microwave techniques to help solve future energy problems," IEEE Int. Microwave Symposium, *73 CHO 736-9 MTT*, pp. 189-191. New York: IEEE, 1973.

[9] CALVIN, M., "Solar energy by photosynthesis," *Science*, vol. 184, pp. 375–381, April 19, 1974.

[10] CHANDRASEKHAR, S., *Radiative Transfer.* New York: Dover, 1960.

[11] CLEWS, H., *Power from the Wind.* East Holden, ME: Solar Wind Co., 1973.

[12] DANIELS, F., *Direct Use of the Sun's Energy.* New Haven: Yale University Press, 1964.

[13] DRUMMOND, A. J., and M. P. THEKAEKARA (eds.), *The Extraterrestrial Solar Spectrum.* Mount Prospect, IL: Institute of Environmental Sciences, 1973.

[14] DUFFIE, J. A., and W. A. BECKMAN, *Solar Energy Thermal Processes.* New York: Wiley, 1974.

[15] DUGUAY, M. A., "Solar electricity: the hybrid approach," *American Scientist*, vol. 65, pp. 422–427, July/Aug. 1977.

[16] FARBER, E. A., "Solar energy conversion research and development at the University of Florida," *Building Systems Design*, vol. 71, pp. *PD-195* to *PD-206*, Feb./March 1974.

[17] FINDLAY, J. W., "Radiotelescopes," *IEEE Trans. on Antennas and Propagation*, vol. AP-12, pp. 853–864, Dec. 1964.

[18] GLASER, P. E., "The potential of satellite solar power," *Proc. IEEE*, vol. 65, pp. 1162–1176, Aug. 1977.

[19] GOLDING, E. W., *The Generation of Electricity by Wind Power*. London: E. & F. N. Spon, Ltd., 1976.

[20] HAY, H. R., and J. I. YELLOTT, "A naturally air-conditioned building," *Mechanical Engineering*, vol. 92, pp. 19–25, Jan. 1970.

[21] HUDSON, R. D., Jr., *Infrared System Engineering*. New York: Wiley, 1969.

[22] JOHNSON, F. S. (ed.), *Satellite Environment Handbook*, 2nd ed. Stanford, CA: Stanford University Press, 1965.

[23] KORMONDY, E., *Concepts of Ecology*. Englewood Cliffs, NJ: Prentice-Hall, 1969.

[24] KRAUS, J. S., *Radio Astronomy*. New York: McGraw-Hill, 1966.

[25] KREIDER, J. F., and F. KREITH, *Solar Heating and Cooling*, rev. 1st ed. New York: McGraw-Hill, 1977.

[26] KRENZ, J. S., *Energy Conversion and Utilization*. Boston: Allyn and Bacon, 1975.

[27] KUZAY, T., et al., "Solar One: first results," U.S. Section Annual Meeting. Fort Collins, CO: International Solar Energy Society, Aug. 20–23, 1974.

[28] LONDON, J., and T. SASAMORI, "Radiative energy budget of the atmosphere," *Space Research*, vol. XI, pp. 639–649, 1971.

[29] LOVE, A. W., "Spherical reflecting antennas with corrected line sources," *IRE Trans. on Antennas and Propagation*, vol. AP-10, pp. 529–537, Sept. 1962.

[30] MCVEIGH, J. C., *Sun Power*. Elmsford, NY: Pergamon, 1977.

[31] MEINEL, A. B., and M. P. MEINEL, *Applied Solar Energy*. Reading, MA: Addison-Wesley, 1976.

[32] MSFC, *The Development of a Solar-Powered Residential Heating and Cooling System*. Marshall Space Flight Center, AL: George C. Marshall Space Flight Center, May 10, 1974.

[33] NSF/NASA Solar Energy Panel, *Solar Energy as a National Energy Resource*. College Park, MA: Department of Mechanical Engineering, University of Maryland, Dec. 1972.

[34] ODUM, E. P., and H. T. ODUM, *Fundamentals of Ecology*, 2nd ed. Philadelphia: Saunders, 1959.

[35] PUTNAM, P., *Power from the Wind*. New York: Van Nostrand Reinhold, 1948.

[36] RABINOWITCH, E. I., and GOVINDJEE, *Photosynthesis*. New York: Wiley, 1969.

[37] RABL, A., "Comparison of solar concentrators," *Solar Energy*, vol. 18, pp. 93–112, 1976.

[38] *Science*, vol. 197, "Solar energy research. Making solar after the nuclear model," pp. 241–244, July 15, 1977.

[39] *Science*, vol. 197, "Solar thermal electricity: power tower dominates research," pp. 353–356, July 22, 1977.

[40] *Science*, vol. 197, "Photovoltaics: the semiconductor revolution comes to solar," pp. 445–447, July 29, 1977.

[41] *Science*, vol. 197, "Solar thermal energy: bringing the pieces together," pp. 650–651, Aug. 12, 1977.

[42] *Science*, vol. 197, "Wind energy: large and small systems competing," pp. 971–973, Sept. 2, 1977.

[43] SHURCLIFF, W. A., *Solar Heated Buildings: A Brief Survey*. Cambridge, MA: W. A. Shurcliff, 19 Appleton St., Cambridge, MA 02138. (New editions issued at intervals.)

[44] SIMMONS, D. M., *Wind Power*. Park Ridge, NJ: Noyes Data Corp., 1975.

[45] SMIC (Study of Man's Inpact on Climate), *Inadvertent Climate Modification*. Cambridge, MA: M.I.T. Press, 1971.

[46] THEKAEKARA, M. P. (ed.), *The Energy Crisis and Energy from the Sun*. Mount Prospect, IL: Institute of Environmental Sciences, 1974.

[47] THRELKELD, J. L., *Thermal Environmental Engineering*, 2nd ed. Englewood Cliffs, NJ: Prentice-Hall, 1970.

[48] TLEIMAT, B. W., and E. D. HOWE, "A solar-assisted heat pump for heating and cooling residences," U.S. Section Annual Meeting. Fort Collins, CO: International Solar Energy Society, Aug. 20–23, 1974.

[49] TRYBOUT, R. A., and G. O. G. LÖF, "Solar house heating," *Natural Resources Journal*, vol. 10, pp. 268–326, April 1970.

[50] U.S. Department of Commerce, *Climatic Atlas of the United States*. Washington, D.C.: U.S. Government Printing Office, 1968.

[51] VALLEY, S. L. (ed.), *Air Force Cambridge Research Laboratories, Handbook of Geophysics and Space Environment*. New York: McGraw-Hill, 1965.

[52] WOODWELL, G. M., "Toxic substances and ecological cycles," *Scientific American*, vol. 216, pp. 24–31, March 1967.

[53] WOODWELL, G. M., "The energy cycle of the biosphere," *Scientific American*, vol. 223, pp. 64–74, Sept. 1970.

6 REMOTE SENSING OF THE ENVIRONMENT

6-1 INTRODUCTION; AERIAL PHOTOGRAPHY AND SPACE IMAGERY

The techniques and principles of operation of certain systems utilized for remote sensing and the application of these systems to remote sensing of the atmosphere, sea, and land are considered in this chapter. Aerial photography is discussed briefly, and the subject of space imagery is introduced in this first section. In the following two sections we shall treat radiometric and radar techniques, including those which can be carried out from the ground as well as those that are useful in aircraft and satellites. In the final section we shall consider applications of the techniques discussed in the earlier sections. Aerial photography is a time-honored and important remote sensing technique, and space imagery is a related, newer, exciting technique that has produced excellent results in the Landsat, NOAA, Nimbus, Skylab, and other programs.

Conventional aerial cameras utilize magazines accommodating film that is $9\frac{1}{2}$ in. wide and 200 ft long and that provides 250 exposures, each 9 in. square. The film is held flat during exposure by suction in order to obtain uniformly sharp images. Multiband and panoramic cameras are also utilized in aerial photography. Multiband cameras may have as many as nine lenses and film-filter combinations to cover the range of wavelengths from 0.4 to 0.9 μm, which includes the visible spectrum and part of the near infrared. The panoramic camera utilizes a narrow slit, aligned parallel to the line of flight, in an opaque partition near the focal plane of the camera. The camera also scans perpendicularly to the line of flight to obtain coverage of a large area in a single exposure.

Most aerial photography in the past utilized black and white prints, but color negatives have been used widely more recently. Black and white and color infrared films are also available, Kodak Ektachrome Infrared Aero film, having layers sensitized to green, red, and infrared wavelengths and normally used with a Wratten No. 12 filter to eliminate blue light, is a very useful film. The dyes of the film are such that green objects appear blue (with the exception of vegetation which is highly infrared reflective), red objects appear green, and infrared-reflective objects such as healthy vegetation appear red. The film also has excellent haze penetration when used with a Wratten No. 12 filter. It has been asserted that color photography is a more practical tool than multiband photography because the eye can distinguish subtle differences in color better than it can distinguish differences in black and white photographs.

Aerial photography has remained a useful technique after the advent of satellite imagery because it is often necessary to carry out rather close-range photography and an aircraft is economical to schedule for coverage of a limited area of interest. Satellite imagery has the advantage of covering large areas economically and on a regular schedule and showing features which are not evident in close-range photography.

Since the first Sputnik, launched by the U.S.S.R. on October 4, 1957, and the first American satellite, Explorer 1, launched on January 31, 1958, many earth satellites and space vehicles have been placed in orbit for a variety of purposes. Communication satellites were mentioned in Chap. 2, and brief mention was made in Chap. 3 of research satellites used for studying space and the earth's atmosphere, including the Van Allen radiation belts which were discovered by instrumentation on Explorer 1. In this chapter interest is directed mainly to the earth satellites, which are designated as meteorological or earth resource satellites, especially the NOAA, Nimbus, and Landsat satellites. The meteorological satellites are useful for remotely sensing the earth's surface as well as the atmosphere, and applications to both the earth's atmosphere and surface will be mentioned.

The first meteorological satellite was Tiros-1, which was launched on April 1, 1960 as the first of the Tiros-ESSA-NOAA series. ESSA-1, launched on February 3, 1966, was the first operational weather satellite and initiated the procedure of supplying daily photographs of cloud cover of the type which have become familiar to television viewers. The first satellite of the Nimbus series, Nimbus-1, was launched on August 28, 1964. The Nimbus satellites can be described as research and development, rather than operational, satellites, and the Nimbus program has been concerned with developing and testing instrumentation and techniques. Both the NOAA and Nimbus satellites are in near-polar sun-synchronous orbits and supply remote sensing data to a variety of users.

The Landsat satellites fall in the category of earth resources satellites, and the Landsat program was formerly designated as the ERTS (Earth Resources Technology Satellite) program. Landsat-1 was launched on July 23, 1972. The Landsat satellites provide the highest spatial resolution (100 m) for sensing the earth's surface of any of the satellites mentioned, although low-level military satellites orbiting at altitudes of 95–400 km are said to have resolutions of about 0.6 m [27]. The sensors on the satellites include those utilizing reflected sunlight and those using radiometric and radar techniques. The Landsat satellite uses reflected sunlight, and the NOAA and Nimbus satellites use radiometric and radar techniques. These techniques are considered in Secs. 6-2 and 6-3, and satellite applications thereof are treated in Sec. 6-4, especially Secs. 6-4-2 and 6-4-6. The polar orbit of the Landsat satellites is at an altitude of about 910 km. A Landsat satellite images every location every 18 days, and when Landsat-1 and Landsat-2 were both in orbit a 9-day repeat period was provided. Each Landsat image covers an area of 185×185 km, and a multispectral scanning system (MSS) operates in four wavelength bands. The bands are 0.5–0.6 μm (green), 0.6–0.7 μm (red), 0.7–0.8 μm (near infrared), and 0.8–1.1 μm (near infrared). The images formed appear like photographs. Certain features are obvious upon visual inspection, and the images can also be analyzed by the techniques of multispectral analysis, change recognition analysis, and pattern recognition.

The ATSs (Application Technology Satellites) are synchronous satellites which have been employed for a variety of experiments involving education, medicine, remote sensing, etc. Situated at the earth's equator at a distance of about 35,900 km, as for the INTELSAT communication satellites, the ATSs can image a large portion of the earth's surface at a time but cannot supply the detail and resolution of the meteorological and earth resource satellites in low near-polar orbits. The NASA-NOAA SMS/GOES weather satellites that transmit X-ray, proton, and vector geomagnetic field data to the Space Environment Laboratory (Sec. 3-2-7) are also synchronous satellites. Other satellites which have been used to remotely sense the earth are Skylab, a manned satellite, and the DMSP (formerly DAPP) satellite. DMSP stands for Defense Meteorological Satellite Program.

An application of satellites other than that of remote sensing in the usual sense is that of receiving data from buoys, balloons, remote platforms, and even transmitters mounted on large animals. In such applications it is necessary to know the locations of the remote sources of data. One technique is to record the frequency of transmissions from the remote source and to determine position on the basis of the rate of variation in Doppler shift of the frequency. Balloons released from Christchurch, New Zealand have been tracked by this procedure and have been found to have circled the earth in the southern hemisphere, moving from west to east at the 100-mb level, as many as 25 times.

A very interesting satellite scheduled for launch in 1978 is the Seasat-A satellite. It is designed primarily for remote sensing of the oceans and seas and is mentioned further in Sec. 6-4-5.

The subject of remote sensing is large, and only a brief introduction can be provided here. A thorough, two-volume treatment of the topic has been edited by Robert Reeves and published by the American Society of Photogrammetry [63]. The annual Symposia on Remote Sensing of Environment, organized by the Environmental Research Institute of Michigan, Ann Arbor, Michigan, and the published proceedings of these symposia are good sources of information on remote sensing developments and interests, and the periodicals *Photogrammetric Engineering and Remote Sensing* and *Remote Sensing of Environment* are devoted to the subject. The practice of listing cited references at the end of the chapter is observed for this final chapter, and following that a separate selected bibliography on remote sensing, emphasizing applications to the land, is included in addition.

6-2 RADIOMETRIC TECHNIQUES

Foundations of the radiometric method were presented in Chap. 5 where the concepts of antenna temperature and blackbody radiation were developed. Radiometric techniques for remote sensing may be used in the microwave and infrared frequency ranges. A microwave or infrared radiometer can be calibrated to measure average or effective brightness temperatures of the areas of the earth within their fields of view from an aircraft or satellite. In remote sensing of the earth's environment, more thermal energy is available in the infrared range than in the microwave range, but sensitivities of microwave detectors are higher than those for infrared, and overall performance in the two frequency ranges may be comparable. In contrast to systems using reflected sunlight, such as the MSS of the Landsat Satellites, radiometric systems, such as the VHRR (very high resolution radiometer) of the NOAA satellites, can utilize thermal radiation and have the potential for obtaining information in darkness. Microwave radiometers, such as the ESMR (electrically scanning microwave radiometer) of the Nimbus satellites, furthermore, can penetrate overcast.

Microwave radiometers consist of an antenna and a sensitive receiver which provides sufficient amplification to drive some type of indicating system. Antenna input power is proportional to the average brightness temperature that the antenna "sees." Radiometers can be constructed to record thermal radiation covering a relatively broad frequency spectrum or to record radiation at discrete frequencies such as that at 1410 MHz which is caused by interstellar hydrogen. Radiometer receivers for environmental monitoring have much in common with those designed for radio astronomy.

The antenna for a microwave radiometer commonly consists of a paraboloidal reflector and a "feed" located at the focal point of the reflector. Cassegrain antennas, which have the feed horn located behind the main reflector and use a hyperboloidal subreflector, however, have the advantages of lower line losses and less spillover. Other multiple-reflector antennas, horn reflector antennas, and slotted-waveguide antennas have also been utilized in radiometer applications. In general, radiometer antenna types are the same as those utilized in other microwave applications.

Radiometer receivers are also similar to microwave communications receivers in many respects but have high sensitivity and gain and must be calibrated accurately so as to measure small temperature differences. Radiometer receivers usually have a square-law detector followed by a long-time-constant *RC* network which integrates the signal and effectively provides a very small postdetection bandwidth. The minimum variation or change in temperature ΔT_{min} which can be detected is an important characteristic of a radiometer receiver and is given by

$$\Delta T_{min} = \frac{T_{sys}}{\sqrt{\Delta f t}} \tag{6-1}$$

where T_{sys} is the system noise temperature defined by Eq. (2-16), Δf is the receiver predetection bandwidth (essentially the IF bandwidth), and t is the postdetection integration time. A low value of ΔT_{min} represents high sensitivity. The expression for ΔT_{min} appears reasonable if it is noted that the measurement of T_{sys} is like measuring the noise output of a resistor at the same temperature and that measuring the noise output of a resistor is equivalent to measuring the number of independent noise pulses per second appearing at the resistor terminals. This measurement can be made with a standard deviation equal to the mean value divided by $\sqrt{n}$, where n is the total number of individual noise pulses or contributions to the noise signal. n is equal to the product of Δf, the receiver bandwidth, or the number of independent noise-pulse contributions which can be measured per second, and the integration time t. Only ΔT values which are equal to or greater than the standard deviation of T_{sys} can be detected.

The above discussion assumes a perfectly stable receiver. However, any receiver is subject to some degree of gain instability, and this is a serious matter in the case of high-gain, highly sensitive receivers. The problem of gain instability can be largely overcome by switching the receiver input between the antenna and a comparison noise source at a sufficiently high rate, perhaps between 10 and 1000 Hz. This technique was developed by Dicke [20], and a receiver employing this principle is called a Dicke receiver.

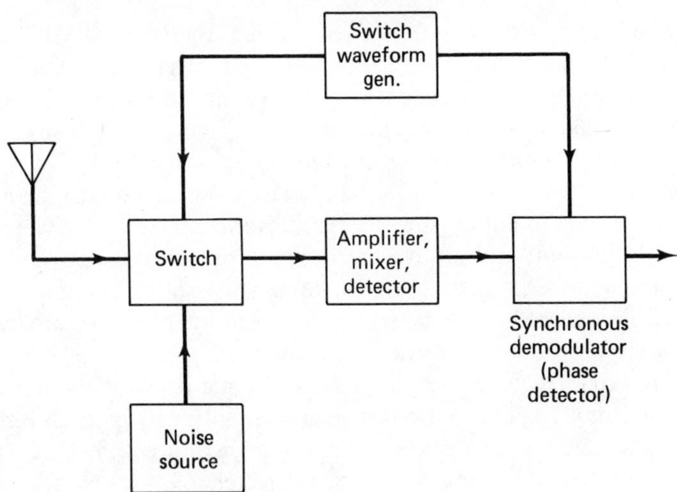

FIGURE 6-1. Dicke or switched receiver.

Such a receiver (Fig. 6-1) includes a synchronous demodulator which receives the amplified detected switched signal. The overall effect when narrow-band amplification is inserted between the detector and the demodulator in a Dicke receiver is that the value of $\Delta T_{\min}$ is given by

$$\Delta T_{\min} = \frac{\pi}{\sqrt{2}} \frac{T_{\text{sys}}}{\sqrt{\Delta f t}} \tag{6-2}$$

Even with the arrangement described, gain variations still affect the measurement of the difference between the reference temperature and the antenna temperature. The several techniques developed to further overcome this problem include adjusting the reference temperature to be close to the antenna temperature; injecting noise in the signal channel until $T_{\text{sys}} + T_i$, where T_i is the injected noise temperature, equals T_R, the reference temperature; and the use of gain modulation.

Calibration of a radiometer is necessary, and this can be accomplished by injection of a calibration signal into either the signal or reference channels. $\Delta T_{\min}$ values as low as 0.05 K rms have been quoted for modern broadband microwave radiometers.

Many of the concepts and techniques discussed with relation to microwave radiometers apply also to infrared radiometers. Corresponding to the microwave antenna is an optical system. Actually Cassegrain systems are used at both microwave and infrared frequencies, and systems for the two frequency ranges may thus be similar in that respect. Nothing comparable to the superheterodyne receivers that are used at microwave frequencies, however, is available in the infrared range. The detectors which are placed at the foci of the optical system in the infrared range are of two types, namely thermal detectors and quantum detectors. Examples of the former are bolometers, thermocouples, and Golay cells (which operate on the principle of a gas thermometer). The quantum detectors include the photoelectric, photovoltaic, photoconductive, and photoelectromagnetic varieties. Lead sulfide (PbS) is a commonly used photoconductive detector, and indium arsenide (InAs) is a common photovoltaic detector [47]. Another feature of infrared systems is that the radiation may pass through an optical modulator before reaching the detector. Optical modulators, also known as reticles, are useful for detecting a small discrete source in the presence of background radiation.

Images can be generated directly (on film in a camera) in the optical and near-infrared frequency ranges, but such direct image formation is not possible at far-infrared and microwave frequencies. Radiometers operating at these frequencies, however, may have sufficiently narrow beamwidths to allow the construction of images by angular scanning. Scanning radiometers have been developed in both the infrared and microwave frequency ranges.

Environmental applications of radiometers can be classified roughly into two categories, the determination of atmospheric parameters and the measurement of surface characteristics.

6-3 RADAR TECHNIQUES

6-3-1 Introduction: Range and Velocity Measurements

The ionosonde, developed in 1925–1926 and used for monitoring the ionosphere since, was the first practical radar system of significance to be put into service. Operating in the HF frequency range and developed some years prior to surveillance radars, before the term radar (for radio detection and ranging) was introduced, the ionosonde may not be thought of by some as a radar, but it clearly is a special-purpose radar system. Surveillance radars were developed not long before World War II. A British 25-MHz chain of radars began operation in 1935, and 200-MHz radars were used by the U.S. Navy by 1938. It was during the war, however, that efficient microwave radars were developed in Great Britain and the United States. The Radiation Laboratory of the Massachusetts Institute of Technology was one of the leading laboratories in radar research and development, and the series of 28 volumes written under Radiation Laboratory auspices is a classic in the field [62]. Other useful references concerning radar are those by Barton [8], DiFranco and Rubin [21], Nathanson [59], and Skolnik [70, 71].

A common type of radar is the pulsed search or surveillance radar which radiates short pulses of electromagnetic energy at a certain repetition rate and which uses a narrow-beamwidth antenna that rotates continuously in azimuth. The distance R to a target from which an echo is received is determined by measuring the time t for a pulse to travel from the radar to the target and back. Direction to the target is given by the direction the antenna is pointing when the echo is received. The repetition period of the pulses ($1/f_r$, where f_r is the repetition rate, a typical value being 1000 pulses/s) is extremely short compared with the period of rotation of the antenna, and a number of echo pulses is normally received from each individual target as the antenna rotates. This number is a function of the antenna beamwidth, antenna rotation rate, and the pulse repetition rate. Resolution in azimuth is determined by the antenna beamwidth, which might be on the order of a few degrees or less between half-power angles in the case of a microwave radar. In terms of linear distance in the azimuthal direction, the resolution is about $R\phi_{\mathrm{HP}}$, where R is the distance to the target and ϕ_{HP} is the half-power beamwidth in the horizontal or azimuthal direction.

The echoes received by a surveillance radar are commonly displayed on a PPI (plan position indicator) display which presents a radar map of the surroundings in terms of polar coordinates (range and azimuth), with echoes showing as bright areas on a cathode-ray tube. Such surveillance radars commonly have a relatively narrow horizontal beamwidth and a wider vertical beamwidth, but some surveillance radars use "pencil" beams which are equally narrow in both directions. Many other types of radars and modes of operation are utilized. A slight variation of a surveillance radar is one which has the capability of sector scanning as well as continuous rotation. Tracking radars have the ability to lock onto and follow individual targets, providing information on azimuth angle, elevation angle, and range in the process. If highly precise elevation angles and ranges are desired, it may be necessary to correct for atmospheric refraction and time delay. RHI (radar height indicator) systems scan up and down in elevation angle and tend to have narrow vertical beamwidths and relatively broad horizontal beamwidths. Other types of radars are discussed in Secs. 6-3-3 and 6-3-4.

Returning to the measurement of the range R to the target, $2R = ct$ where $c = 3 \times 10^8$ m/s and

$$R = \frac{ct}{2} \qquad (6\text{-}3)$$

The range resolution ΔR, the minimum distance by which two targets can be separated in range and be distinguished, is given by

$$\Delta R = \frac{c\tau}{2} \qquad (6\text{-}4)$$

where τ, the width of the transmitted pulse, has been substituted for t in Eq. (6-3). A commonly used width is 1 μs, and the corresponding range resolution is 150 m.

To detect a pulse of width τ in the presence of noise a bandwidth B approximately equal to $1/\tau$ is optimum for the conventional radar employing a superheterodyne system for signal amplification. If a bandwidth of $1/\tau$ is utilized, ΔR can be expressed as

$$\Delta R = \frac{c}{2B} = \frac{v}{B} \tag{6-5}$$

where v is an effective velocity. A radar actually measures time intervals, and the range resolution ΔR corresponds to a time resolution Δt, where

$$\Delta t = \frac{1}{B} \tag{6-6}$$

Thus the maximum time resolution and therefore the maximum range resolution are determined by available bandwidth, which is usually the bandwidth of the IF (intermediate-frequency) amplifier.

Consider that a radar system is receiving an echo from an object moving toward the radar with a velocity v_R m/s. In time t the phase of the received signal will advance by an amount $\Delta\phi$, where

$$\Delta\phi = 2kv_R t = 2k\Delta d \tag{6-7}$$

$k = 2\pi/\lambda$ and is the propagation constant, and Δd is the distance moved by the object in time t. The factor of 2 accounts for the fact that the electromagnetic wave traverses the distance between the radar and the object twice, once in each direction. Because the phase of the received signal is changing, the frequency of the received signal differs from the transmitted frequency by the Doppler frequency f_D, where

$$f_D = \frac{1}{2\pi}\frac{d\phi}{dt} \tag{6-8}$$

Thus

$$f_D = \frac{2v_R}{\lambda} \tag{6-9}$$

More generally when the object is moving in a direction θ measured from the radial direction,

$$f_D = \frac{2v\cos\theta}{\lambda} \tag{6-10}$$

If the object is moving but has no radial velocity, $\theta = 90°$ and $f_D = 0$. The Doppler shift f_D is usually small compared with the IF bandwidth, and the Doppler-shifted signal can be received and displayed in the same way as the echo from a stationary target. Indeed the conventional noncoherent pulse radar without MTI or Doppler capability cannot measure Doppler frequency or distinguish between stationary and moving targets except by noting the change of position with time. Coherent pulse radars, or pulse-Doppler radars, and certain modulated CW (continuous-wave) radars, however, can measure and record both range and velocity values. Some CW radars can measure velocity but not range.

The resolution to which the Doppler frequency f_D can be measured is $1/T'$, where T' is the observation or dwell time. For the case where a single pulse of width τ is transmitted,

$$\Delta f_D = \frac{2\Delta v_R}{\lambda} = \frac{1}{\tau} \tag{6-11}$$

and

$$\Delta v_R = \frac{\lambda}{2\tau} \tag{6-12}$$

where Δf_D and Δv_R are resolutions, namely the smallest variations in Doppler frequency and radial velocity that can be distinguished. Note that the product of ΔR, given by Eq. (6-4), and Δv_R, given by Eq. (6-12), has the value $c\lambda/4$ and is thus independent of pulse length.

In practical radar operations T' is determined by the characteristics of the radar and the mode of operation. If the radar is a continuously rotating surveillance radar, T' may be determined by the antenna beamwidth and the rate of rotation. If the antenna can be pointed continuously at a given target, considerable latitude in the choice of T' may be available. In some applications it may be convenient to define T' by $T' = NT$, where $T = 1/f_r$ and is the repetition period and N is the number of sweeps (same as the number of transmitted pulses) over which the measurement is conducted. If $T' = NT$,

$$\Delta v_R = \frac{\lambda}{2NT} \tag{6-13}$$

The possibility of an ambiguous range arises in the case of an echo from a target at a range greater than that corresponding to the period T between consecutive pulses, where $T = 1/f_r$. Thus R_m, the maximum unambiguous range, is given by

$$R_m = \frac{cT}{2} = \frac{c}{2f_r} \tag{6-14}$$

A maximum unambiguous velocity v_{Rm} also exists, as can be seen by recalling the sampling theorem of communication theory. To sample (or reproduce or represent) a frequency f, one must sample at a rate f_s or greater, where $f_s = 2f$. Thus the maximum unambiguous Doppler frequency f_{Dm} that can be recorded is $f_r/2$, and

$$f_{Dm} = f_r/2 = \frac{2v_{Rm}}{\lambda} \tag{6-15}$$

and

$$v_{Rm} = \frac{\lambda f_r}{4} \tag{6-16}$$

Also

$$R_m v_{Rm} = \frac{c\lambda}{8} \tag{6-17}$$

Equation (6-17) shows that the values of R_m and v_{Rm} cannot both be chosen arbitrarily. Increasing R_m by decreasing f_r, for example, results in a decrease in v_{Rm} and vice versa.

Targets which have ranges greater than R_m may be confused with targets at smaller ranges, and targets which have velocities greater than v_{Rm} may be confused with targets having lower velocities and vice versa. In many cases the ranges and velocities of targets may be known to fall within certain limits, and no problem of ambiguity may develop. It is advisable to be alert, however, to the possibility of ambiguity, and it should be kept in mind that designing a radar system to have certain values of R_m and v_{Rm} provides no guarantee that echoes with higher values will not be encountered.

The ability of a radar system to measure range and velocity, to distinguish between closely spaced ranges and velocities, and to distinguish targets of given ranges and velocities from other targets without ambiguity can be described by the radar ambiguity function introduced by Woodward in 1967. This function can be described as a radar response pattern, similar to an antenna pattern except that for antennas two orthogonal directions in space may be considered, whereas the two "directions" or "dimensions" of the ambiguity function are range and velocity. Complete radar response and antenna patterns are three dimensional, but two-dimensional presentations can be utilized, either plan views showing contours of constant response or separate plots of range response versus range or velocity response versus velocity. Figure 6-2 shows two-dimensional representations of ambiguity functions for single long and short pulses. As illustrated in the figure, a long pulse tends to provide low range resolution but high Doppler-frequency and velocity resolution, whereas a short pulse tends to provide high range resolution but low Doppler-frequency resolution. For trains of pulses, the central part of the response pattern is as suggested by the two-dimensional drawing in Fig. 6-3. The patterns or ambiguity functions are centered or maximized on a particular target of given range and velocity and show the response to targets having other ranges and velocities. Considering the response as a function of range, zero reponse is recorded between the given range and

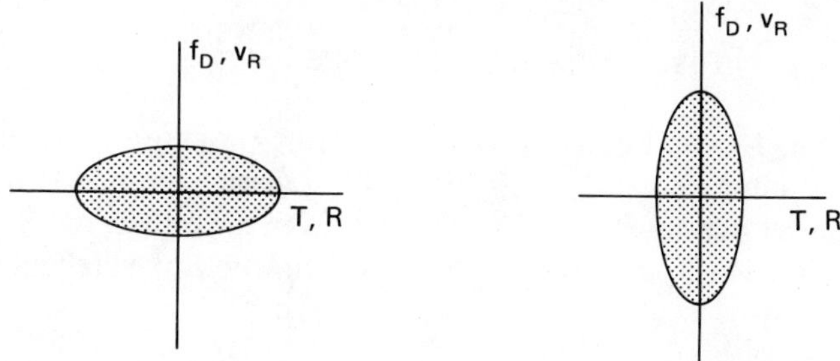

FIGURE 6-2. Two-dimensional ambiguity diagrams for a single pulse: (a) long pulse, (b) short pulse.

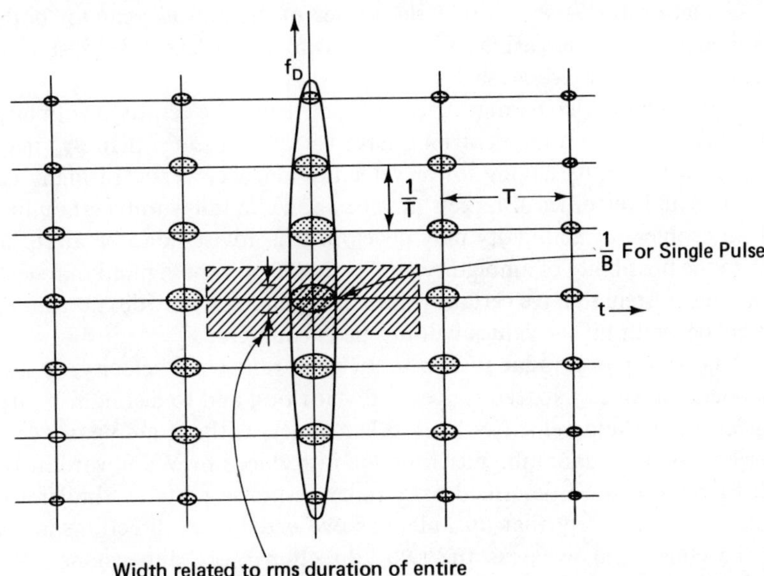

FIGURE 6-3. Central part of the ambiguity function for a limited Gaussian pulse train. If operation is restricted to the rectangular crosshatched region, no problem of ambiguity is encountered. (Adapted from W. S. Burdic, *Radar Signal Analysis*, Prentice-Hall, Inc., 1968.)

ranges differing by multiples of $cT/2$. In the case of velocity, the correct Doppler frequency differs from its nearest alias by $f_r = 1/T$. For a train of Gaussian pulses, for which pulse amplitude varies as exp $(-Ct^2/\tau)$, where C is a constant, t is time, and τ is half-power pulse width, there is zero response between the proper velocity and its aliases also, as in Fig. 6-3. For other pulse shapes, including the rectangular shape, a "velocity side-lobe" structure occurs along the f_D axis around the responses shown in Fig. 6-3, but the situation is not very significantly different from that shown in the figure.

6-3-2 Pulse-Radar Performance

Radar Range Equation for Discrete Targets. Consider a pulsed radar transmitter radiating a power of W_t and having an antenna gain of G_t. The power density P, at a distance of R from the transmitter in the direction the gain is specified (normally in the direction of maximum gain), is given by

$$P = \frac{W_t G_t}{4\pi R^2} \quad \text{W/m}^2 \tag{6-18}$$

If at this distance some of the radiated power is intercepted by a target having a radar cross section of σ m^2 and some of the intercepted power is scattered back to the radar location, the backscattered power density at the radar is given by

$$P_r = \frac{W_t G_t \sigma}{(4\pi R^2)^2} \quad \text{W/m}^2 \tag{6-19}$$

Note that, although most targets do not scatter isotropically, the value of the radar cross section σ is that corresponding to an equivalent isotropic scatterer, which would return the same amount of power to the receiver as the actual scatterer. Given the above power density, the radar receiver has a power input W_r that is the product of P_r and the effective area for calculating gain A_r of the receiving antenna so that

$$W_r = \frac{W_t G_t A_r \sigma}{(4\pi R^2)^2} \quad \text{W} \tag{6-20}$$

The effective area for gain A is related to gain by $G = 4\pi A/\lambda^2$. By using this relation and considering a monostatic radar, employing the same antenna for transmission and reception, so that subscripts are not needed for antenna gain or area, W_r is also given by

$$W_r = \frac{W_t G^2 \lambda^2 \sigma}{(4\pi)^3 R^4} = \frac{W_t A^2 \sigma}{4\pi \lambda^2 R^4} \tag{6-21}$$

The effective area for gain is less than the aperture area and is commonly taken as 0.54 times the aperture area for an ordinary antenna. (See Sec. 2-4, where the effective area for gain is designated as A_{eff}, for further discussion.) Some minimum value of W_r is required for an echo to be recognized or distinguished from noise. Thus there is a maximum range $R_{\max}$ at which a target of cross section σ can be detected. This range can be found by setting W_r equal to $S_{\min}$, the minimum usable signal power, and solving for $R_{\max}$. Using the first of the above expressions for W_r results in

$$R_{\max} = \left[\frac{W_t G A \sigma}{(4\pi)^2 S_{\min}}\right]^{1/4} \quad \text{m} \tag{6-22}$$

The value of $S_{\min}$ is determined by the receiver noise power N and the minimum signal-to-noise ratio $(S/N)_{\min}$ which can be tolerated. That is,

$$S_{\min} = N\left(\frac{S}{N}\right)_{\min} \tag{6-23}$$

where $N = kTB_nF$. k is Boltzmann's constant, T is absolute temperature, B_n is the noise bandwidth, and F is the receiver noise figure. Substituting the expression for $S_{\min}$ into Eq. (6-22) and introducting a loss factor L, we obtain

$$R_{\max} = \left[\frac{L W_t G A \sigma}{(4\pi)^2 k T B_n F_n (S/N)_{\min}}\right]^{1/4} \tag{6-24}$$

In practice, the noise bandwidth B_n is usually determined by the receiver

IF bandwidth, which is designed to be equal to about $1/\tau$, where τ is the pulse width. In this case $1/\tau$ can be substituted for B_n, resulting in

$$R_{\max} = \left[\frac{LW_t\tau GA\sigma}{(4\pi)^2 kTF_n(S/N)_{\min}}\right]^{1/4} \tag{6-25}$$

The product $W_t\tau$ is the energy of the transmitted pulse, and this pulse energy is a more fundamental quantity than pulse power. By designating this quantity by E_t and also noting that $(S/N)_{\min}$ in the denominator can be written as $(E/N_0)_{\min}$ (where E is received energy per pulse and $N_0 = N/B_n = kTF_n$),

$$R_{\max} = \left[\frac{LE_tGA\sigma}{(4\pi)^2 kTF_n(E/N_0)_{\min}}\right]^{1/4} \tag{6-26}$$

The loss factor L can be used to represent various system losses which have not yet been taken into account, such as attenuation in the transmitting medium and losses in waveguides, transmission lines, or RF plumbing. The pulse energy, rather than power, could have been utilized at the beginning, in Eq. (6-18), etc. To express Eqs. (6-18)–(6-21) in terms of pulse energy, for example, it is only necessary to multiply both sides by the pulse width τ. The value of $(S/N)_{\min}$ to be used depends on the application and may be 1 to 2 in the case of a visual observer watching a PPI scope.

Aircraft present discrete radar targets, and air-traffic control is an important area of application for radar. In the United States, the Federal Aviation Administration (FAA) operates a network of Air Route Surveillance Radars (ARSRs) and Airport Surveillance Radars (ASRs) such that aircraft flying at the altitudes of commercial aircraft are essentially always under surveillance over the continental United States. The signals from ARSRs are transmitted to 21 Air Route Traffic Control Centers in the continental United States where they are processed by computer so that the air-traffic controllers work with computer-generated displays, with no rotating sweep apparent as on the PPI displays formerly used. Each center receives signals from up to about 10 ARSRs or similar radars. Transponder signals are normally employed in the case of commercial aircraft, and information on aircraft altitude obtained by an airborne altimeter is transmitted to the centers along with the transponder signal that indicates range and azimuth. (The broad vertical beamwidth of the ARSRs does not allow them to obtain height information directly.) Representative system parameters (some variation occurs from model to model) are as follows. The ARSRs operate in the 1280- to 1350-MHz range (L band), with peak powers of 4.2 MW, a pulse length of 2 μs, and a pulse repetition rate of 360 pps. The horizontal beamwidth is 1.35°, the vertical beamwidth is 6.2°, and the antenna rotates at 6 r/min. The radars have the ability to detect a target of 0.01-m² cross section at 42 nmi (77 km), and the maximum sweep length is 200 nmi (366 km). The noise figure of the receiver when using a parametric amplifier input is 4.9 dB. The

ASRs operate in the 2700- to 2900-MHz range (S band), with peak powers of 425 kW, a pulse length of 0.833 μs, and pulse repetition rates between 1000 and 1200 pps. The horizontal beamwidth is 1.5°, and the vertical beamwidth is 5°, but with a $\csc^2$ response to 30°. The antenna rotation rate is 15 r/min, and the maximum sweep length is 60 nmi (110 km). The minimum detectable signal is 109 dBm with normal video. At an opposite extreme from the high-power ARSRs are the small X-band (9000- to 10,000-MHz) shipborne marine radars having peak powers from about 5 to 25 kW, pulse lengths between 0.05 and 1 μs, and sweep ranges from $\frac{3}{4}$ nmi (1.37 km) to 48 nmi (87.8 km). The shortest pulse length is used on the shortest ranges. They usually employ slotted-waveguide antennas up to 3.7 m (12 ft) in length that provide a narrow horizontal beam and a broad vertical beam.

Radar Range Equation for Distributed Targets. Suppose that the radar receives echoes from objects which are distributed with a certain density throughout the volume of space which the radar interrogates, instead of receiving an echo from a single discrete target. The drops in a region of uniform rainfall can serve as an important example. In such a case the individual objects backscatter incoherently or independently and the backscattered powers add. (In coherent reflection from a number of objects the backscattered field intensities add. A much larger effect would be observed in that case as power is proportional to field intensity squared.) The radar cross section per unit volume is σN for independent backscatter, where σ is the cross section of a single object and N is the number of objects per unit volume. The volume interrogated by a radar at a particular instant is given approximately by

$$V = \frac{\pi}{4} R^2 \theta_{\text{HP}} \phi_{\text{HP}} \frac{c\tau}{2} \tag{6-27}$$

where θ_{HP} and ϕ_{HP} are the half-power beamwidths and $c\tau/2$ is the length of the pulse in space. [$c\tau/2$ is also the range resolution as given by Eq. (6-4).] The expression for V has been put into a different form by noting that $4\pi/\theta_{\text{HP}}\phi_{\text{HP}} \approx 4\pi A/\lambda^2$. Thus $\theta_{\text{HP}}\phi_{\text{HP}} \approx \lambda^2/A$, providing a basis for the approximate expression

$$V = \frac{\pi R^2 \lambda^2 c\tau}{8A} \tag{6-28}$$

If $\sigma N V$ is now substituted in place of σ alone in Eq. (6-21),

$$W_r = \frac{W_t A c\tau}{32R^2} \sigma N \tag{6-29}$$

and

$$R_{\max} = \left(\frac{W_t A c\tau \sigma N}{32 S_{\min}}\right)^{1/2} \tag{6-30}$$

The important point to notice is that the received power now varies inversely as R^2 instead of inversely as R^4. Equations (6-29) and (6-30) can of course be manipulated into other forms, and substitution can be made for $S_{\min}$ as in the previous treatment for a discrete reflecting object. In some applications a cross section per unit solid angle σ' is used in place of σ. The numerical relation between these two quantities is $\sigma = \sigma' 4\pi$. In general a loss factor L may need to be taken into account by introducing it in the numerator of Eq. (6-30).

BRAGG REFLECTION. When monostatic-pulse-radar echoes are received from regions having a spatial spectrum of irregularities, it is the irregularities having periodicities near $\lambda/2$ (along the direction of wave propagation) that are primarily responsible for useful echoes. It is these irregularities which cause constructive interference of backscattered waves, much as in the case of Bragg scattering of X-rays from crystal structures. Situations where the Bragg reflection mechanism is effective include the backscatter of HF waves from ocean waves and backscatter of radio waves and acoustic waves from turbulent regions of the atmosphere. Equation (4-126) for the volume reflection coefficient in a turbulent region is consistent with this statement because it is the refractive index power density corresponding to $\lambda/2$ that causes the finite reflectivity. [$k = 4\pi/\lambda$ in Eq. (4-126), where λ is radar wavelength.] To obtain information about more than one size of irregularity in such cases, multifrequency radar capability is needed.

6-3-3 FM-CW, MTI, and Pulse-Doppler Radars

The Doppler effect can be used in a simple unmodulated CW radar system to detect moving objects. Echoes from stationary objects exhibit no Doppler shift, but echoes from objects with a sufficient radial velocity have a measurable Doppler shift. The use of a directional antenna allows the determination of direction. Such a Doppler radar, however, can give no information about target range.

If the phase or frequency of a CW radar output is modulated, however, range can be determined. Consider a repetitive waveform that increases linearly with time over an interval T as shown in Fig. 6-4. If an echo is received from an object, the plot of the echo frequency versus time will be as shown by the dotted line, which is delayed by a time interval Δt from the transmitted signal, where $\Delta t = 2R/c$ and $R = \Delta tc/2$, consistent with Eq. (6-3). However, the difference frequency f_B is what is measured, and

$$f_B = \frac{\Delta t B}{T} = \frac{2RB}{cT} \tag{6-31}$$

or

$$\Delta t = \frac{f_B T}{B}$$

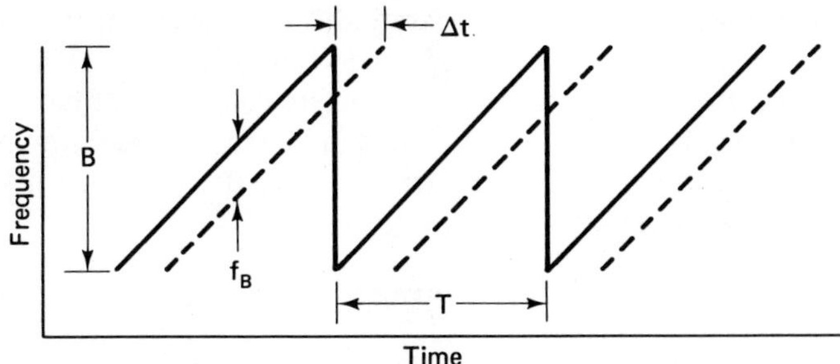

FIGURE 6-4. Waveform of FM-CW radar.

where B is the total frequency deviation and T is the time during which the frequency is swept over the range B. Thus the distance to the target can be determined from

$$R = \frac{c}{2}\frac{f_B T}{B} \tag{6-32}$$

Radars using the principle discussed are called FM-CW radars. They have the advantages of a low minimum range, allowing them to record closer echoes than pulse radars, and high resolution. The Naval Electronics Laboratory (NEL), San Diego, pioneered in the use of FM-CW radars for studying the lower troposphere [64], and the Institute of Telecommunication Sciences, Boulder, began using a similar radar a short time later. At present NEL and the Wave Propagation Laboratories, NOAA, Boulder, are using FM-CW radars to study phenomena of the troposphere. Range resolution in an FM-CW radar is determined by the fact that the beat frequency f_B is sampled for a time interval T. The resolution of the frequency measurement is thus limited to $1/T$. If two frequencies separated by $1/T$ can be resolved, or $\Delta f_B = 1/T$, then Eq. (6-32) leads to

$$\Delta R = \frac{c}{2B}$$

which is the same expression as Eq. (6-5).

As the beat frequency is used in FM-CW radars for a determination of range, a potential problem exists for moving targets which cause Doppler frequency shifts because of their motion. For a point target this problem can be solved by using a triangular sweep with symmetrical up-frequency and down-frequency portions of the sweep waveform. By taking the average of the up-frequency and down-frequency beat frequencies, the effect of target motion can be eliminated, and the correct range can be obtained. Also target velocity can be determined by taking half of the difference between the up-frequency and down-frequency beat frequencies.

For distributed targets, for example for turbulent regions of the atmosphere, the procedure mentioned above cannot be used, and it has been assumed that it was impossible to separate the components of the difference or beat frequency that are due to range and velocity. Strauch [75], however, has shown that the measurement of signal phase can be employed to obtain target velocity, while target range can be obtained in the usual way but with a correction, if need be, for velocity, which is also determined. An accumulated phase-change term of

$$\Delta\phi = 4\pi n f_0 \frac{vT}{c} \tag{6-33}$$

where n is the number of sweeps of duration T, f_0 is the frequency at the start of the sweep, c is 3×10^8 m/s, and v is the desired velocity, is what provides the velocity information. This expression for $\Delta\phi$ is essentially a restatement of Eq. (6-7), with Δd replaced by vnT and the λ of $k = 2\pi/\lambda$ replaced by c/f_0.

An MTI radar is a pulsed radar with an additional feature that tends to eliminate stationary targets and show only moving targets. The discrimination is based on the Doppler effect. The concept of MTI is demonstrated by a simple system in which the signal from a stable reference is mixed with the returned echo signal. Whereas the signal from the reference oscillator is represented by

$$V_{\text{ref}} = A \sin 2\pi f t \tag{6-34}$$

that of the echo is given by

$$V_{\text{echo}} = A_1 \sin \left[2\pi(f \pm f_D)t - \frac{4\pi f R}{c} \right] \tag{6-35}$$

and the beat or difference frequency signal is

$$V_b = A_2 \sin \left(2\pi f_D t - \frac{4\pi f R}{c} \right) \tag{6-36}$$

A bipolar video signal can be formed from V_b. Echoes will also be received from stationary objects, but f_D will be zero in that case, and the echo intensity will remain essentially constant from pulse to pulse. If the bipolar signals are viewed on an A scope, a presentation like that shown in Fig. 6-5 may result. The echoes from the stationary objects or background remain constant, but the echo from the moving object varies as $\sin 2\pi f_D t$. To eliminate the stationary echoes a system like that shown in Fig. 6-6 can be used. The signals are fed through a delay line giving a delay of T, where T is the repe-

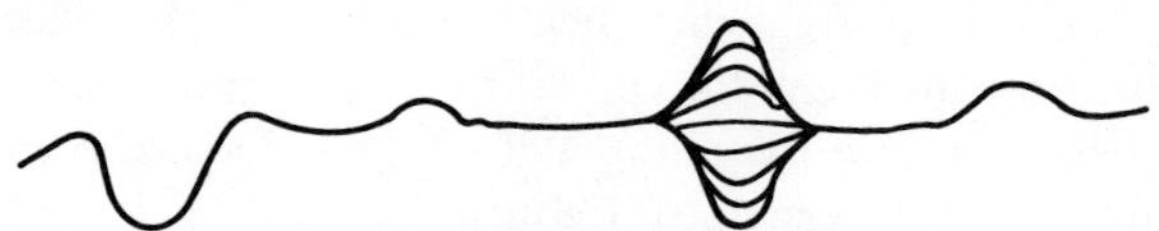

FIGURE 6-5. Bipolar video signal.

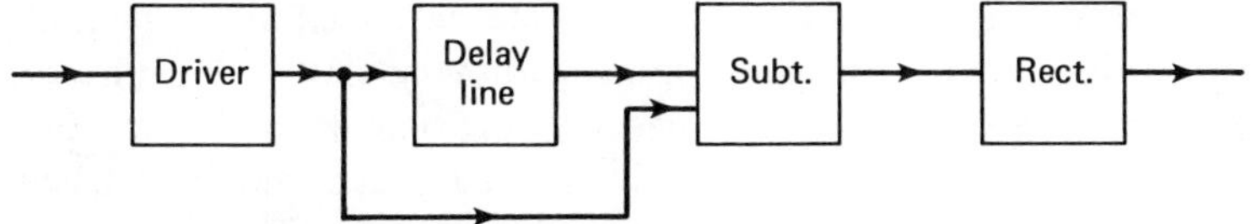

FIGURE 6-6. Block diagram of delay-line portion of MTI system.

tition period. Then the delayed signals are compared with undelayed signals in a subtraction circuit. As the stationary echoes remain constant but the moving echoes do not, only the moving echoes are retained. They are then rectified or put into unipolar form and can be applied to a PPI. Actually there are some blind speeds which give zero output from the subtraction process, as do echoes from stationary targets (see Prob. 6-6).

The waveforms of Eqs. (6-34) and (6-35) are IF rather than RF waveforms in practical MTI systems. In radars using magnetrons as transmitting tubes, the transmitted waveform is not coherent from pulse to pulse, but a reference IF oscillator, called the COHO, is synchronized with a waveform derived by mixing an attenuated replica of each transmitted pulse with the output of a stable local oscillator, called the STALO. Being resynchronized in this way by each transmitted pulse, the COHO is thus able to provide a suitable IF reference signal to allow the process described above to be carried out. The elements of such a system are shown in Fig. 6-7.

Another form of MTI, which can be designated as amplitude-processing

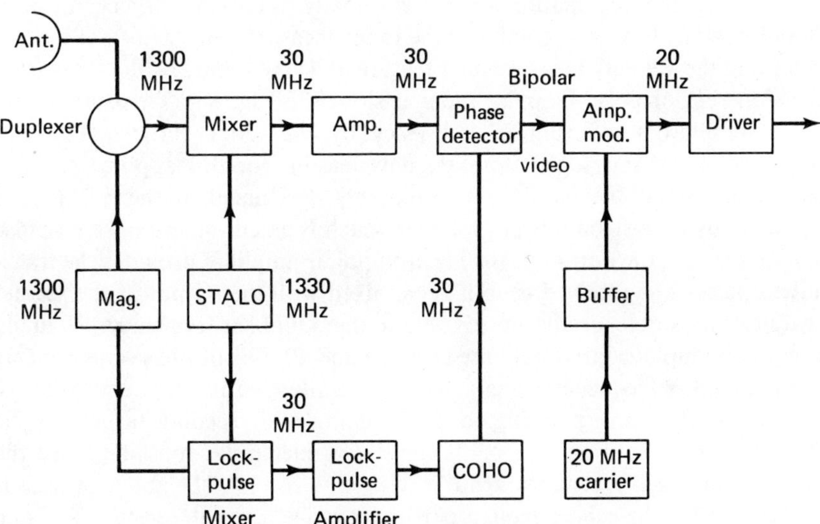

FIGURE 6-7. Block diagram of RF and IF portions of MTI system.

MTI in contrast to the phase-processing MTI described above [59], detects moving objects only in the presence of stationary clutter. No phase comparison is carried out, and performance is dependent on the fact that the echo itself from a region where both stationary and moving objects are present fluctuates in amplitude as suggested in Fig. 6-8. The signal E_T is the sum of the echo from the stationary object, E_s, and the echo from the moving object, E_m. The phase of E_s is constant, whereas the phase of E_m varies continuously. The sum of E_s and E_m thus varies in amplitude, as shown by the examples in Figs. 6-8(a) and (b). In a radar using this technique, an ac amplifier suffices to amplify the signals from moving targets and reject the echoes from stationary targets.

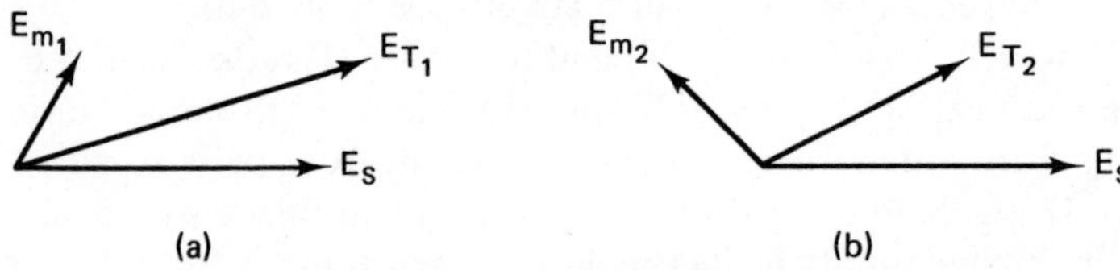

FIGURE 6-8. Concept of amplitude-processing MTI.

Most FAA and Air Force search radars have MTI capability. Either MTI or normal video signals can be selected for display on their PPI or storage-tube displays.

Although an MTI radar eliminates stationary targets or tends to, it does not provide information about the velocity of moving targets. A pulse-Doppler radar, however, combines the range measurement feature of a pulse radar and the velocity measurement feature of a CW radar. A block diagram of a coherent or pulse-Doppler radar is shown in Fig. 6-9. Two stable low-power oscillators differing in frequency by the radar IF frequency are employed. Solid-state oscillators are now feasible for this application. The output of one of the oscillators, commonly designated at the STALO, is amplified by a klystron power amplifier which is gated on and off by a pulse modulator, and the output of the klystron power amplifier provides the transmitted pulse. The received echo is then mixed with the output of the second or reference stable oscillator to obtain the Doppler frequency. Actually circuitry is employed to obtain the in-phase and 90° out-of-phase components of the Doppler frequency signal, this feature allowing positive and negative Doppler shifts, corresponding to approaching and receding targets, to be resolved. The repetition rate of the pulse-Doppler radar must be twice the highest Doppler frequency which it is desired to record. The above process is carried out for the echoes from a particular range interval, selected by using a range "gate." As many different range gates as are desirable and practical can be utilized.

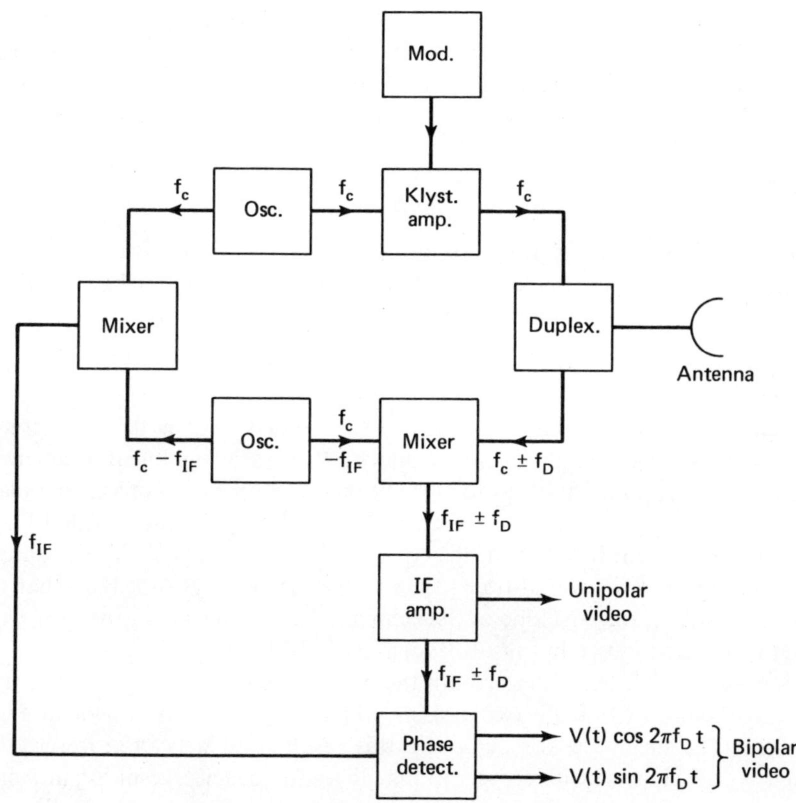

FIGURE 6-9. Block diagram of pulse-Doppler radar.

6-3-4 Chirp and Side-Looking Radars

A basic limitation of conventional pulsed radars is that if short pulses are used to obtain high range resolution, reduced sensitivity for distant targets results, assuming the peak transmitter power remains the same. Also wide receiver bandwidths must be used with short pulses, and receiver noise is proportional to bandwidth. Sensitivity can be improved by increasing the pulse length and thus increasing average signal power and decreasing noise power, if the receiver bandwidth is narrowed accordingly, but this action is taken at the expense of range resolution. A possible solution to this problem, however, lies in the use of pulse compression techniques. Basically these involve increasing the bandwidth of the transmitted pulse by some form of modulation or coding. The transmitted pulse in this case may be made as long as necessary to obtain sufficient sensitivity, and the received pulse is subsequently compressed in time to provide the range resolution that is needed. The point is that a certain bandwidth is needed to obtain a certain range resolution and there is more than one way of obtaining the bandwidth. One may use short pulses, as discussed previously, or one may employ frequency modulation within a longer pulse. In either case the needed bandwidth B is given by

$$B = \frac{c}{2\Delta R} = \frac{v}{\Delta R} \tag{6-37}$$

where ΔR is the desired range resolution and v is an effective velocity. A measurement of range involves a measurement of time, and the bandwidth B needed for a time resolution of Δt is given by

$$B = \frac{1}{\Delta t}$$

Linear frequency modulation within the transmitted pulse is the most commonly used pulse compression technique. Radars using linear frequency modulation are commonly called chirp radars. The operation of a chirp radar is as suggested in Fig. 6-10. In Figs. 6-10(a) and (b), an unmodulated long RF pulse is shown. If pulses of this type impinge on two small closely spaced targets such that the round-travel time to the second is greater than that to the first by Δt s, the envelope of the echo and the video signal generated will be somewhat as shown in Figs. 6-10(c) and (d). If, however, frequency modulation over a sufficient range within the pulse is used, as in Fig. 6-10(f), the returned echoes from the two objects will be separated as shown in Fig. 6-10(h). The technique for achieving this result is to delay the lower frequencies of the pulse with respect to the higher frequencies, resulting in pulse compression and separation as shown. Thus desired range resolution is obtained without sacrificing average power.

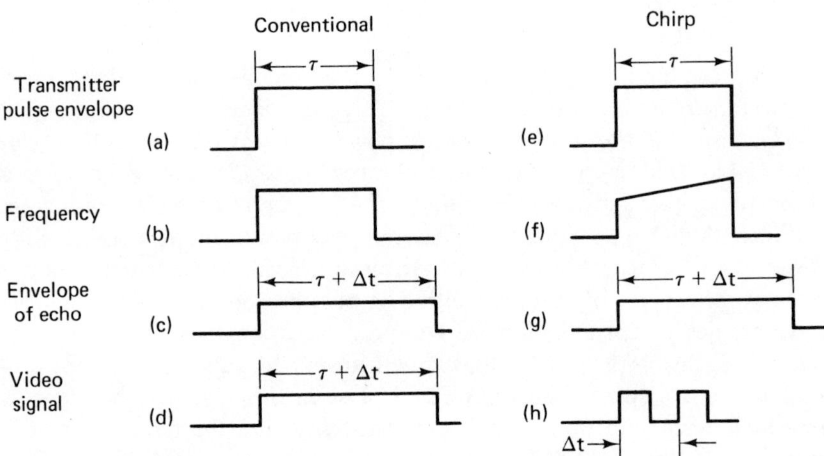

FIGURE 6-10. Waveforms illustrating principle of chirp radar.

The side-looking radar is grouped with the chirp radar because of certain similarities, but in other ways the side-looking radar is distinctly different. It is used in aircraft or satellites for forming an image of the terrain below it, and its beam is directed perpendicular to the direction the vehicle is traveling. Two types of side-looking airborne radars (SLARs) can be distinguished, those using real apertures and synthetic aperture radars (SARs). The first type may employ a slotted-waveguide type of antenna which is long in one dimension and narrow in the other. (The antennas used for marine radars are commonly of this type.) This antenna is mounted on an aircraft to provide a broad vertical beam and a narrow beam in the along-track direction, namely the direction the aircraft is flying. Real aperture radars can provide good along-track resolution when used in aircraft that fly at rather low altitudes, but for precision imagery from higher altitudes or, especially, from satellites it is necessary to use synthetic apertures. Both types of SLARs can achieve good range resolution, perpendicular to the track of the aircraft, by employing chirp techniques. Additional information about SLARs and the entire range of microwave techniques for remote sensing has been provided by Moore [58].

We now explain how synthetic-aperture radars achieve their high along-track resolution. For analysis purposes one can consider that the vehicle is stationary and that the target is moving as in Fig. 6-11. The distance to the moving target is

$$R^2(t) = R_0^2 + (vt)^2 \tag{6-38}$$

where t can be positive or negative and is measured from the time when $R(t) = R_0$. In practical application the echo from the target is received over a rather small angle such that

$$R(t) \simeq R_0 + \frac{v^2t^2}{2R_0} \tag{6-39}$$

Under this condition the phase of the received echo varies as

$$\phi(t) = 2kR_0 + \frac{kv^2t^2}{R_0} \tag{6-40}$$

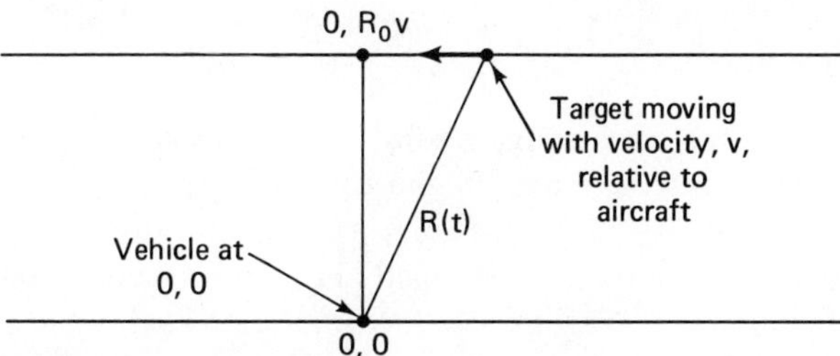

FIGURE 6-11. Geometry for discussing along-track resolution of synthetic-aperture radar.

where $k = 2\pi/\lambda$ is the propagation constant. Doppler frequency f_D is related to phase by

$$f_D = \frac{1}{2\pi}\frac{d\phi}{dt} \tag{6-41}$$

Therefore,

$$f_D = \frac{2v^2t}{R_0\lambda}$$

or

$$\Delta f_D = \frac{2v^2T}{R_0\lambda} \tag{6-42}$$

where T is the total interval over which data are received and Δf_D is the total variation in Doppler frequency. This value of Δf_D is also the bandwidth which is available to provide azimuthal or along-track resolution. That is, Δf_D corresponds to B of Eq. (6-37) insofar as considering along-track range resolution is concerned. T, the interval over which an echo is received from a particular target, is given by

$$vT = R_0\phi_{HP} \tag{6-43}$$

where v is the velocity of the aircraft, R_0 is the shortest distance to the target, and ϕ_{HP} is the beamwidth between half-power points of the radar antenna. Substituting $T = R_0\phi_{HP}/v$ into Eq. (6-42) results in

$$\Delta f_D = \frac{2v\phi_{HP}}{\lambda} \tag{6-44}$$

But the beamwidth ϕ_{HP} is given approximately by

$$\phi_{HP} = \frac{\lambda}{d} \tag{6-45}$$

where d is the antenna diameter. Thus

$$\Delta f_D = \frac{2v}{d} \tag{6-46}$$

Now by using Eq. (6-37) with Δf_D substituted for B and considering the aircraft to be flying in the x direction,

$$\Delta x = \frac{v}{\Delta f_D} = \frac{d}{2} \tag{6-47}$$

where Δx is the along-track range resolution. Note that v is the velocity of the aircraft in this case, whereas in the original application of Eq. (6-37) v was $c/2$.

The along-track resolution obtained above can be compared with that for a conventional or real aperture side-looking radar having the same beamwidth, $\phi_{HP} = \lambda/d$. The resolution in this case is given approximately by

$$\Delta x = \frac{\lambda R}{d}$$

The latter value for Δx will be much larger than the value given by Eq. (6-47) when R is very large. Thus the SAR can provide much higher along-track resolution than a real aperture radar. SARs are coherent radars; the echo pulses received by an SAR are compared in phase with the stable frequency source from which the transmitter pulses are derived. The output of the phase comparator is usually recorded on photographic data film which is advanced in synchronism with the aircraft motion. The data on the film constitute a holographic record, and the type of radar described can be called a holographic radar. Film can record only variations in density, but, as in any hologram, the pattern of interference between the reference source and the signal contains both amplitude and phase information. By passing coherent light through the data film in an optical processor, an image of the terrain is obtained and can be recorded on an image film.

The simplified analysis given above illustrates how the synthetic-aperture radar can acheive greater along-track resolution than a real-aperture radar and shows that the along-track resolution tends to be independent of range and wavelength. It is only practical, however, to operate such systems with a beamwidth ϕ_{HP} of limited extent, and an improved along-track resolution cannot be obtained indefinitely by making d smaller and smaller as casual consideration of Eq. (6-47) might appear to imply. The complexities of SAR systems are such, furthermore, than an along-track resolution as good as that indicated by Eq. (6-47) should not be expected in practice, and along-track resolution will not be completely independent of range or wavelength.

The SAR radar developed by Goodyear Aerospace Corp. and used extensively in South America [69] is an X-band radar utilizing a bandwidth of 15 MHz and an antenna 1.2 m in length and 0.15 m high. A range resolution of approximately 12 m is provided in the across-track direction, and a resolution of 10 m is provided in the along-track direction. A swath 37 km in width is imaged. It is necessary to give careful attention to maintaining stable, level flight and to recording precise information on aircraft position when using SAR radar. In SAR operations in Brazil, 30 ground locations were positioned to an accuracy of 10 m using the Transit Satellite Positioning System, and transponders for use with Shoran radio-positioning equipment were installed at each location. By simultaneously receiving transmissions from two transponders, position was recorded to an accuracy of 75 m. Altimeter data on altitude are also recorded continuously in SAR operations. Goodyear has employed a Caravelle jet to carry the airborne SAR equipment.

6-3-5 Lidar

Lidar or laser radar operates on the same principle as conventional radar. Short pulses are transmitted, and the range to the region from which echoes are received is determined by measuring the time delay of the echoes.

The atmosphere tends to attenuate and backscatter radiation more intensively, however, at the frequencies employed for lidar than at microwave radar frequencies. Ruby lasers operating at a wavelength of 0.6943 μm, neodymium lasers operating at 1.06 μm, and nitrogen lasers operating at 0.3771 μm are examples of the sources used in lidar systems.

The same range equations developed for conventional radar in Sec. 6-3-2 can also be applied to lidar. At optical frequencies it is common practice, however, to specify beamwidth rather than gain. If $4\pi/\theta_t^2$ is substituted for G_t in Eq. (6-20) and if a loss factor L is introduced, it becomes

$$W_r = \frac{W_t A \sigma L}{4\pi R^4 \theta_t^2} \tag{6-48}$$

L can be used to account for losses in the transmitting and receiving optics and in the propagation medium. Unlike radar, which commonly uses the same antenna for transmission and reception, with a TR system to protect the receiver during the transmitted pulse, the transmitter and receiver mirrors in a lidar are physically separate. They can be placed close together, however, to achieve monostatic operation.

Lidars have been used for various applications, including the docking of space craft. In remote sensing of the troposphere, Eq. (6-29), which was developed for the case of distributed targets, is applicable. It is common practice, however, to use a cross section per unit solid angle in lidar applications. Designating the latter cross section by σ' and substituting $\sigma' 4\pi$ for σ in Eq. (6-29) and also introducing the loss factor L, we obtain

$$W_r = \frac{W_t A c \tau \sigma' N \pi L}{8R^2} \tag{6-49}$$

Also in calculating the scattering volume, the $\pi/4$ factor of Eq. (6-27) is usually omitted in the case of the narrow beamwidths employed in lidar. Finally the lidar receiver can be gated on for any length l, rather than specifically for the distance $c\tau/2$, and the factor L can be expressed more precisely as $\eta_t \eta_r \eta_q \exp\left[-\int_0^R (\alpha_t + \alpha_r)\, ds\right]$, where η_t is the efficiency of the transmitting optics, η_r is the efficiency of the receiving optics, η_q is the efficiency of the detector, and α_t and α_r are the power attenuation or extinction coefficients along the path. The received signal is at a frequency different from the transmitted signal if Raman backscatter is observed, so two different extinction coefficients are shown. By taking these points into account, Eq. (6-49) becomes

$$W_r = \frac{W_t}{R^2} A l \sigma' N \eta_t \eta_r \eta_q \exp\left[-\int_0^R (\alpha_t + \alpha_r)\, ds\right] \tag{6-50}$$

Still another variation is to write the equation in terms of the number of photons transmitted and received per pulse, making use of the relation that

$$N' = \frac{W\tau}{h\nu} \tag{6-51}$$

where τ is pulse length, h is Planck's constant, and ν is frequency. In terms of N' the equation becomes

$$N_r' = \frac{N_t' Al\sigma' N\eta_t\eta_r\eta_q \exp\left[-\int_0^R (\alpha_t + \alpha_r)\, ds\right]}{R^2} \tag{6-52}$$

In some cases, as when an intensely scattering layer is encountered in an otherwise clean sky, effects due to variations in σ' can be readily identified and separated from effects associated with α_t and α_r. In other cases when α_t and α_r have large values it may be difficult to identify atmospheric structure by using lidars.

Two important types of noise affecting lidar are detector noise and sky background noise. The detector or dark noise can be expressed in photons per second and designated as N_D'. The background count is given by

$$N_B' = \eta_q\eta_r \frac{B_\lambda \Omega\, \Delta\lambda A}{h\nu} \tag{6-53}$$

where B_λ is the brightness, Ω is the solid angle of the receiver, and $\Delta\lambda$ is the bandwidth of the receiver. Then the total number of noise counts per transmitted pulse, N_n', is given by $N_n' = (2l/c)(N_D' + N_B')$, and the number of noise counts per second is given by ZN_n', where Z is the pulse repetition frequency. Strauch and Cohen [76] state that the signal-to-noise ratio for 1-s averaging is

$$\frac{ZN_r'}{\sqrt{Z(N_r' + 2N_n')}} = \frac{\sqrt{ZN_r'}}{\sqrt{(N_r' + 2N_n')}}$$

and for averaging over T s,

$$\frac{S}{N} = \frac{\sqrt{ZT}\, N_r'}{\sqrt{N_r' + 2N_n'}} \tag{6-54}$$

These expressions apply when noise count is not accurately measured, and it is stated that the $2N_n'$ in the denominator can be replaced by N_n' if the noise is accurately determined.

6-3-6 Acoustic Sounders

The acoustic sounder or acoustic radar equation for distributed targets has the same general form as for microwave radar and lidar. The cross section per unit volume per unit solid angle σ_v' is commonly used. The loss factor L can be expressed as $\eta_t T^2$, where η_t is antenna efficiency and T is the atmospheric transmission factor. As in the case of the lidar equation, the $\pi/4$ factor in the expression for pulse volume [see Eq. (6-27)] is generally omitted. Thus the equation takes the form

$$W_r = \frac{W_t \sigma'_v v_s \tau A \eta_t T^2}{2R^2} \tag{6-55}$$

Note that $v_s\tau/2$ (corresponding to $c\tau/2$) has been used in place of the l of Eq. (6-50), where v_s is the acoustic velocity. The attenuation of acoustic or sound waves in the earth's atmosphere limits acoustic sounders to rather short ranges, but the high sensitivity of acoustic waves to temperature and wind structure makes the acoustic sounder highly useful for studying the lower atmosphere. The acoustic power attenuation coefficient α is composed of three components so that

$$\alpha = \alpha_c + \alpha_m + \alpha_s$$

where α_c is the classical coefficient associated with heating of air, α_m is the molecular coefficient that is strongly affected by humidity, and α_s is due to scattering. The molecular coefficient for air at 20°C and 15% humidity at a frequency of 2 kHz is 35 times the value of the classical coefficient. Values for the scattering coefficient are not well established.

The cross section σ'_v has been shown [40] to be

$$\sigma'_v = \frac{32\pi^5 \cos^2\theta}{\lambda}\left\{\frac{\Phi(v)[(4\pi/\lambda)\sin(\theta/2)]\cos^2(\theta/2)}{v_s^2} + \frac{\Phi(T)[(4\pi/\lambda)\sin(\theta/2)]}{4T^2}\right\} \tag{6-56}$$

where θ is the scattering angle measured from the direction of propagation and $\Phi(v)$ and $\Phi(T)$ are the three-dimensional spectral densities of fluctuations in wind velocity and temperature corresponding to a spatial scale λ', where

$$\lambda' = \frac{\lambda}{2\sin(\theta/2)} \tag{6-57}$$

and λ is the acoustic wavelength. For $\theta = 180°$, $\lambda' = \lambda/2$. These relations are consistent with the requirement for Bragg scattering. It can be noted that Eq. (6-56) predicts backscatter ($\theta = 180°$) from temperature structure but not from velocity or wind structure.

Acoustic sounders can also record echoes from gradients in temperature, but the magnitudes of the actual gradients in the atmosphere are poorly known, and it is difficult to predict the magnitudes of the corresponding reflection coefficients.

Although velocity structure itself does not contribute to backscatter, wind velocities can be measured by recording Doppler shifts of the signals backscattered from temperature structure.

The lowest theoretical limit of acoustic noise is provided by the random thermal motion of the atmospheric molecules. This value, as given by Little

[55], is 4×10^{-19} W for a standard atmosphere and a 100-Hz receiver bandwidth. Other noise sources, such as wind, insects, birds, and human activity, usually predominate. Signal levels are stated by Hall [40] to range from near the theoretical noise limit for 20 W of transmitted power and ranges of 100 m to 1 km in nonturbulent regions to values 60 or 70 dB above the noise limit at a range of 50 m in extremely turbulent regions.

6-4 ENVIRONMENTAL CONSIDERATIONS AND APPLICATIONS

6-4-1 Introduction

Applications of and environmental considerations concerning some of the previously mentioned remote sensing systems are considered in this section. Applications of radiometric techniques to remote sensing are mentioned in Sec. 6-4-2. Environmental effects on radar systems and applications of radar systems to sensing the atmosphere are treated in Secs. 6-4-3 and 6-4-4. Sections on remote sensing of sea and land (Secs. 6-4-5 and 6-4-6) and a section on the ionosphere (Sec. 6-4-7) complete Sec. 6-4.

Although lidars and acoustic sounders were mentioned briefly in Sec. 6-3, the term *radar* itself should only be applied to sounders using microwaves or lower-frequency radio waves and is so applied in this section. Radar is the most highly developed of the three types of systems and has the advantage of being less affected by the transmitting medium than lidars and acoustic sounders [19]. If one desires to study certain characteristics of the atmosphere, however, lidars and acoustic sounders may have advantages. Lidars can detect, and to some degree identify, atmospheric pollutants of a particulate and gaseous nature. Acoustic sounders have much greater sensitivity to atmospheric temperature flutuations than radars. Practical considerations concerning the scope of this volume, however, suggested that attention to lidars and acoustic sounders be largely limited to the discussion of the principles of operation already presented. The laser profilometer is mentioned, however, in the section on remote sensing of the oceans.

6-4-2 Applications of Radiometry to Remote Sensing

Passive Microwave Imagery of the Earth. The scanning microwave spectrometer (SCAMS) on Nimbus 6 has produced the first passive microwave spectral images of the earth and has produced maps showing temperatures of three atmospheric layers below 20 km, the water vapor and liquid water distributions over the oceans, and coverage and type of ice and snow [73]. SCAMS operates at frequencies centered on 22.235, 31.65, 52.85, 53.85, and 55.45 GHz and is essentially an improved scanning version of the NEMS system on Nimbus 5. Some points of interest about the results for ice and snow are as follows. Permanent snow and ice cover over Geenland and Antarctica is characterized by the very low values for T_B of 140 to 160 K at 31.65 GHz. A fine structure in the temperature pattern occurs over Antarctica in its winter. Utilizing the difference between values for T_B at 22.235 and 31.65 GHz for sea ice and seasonal snow cover over land allows distinguishing different types of ice and snow. Results are mentioned further under the headings of atmospheric profiles and applications to precipitation and water vapor content.

Atmospheric Profiles. In Sec. 3-2-3, the formation of ionospheric layers, by the absorption of incident solar radiation, was analyzed. Above the peak of a layer the density of the constituent that is ionized by radiation in a particular frequency range is relatively low, and therefore the pertinent power attenuation coefficient α_p is relatively low and relatively little ionization is produced. Below the peak of the layer the density of the constituent and the value of α_p may be high, but the solar radiation intensity P_0 incident upon the region of interest has been attenuated to such an extent that little additional ionization can be produced. Thus it is the product $P_0 e^{-\alpha_p dz}$ that determines the ionization produced in a layer of thickness dz. Somewhat the same situation prevails if one considers the absorption of radiation near 60 GHz by the oxygen molecules of the troposphere, although the absorption in this case does not result in ionization. Maximum absorption by oxygen (O_2) as a function of frequency occurs near 60 GHz. Peak absorption by O_2 as a function of altitude therefore occurs at a higher altitude at 60 GHz than at 50 GHz, in the case of radiation incident from above. Curves showing absorption as a function of altitude can be drawn for given frequencies.

In the application of radiometry one is concerned most directly with emission rather than absorption, but it can be recalled that absorption and emission coefficients have identical values for a given material or surface. It develops that if one records emission from tropospheric oxygen from a satellite, for example, the situation is similar to that discussed for absorption in that the emission recorded at a given frequency will emanate primarily from a limited altitudinal range and the altitudinal range will vary with frequency. Intense emission will take place at lower altitudes, but this emission will be strongly attenuated before reaching the satellite. Emission from higher altitudes will be less intense because of reduced density of oxygen. Figure 6-12 shows plots of received emission versus source altitude for several frequencies for the case of a downward-pointing radiometer in a satellite. The curves shown are commonly referred to as weighting functions. The situation for emission differs from that for absorption in that the curves of Fig. 3-4

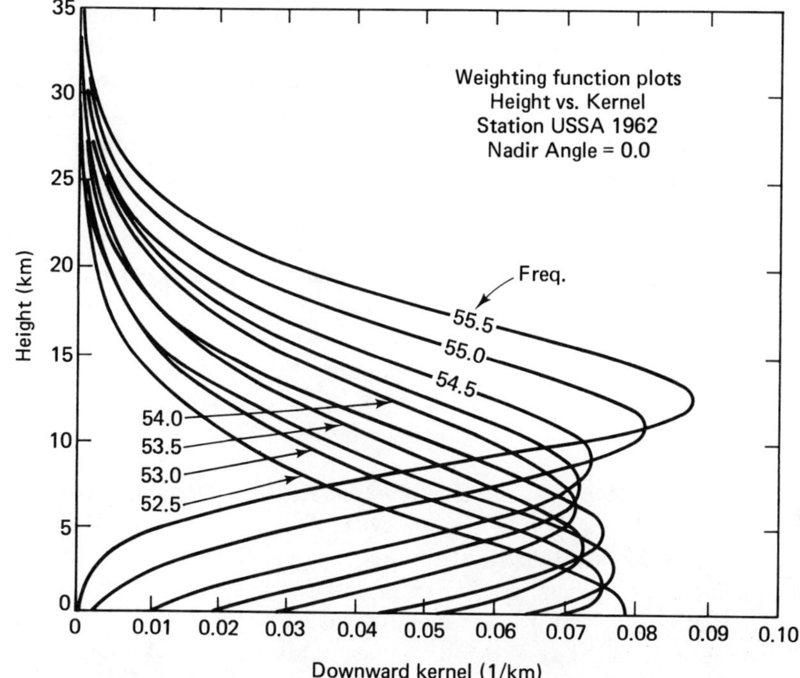

FIGURE 6-12. Weighting functions for downward-pointing multispectral radiometer. (From Westwater and Strand, "Inversion Techniques," in Derr, *Remote Sensing of the Troposphere*, 1972 [86].)

represent true absorption as a function of height, whereas the curves of Fig. 6-12 represent apparent or effective emission (emission modified by absorption).

If radiometer measurements are made from the earth's surface, somewhat similar effects are encountered except that in this case the region of greatest absorption and emission is that closest to the radiometer. The weighting functions therefore have different shapes as compared with the case when the radiometer is above the atmosphere. The net effect is the same as when making measurements from a satellite; emission from different altitudinal ranges can be observed by varying the frequency. Another possibility is to vary the angle at which the radiometer is pointed. A vertically pointing radiometer will respond to conditions at a higher altitude than one pointing away from the zenith.

Now that this background discussion has been presented, it is possible to describe how radiometers operating near 60 GHz can, by varying either the frequency or angle of view, determine atmospheric temperature profiles.

First, it can be assumed that the temperature of the oxygen component is the temperature of the atmosphere, as oxygen is well mixed with, and in constant ratio to, nitrogen throughout the troposphere. Thus oxygen temperature and atmospheric temperature are one and the same. Second, it has been explained that a radiometer operating at a given frequency and angle responds primarily to the emission coming from a given altitudinal range. The radiometer antenna temperature is therefore primarily determined by the temperature in the given altitudinal range. However, the weighting functions as shown in Fig. 6-12 are rather broad and overlap, and a measurement at one frequency does not uniquely determine the temperature at a particular altitude. To calculate temperature profiles from the data, it is necessary to use inversion techniques. The relation giving the observed brightness temperature at a given frequency ν for a downward-pointing radiometer is

$$T_B(\nu) = \epsilon' T_s e^{-\tau_\nu} + \int_0^{h_s} T(h) W(h, \nu)\, dh \tag{6-58}$$

The first term on the right-hand side represents emission from the surface as modified by the atmosphere. ϵ' is emissivity, T_s is surface temperature, and τ_ν is atmospheric optical depth at the frequency ν. $W(h, \nu)$ is the weighting function for the frequency ν. $T(h)$ represents the temperature expressed as a function of height, and it is the quantity that it is desired to determine by inversion techniques. The $\epsilon' T_s e^{-\tau_\nu}$ term must be taken into account in general. Use of a frequency falling in an atmospheric window and thus affected little if at all by the atmosphere facilitates the determination of the magnitude of this term.

The technique described has been used at the Wave Propagation Laboratory, NOAA, Boulder, by Snider and Westwater [72]. Parameters of the ground-based system employed are shown in Table 6-1. Appropriate inversion techniques have been treated by Westwater and Strand [86]. SCAMS on Nimbus 6 has utilized its 50-GHz frequencies for determining temperature profiles or actually maps of temperatures in the 1000- to 500-mb, 500- to 250-mb, and 250- to 100-mb layers, all below a height of 20 km [73]. Nimbus 6 also has an infrared radiometer system for obtaining temperature profiles.

Table 6-1 Multifrequency Radiometer Characteristics.

Operating frequency	52.5, 53.5, 54.5, 55.5 GHz
Antenna characteristics	1.2-m-diameter conical horn, 3-dB beamwidth 0.3°
Receiver type	Dicke switching radiometer, dual-conversion superheterodyne: 1st IF: 4–8 GHz; 2nd IF: 60-MHz center frequency, noise bandwidth approx. 30 MHz
Sensitivity	Approx. 1 K for 60-s integration time

Its high resolution infrared radiation sounder (HIRS) utilizes a 13.4- to 15.0-μm band to obtain temperatures in the relatively cold regions of the atmosphere on the basis of absorption, primarily by CO_2. The 4.24- to 4.57-μm band is used to obtain temperatures in the warm regions of the atmosphere where CO_2 and N_2O are the principal absorbing constituents. Windows at 0.69, 3.71, and 11.0 μm allow determination of surface brightness temperatures and the detection of clouds.

APPLICATIONS TO PRECIPITATION AND WATER VAPOR CONTENT. Attenuation due to precipitation is an important factor limiting the use of millimeter waves, as between a satellite and a ground station, and it is highly desirable to be able to know approximately the magnitude of the attenuation to be expected in various areas under a variety of meteorological conditions. Also, water vapor content is a major factor in causing range errors when distances are measured by using electromagnetic waves, and the ability to measure this content remotely would be advantageous. Procedures for accomplishing these purposes are based on Eq. (5-17), which is repeated below for convenience:

$$T_b = T_s e^{-\tau} + T_i(1 - e^{-\tau}) \tag{5-17}$$

Consider first the measurement of attenuation due to precipitation. Two procedures were used by Wulfsberg and Altshuler [89]. One involved measuring T_b, the brightness temperature recorded by a radiometer, first using the sun as a source and then pointing the radiometer slightly away from the sun and making the measurement again. The difference between the two values obtained during a period of rather uniform widespread rainfall gave $T_s e^{-\tau}$. T_s itself was determined by measuring T_b when pointing at the sun with no rain falling and by correcting for residual attenuation by using the following empirical relations:

$$A(\phi)_{dB} = 0.055 + 0.004e \sec \chi \qquad (15 \text{ GHz}) \tag{6-59}$$

$$A(\phi)_{dB} = 0.17 + 0.013e \sec \chi \qquad (35 \text{ GHz}) \tag{6-60}$$

where χ is zenith angle and e is absolute humidity in g/m^3. Once T_s was known, then $e^{-\tau}$, the attenuation which it was desired to determine, was also known.

The second procedure involved measuring only the second term of Eq. (5-17), that is, measuring sky temperature in the absence of a source. Under these conditions

$$T_b = T_i(1 - e^{-\tau}) \tag{6-61}$$

and it is necessary to determine T_i in order to know $e^{-\tau}$. Wulfsburg and Altshuler found that good results were obtained in Hawaii by using $T_i =$ 284 K. They were able to show that their second method gave good results by using the results from the first method for comparison. Also the first method could give a value of T_i to be used when later using the second method. The second method is in general the most convenient of the two as it can be made at any time of day or night and along a path oriented in any direction. In applying this method it was assumed that scattering was negligible compared to absorption, which tends to be true in the case of the small drops encountered in Hawaii. Other investigators have also found the sky-noise-temperature radiometric method to be very satisfactory.

The correction of microwave atmospheric ranges also involves the use of Eq. (6-61). In this case the factor $e^{-\tau}$ is due to water vapor, and knowledge of $e^{-\tau}$ allows the determination of mean values of refractivity and water vapor content for a path, as shown by Westwater [85]. Frequencies near the water vapor absorption line at 22.235 GHz were used.

The Nimbus-E Microwave Spectrometer (NEMS) on Nimbus 5 consists of five separate microwave radiometers operating in the oxygen band (53.65, 54.90, 58.80 GHz) and water vapor region (22.235, 31.40 GHz). SCAMS on Nimbus 6 is essentially an improved version of NEMS and has scanning capability allowing it to obtain passive microwave images of the earth. Investigators using the two systems have found the 22.235-GHz channel to be more sensitive to water vapor than the 31.4- or 31.65-GHz channels, with the situation reversed for the case of liquid water [39,73]. Specifically the 22.235-GHz channel is more sensitive to water vapor by a factor of 2.5, while the 31.65-GHz channel is more sensitive to liquid water by a factor of 2 [73]. Use of the two frequencies allows the determination of both water vapor and liquid water content.

6-4-3 Environmental Considerations Concerning Radar Systems

Ideally radars display echoes from objects from which it is desired to receive echoes and have a perfectly clean radar screen otherwise. In practice, however, a certain amount of clutter is always present. The term clutter refers to undesired echoes and can be regarded as noise. What is clutter for one party may be the desired signal for another, however, and vice versa. For example, if the function of radar is to detect aircraft, weather or bird echoes might be regarded as clutter, but some radars are operated specifically for the purpose of obtaining weather information. In practice most radars are operated for a single purpose, but in a number of cases they could well function on a multipurpose basis. One commendable example of operation for more than one purpose is the use of FAA radars by the Weather Service at certain of the FAA centers. It is not true, of course, that any two or more radar functions can always be carried out efficiently by the same radar system, but it appears that too little attention has been given in the past to multipurpose operation.

Radar echoes can be received from the following classes of objects or features:

1. Man-made objects, such as aircraft, ships, satellites, missiles, and automobiles
2. Weather features and structure of the lower atmosphere
3. Ionization, as in the earth's ionosphere
4. Birds, insects, animals, humans
5. Land and water surfaces
6. Extraterrestrial bodies—the moon, planets, and sun

Radars designed for aircraft surveillance, for example, may experience undesired echoes or clutter due to weather, land and water surfaces, birds, etc. Echoes from precipitation may be intense and widespread and may interfere with the detection of aircraft. Energy is often backscattered to the radar from the surrounding land or water, particularly during temperature-inversion conditions, which cause radio waves to bend downward. The resulting clutter may be serious. Radar personnel commonly refer to the effect as AP (for anomalous propagation). Birds are a common source of radar echoes and may cause considerable clutter, especially during the spring and fall migration seasons. MTI circuitry tends to eliminate weather, sea, and ground clutter but is not able to do so completely. A major reason for MTI not being completely effective is the fact that the clutter echoes are commonly Doppler shifted. Water drops falling as rain tend to have components of velocity toward or away from the radar, the sea surface is in continuous motion, and vegetation moves under the influence of wind. Some bird echoes are eliminated by MTI, in the case where the birds have insufficient radial velocity. If most of the birds in the region are traveling in the same direction, for example, the birds in two angular sectors where the radial velocity is small may be eliminated. The echoes from birds in the rest of the display are little affected, however, and MTI video is often to be preferred when it is desired to record bird echoes. STC (sensitivity-time-control) circuitry, which reduces receiver gain for close-in targets, helps to minimize close-in ground clutter and clutter of other types and is needed in most radar operations. It is only partially successful in combating clutter due to birds, however, and some STC action is generally helpful when recording bird echoes. Echoes from large flocks of birds can be as intense as those from large aircraft, and if all bird echoes are eliminated on the basis of amplitude alone, all aircraft echoes may be eliminated as well. The most effective procedure which an aircraft surveillance radar can use, to detect cooperating aircraft, is to use airborne beacons or transponders. Clutter is automatically eliminated if only transponder signals are used, and greater sensitivity and ease of recognition are also achieved. Operation using transponder signals alone, however, defeats one of the purposes of using radar. Small aircraft not equipped with transponders will not be detected, and useful information about weather and bird hazards to aircraft will be lost. Military radar systems, furthermore, must obviously retain the ability to detect noncooperating targets. The trained human eye has a rather good ability to detect targets of interest in the presence of clutter, and the clutter problem tends to be more difficult when digital automatic data-processing and data-transmission systems are employed. Precipitation may cause attenuation of desired radar signals (mostly at X and higher-frequency bands) as well as clutter. Environmental effects are considered by Nathanson [59], who suggests using ambiguity functions and clutter diagrams (showing the distribution of clutter with range and velocity) in conjunction.

6-4-4 Applications of Radar to Atmospheric Remote Sensing

The discussion in the previous section has emphasized the effects of the environment on radars, such as those operated for the purpose of aircraft surveillance. Radar can also be used to monitor and study the environment. This section treats the application of radar to study of the atmospheric phenomena of precipitation, wind, atmospheric structure, and birds and insects.

PRECIPITATION. The application of radar to the weather, especially to observations of precipitation, is one of the oldest and most important environmental applications of microwave radar. Radar meteorology embraces the topics of wind and clear air radar echoes, treated in later paragraphs, as well as precipitation and cloud physics. Important references on radar meteorology include Battan [10], Atlas [2], and the proceedings of the weather radar conferences.

Echoes from precipitation are readily observed on microwave radar screens, and the utility of radar for precipitation and storm detection was recognized early in World War II. The tracking of a rain shower for a distance of 7 miles by a 10-cm radar located on the English coast on February 20, 1941 is credited as marking the birth of radar meteorology [54], and Atlas [2] reported that the first operational radar storm detector was set up in Panama in June 1943.

The horizontal and vertical extent and the pattern of relative intensity of rainfall are the most obvious weather features that can be determined by radar. The horizontal extent and pattern can be shown on PPI displays. If the antenna vertical beamwidth is sufficiently narrow, the vertical extent can be determined by scanning in elevation angle, manually or automatically as in the RHI mode of operation. Interest in the use of millimeter waves on earth-satellite paths has caused emphasis to be given to the spatial and temporal structure of rainfall. The intensity of rain varies with spatial coordinates and time, and a major difficulty in predicting the effect of rain on millimeter propagation is the lack of sufficient rainfall data of a statistical nature. It is known that rain tends to occur in cells that are 2 to 6 km across and are embedded in larger regions having dimensions measured in tens of kilometers [17]. Radar has the capability of obtaining the needed statistical information but as of 1977 had not been applied to the purpose to a sufficient extent.

Considerable attention has been devoted to the application of radar to the measurement of precipitation. The radar cross section of a water drop can be calculated by recognizing that

$$\sigma = 4\pi r^2 \frac{P(\pi)}{P_{\text{inc}}} \tag{6-62}$$

where σ is the desired cross section, r is the distance from the drop in question, $P(\pi)$ is the power density scattered in the direction $\theta = \pi$ (namely backwards), and P_{inc} is the incident power density. To obtain the value of the cross section σ, it may be necessary to use the Mie theory, mentioned in Sec. 4-5-3.

If a/λ is very small, however, where a is the radius, a drop acts like an elementary dipole as described by Eq. (4-112) but with n complex so that

$$\frac{\sigma}{\pi a^2} = \frac{64\pi^4}{\lambda^4}\left|\frac{K_c - 1}{K_c + 2}\right|^2 a^4 \tag{6-63}$$

or

$$\sigma = \frac{\pi^5}{\lambda^4}\left|\frac{K_c - 1}{K_c + 2}\right|^2 d^6 \tag{6-64}$$

where d is drop diameter and K_c is the complex relative dielectric constant for the drop. This is the Rayleigh approximation for backscatter, comparable to that developed for attenuation in Sec. 4-5-2. If all drops are small compared to wavelength, the total scattering cross section per unit volume η is given by

$$\eta = \frac{\pi^5}{\lambda^4}\left|\frac{K_c - 1}{K_c + 2}\right|^2 \sum d^6 \tag{6-65}$$

where the summation is carried out for all the drops in the unit volume considered. The summation is often indicated by the symbol Z. In terms of Z, the radar equation for distributed targets [Eq. (6-29)] becomes

$$W_r = \frac{WtAc\tau}{32R^2}\frac{\pi^5}{\lambda^4}\left|\frac{K_c - 1}{K_c + 2}\right|^2 Z \tag{6-66}$$

It has been shown by Probert-Jones [61], however, that this equation overestimates the received power by a factor of 2 ln 2 and the antenna gain by a factor of $16/\pi^2$. Other limitations in applying Eq. (6-66), nevertheless, tend to obscure these correction factors [11], and the equation has been commonly used. A form including the corrections, used by Wilson and Miller

[87], is

$$W_r = \frac{W_t G^2 \theta_{\mathrm{HP}} \phi_{\mathrm{HP}} c\tau}{1024 \ln 2 R^2} \frac{\pi^3}{\lambda^2} \left| \frac{K_c - 1}{K_c + 2} \right|^2 Z \tag{6-67}$$

After determining Z by use of Eq. (6-66) or (6-67), it remains to relate Z to rainfall rate if radar is to be used for precipitation measurements. An empirical relation which has been used for this purpose is

$$Z = 200r^{1.6} \tag{6-68}$$

where Z is in units of $\mathrm{mm}^6/\mathrm{m}^3$ and where r is the rainfall rate in mm/h. This relation is based on the Marshall-Palmer distribution of raindrops with size. This method has provided useful information about rainfall, and considerable effort has been devoted to presenting the data in a convenient form. Methods have been developed, for example, for displaying isoecho contours of radar echo intensity [2]. The radar method of determining rainfall intensity must nevertheless be used with caution because of the empirical nature of the relation between Z and r and because the actual relation is subject to variation with location and type of rainfall. One application is to determine the amount of rainfall over a watershed.

Because of the associated precipitation, radar is able to monitor and study thunderstorms, tornadoes, and hurricanes. Precipitation echoes from convective clouds usually have a distinctive appearance, and when their echo tops grow rapidly to 7600–9200 m they usually are or soon will become thunderstorms [10]. Tornadoes usually develop from severe thunderstorms. There is considerable interest in methods of determining in advance which thunderstorms are most likely to actually develop into tornadoes [78]. Hurricanes have a distinctive echo pattern of spiral or circular arms.

A precipitation phenomenon of interest is that of the bright band, which may show as a prominent echo of considerable horizontal extent on radar screens. This echo results when falling snow melts, the increase in dielectric constant of water with respect to snow apparently being a major factor in the increase in intensity. Below the bright band the echo intensity decreases because of the increased fall velocity, which results in a decrease in drop density. Other topics in cloud physics for which radar may be useful include precipitation in and growth of convective clouds and cloud seeding and hail suppression studies. Echoes from hail cannot always be readily distinguished from echoes from rain, but large hail is a more effective backscatterer than large raindrops of comparable size [10].

CLEAR AIR RADAR ECHOES, WIND. During World War II, radar echoes of unknown origin from apparently clear regions of the atmosphere were observed. Because of their mysterious nature, they were given the name of radar angel. Also during the same period, short-range radar echoes from birds were recorded [51]. Some persons associated these two observations and suggested that radar angels were caused by birds. Others suggested atmospheric inhomogeneities and insects as the cause of the angels, however, and the subject was controversial for many years. Radar echoes from birds had actually been observed in 1939, before World War II, from a U.S. Navy ship off Puerto Rico [12], but this observation did not become widely known until considerably later. In spite of these early observations of radar echoes from birds, it was not until about 1957 that it was clearly established that radar was useful for studying bird migration. Since that time the subject of radar ornithology has become rather well established [23], although it has not yet been exploited fully.

The subject of radar angels was finally resolved in good measure in about 1966, on the basis of studies utilizing the high-power, multiwavelength radar facilities at Wallops Island, Virginia [44]. These studies showed clearly that echoes can be received from birds, insects, and atmospheric inhomogeneities or structure. Bird echoes can be recorded readily by most operational radars, and insect echoes are apparently not uncommon, especially in the case of radars that use horizontal polarization and are capable of close-in seeing. High sensitivity is required to record noncoherent echoes from atmospheric structure. Radar reflectivities (cross sections) per unit volume in units of cm^2/cm^3 detectable by the Wallops Island radars at wavelengths of 3.2, 10.7, and 71.5 cm are 6×10^{-17}, 4×10^{-18}, and 4.5×10^{-19}, respectively. For comparison, the corresponding value for the Weather Service WSR-57 10-cm radar is 3×10^{-16}. The Wallops Island studies made use of the different wavelength dependence of dot angels (discrete echoes from birds and insects of small spatial extent) and scatter from refractive index fluctuations (atmospheric structure). The dot echoes tend to behave as Rayleigh scatterers having a wavelength dependence of λ^{-4}, as indicated by Eqs. (6-65) and (6-66), whereas the echoes from atmospheric structure behave as $\lambda^{-1/3}$, in accordance with Eq. (4-132).

High-power pulse radars at Wallops Island, Defford, England (10.7 cm), and Westford, Massachusetts (23 cm) and FM-CW radars (10 cm) at San Diego, California and Boulder, Colorado have provided very interesting data about convective cells, atmospheric layers or stratifications, and waves in the clear atmosphere [41,42,64]. The backscatter echoes are due to turbulence associated with convective cells and wind shear occuring at layer boundaries where there are large vertical gradients in index of refraction. FM-CW radars are useful for observing the same general types of phenomena as the pulse radars and are also suitable for close-in seeing and thus for studying the structure of the atmosphere close to the ground.

When a radar scans horizontally in a region containing convective cells, where air is rising, the echoes from the boundaries of the cells have the shapes of doughnuts. Figure 6-13 shows the structure of a convective cell and shows the regions from which echoes are received. Atmospheric gravity

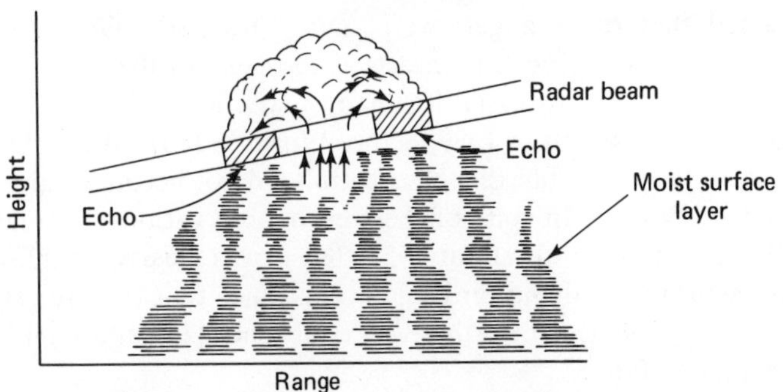

FIGURE 6-13. Illustration of radar detection of thermal cell. (After Hardy and Ottersten, "Radar Investigation of Convective Patterns in the Clear Atmosphere," *Journal of Atmospheric Science*, 1969 [43].)

waves can be detected by radar [46] and are sometimes modified by a Kelvin-Helmholtz instability to form a braided pattern of radar echoes, as suggested by Fig. 6-14. Waves in the lee of the Wales mountains have been recorded by 10.7-cm radar at Defford, England [74]. Insect echoes may provide useful tracers of atmospheric structure. An interesting example of this phenomenon is the observation of atmospheric cellular structure (known as Benard cells) with a 3.2-cm radar at Sudbury, Massachusetts [43]. Measurements of wind by radar means originally involved the use of echoes from hydrometeors (rain, snow, hail), chaff, or balloons. The fact that some echoes from thunderstorms and insects move generally with the wind has been used to obtain estimates of wind direction and velocity. Highly accurate velocities can be obtained by using balloons equipped with transponders which are tracked by precision tracking radars.

Doppler radars are advantageous for wind measurements. One Doppler radar can determine only one component of wind velocity, but two Doppler radars have been used successfully in a technique of coplanar scanning using chaff or hydrometeors [87]. Utilizing this approach, each radar maintains its antenna beam in a common plane, tilted at a given angle above the horizontal, and scans over a common area of the plane. In covering this common area, each radar scans in angle for each of a number of adjacent range gates. It is not practical for each radar to actually point at precisely the same portion of the common area at all times, and the average wind pattern has to be sufficiently stationary for the period over which data are taken to provide good results. In this mode of operation the tilt angle is usually sufficiently small that vertical velocities can be neglected.

In other applications, where the vertical velocity is the quantity of interest, or in the case where three radars are used in an effort to determine all three velocity components, it must be realized that the vertical component

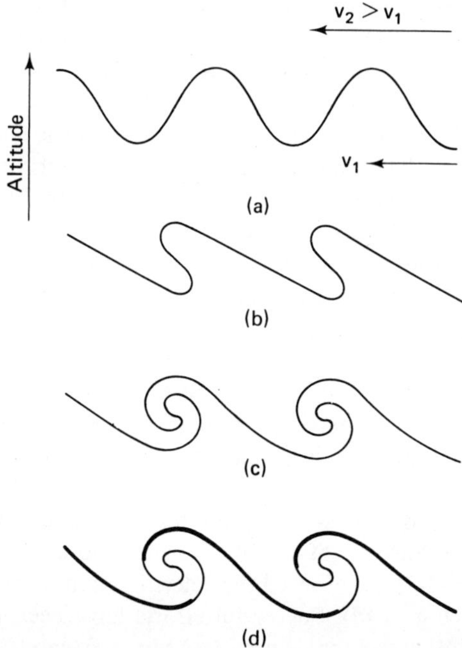

FIGURE 6-14. Sketch of gravitational wave: (a) wave stable, (b) wave becoming unstable, (c) wave breaking and forming vortices, and (d) same as (c) with regions of stronger radar reflectivity blackened. The relative length of the velocity vectors indicates the wind shear. (From Hicks and Angell, "Radar Observations of Breaking Gravitational Waves in the Visually Clear Atmosphere," *Journal of Applied Meteorology*, 1968 [46]. The same type of phenomenon is depicted in [14] and attributed to a Kelvin-Helmholtz instability.)

of target velocity V_f is the sum of the vertical air motion w and the target terminal fall velocity V_t, so that

$$V_f = w + V_t \tag{6-69}$$

The fact that it is difficult to separate w and V_t is one of the important obstacles in Doppler radar research. If w can be neglected, so that $V_f = V_t$, then Doppler frequency measurements can be used for determinations of drop size distributions, as a drop of a given diameter reaches a certain limiting velocity.

The clear air echoes mentioned so far have come from discrete layers, but techniques were developed later which allow, within certain limits, obtaining clear air echoes continuously as a function of time and distance. Such echoes are highly useful for studying winds and turbulence in the atmosphere. The use of coherent radars is essential for exploiting clear air echoes to the fullest, because coherent radars can provide data on velocities and can detect Doppler-shifted echoes which may be obscured by clutter when non-coherent radars are employed. The Sunset VHF meteorological radar, operating at a frequency of 40.6 MHz (wavelength of 7.4 m) near Boulder, Colorado [38], was the first radar built specifically for the purpose of monitoring clear air echoes. The Sunset radar is an outgrowth of experience, with similar echoes, obtained by use of the Jicamarca incoherent-scatter radar near Lima, Peru [88]. It has an antenna area of 7200 m^2, a peak power of 100 kW, pulse lengths greater than 1 μs, a beamwidth of 2.4° in the east-west direction, and a beamwidth of 7.1° in the north-south direction. Spectra of echoes have been obtained up to heights of about 15 km. The Sunset radar can thus measure winds, and it has developed that it can also determine profiles of C_n^2 (Sec. 4-6) [84]. A theoretical model has also been developed to allow the calculation of C_n^2 profiles from radiosonde data. The model involves the theory of turbulence and the fraction of the volume that is turbulent.

The success of the Sunset program in obtaining useful clear air echoes has stimulated other investigations of clear air echoes. The Chatanika 1290-MHz radar (Sec. 6-4-7) has allowed the determination of winds up to a height of about 24 km (Fig. 6-15). Successful efforts have been made also to use an *S*-band FM-CW radar for determining winds from clear air echoes [16]. The power of this CW radar is 200 W, and it utilizes two separate identical antennas 2.44 m in diameter for transmission and reception. The technique for measuring velocity mentioned in Sec. 6-3-3 was utilized. The FM-CW radar has the advantage of being able to "see close in" and has the sensitivity to measure winds continuously to a height of about 2 km. This height is small but complements the capability of the Sunset type of radar, which can record echoes from higher heights but not from below about 1 or 2 km. Heights below 2 km are important to airport operations where wind shear may create hazards for aircraft that are landing or taking off. It has been suggested also that the use of Doppler processing allows the recording of clear air echoes at near ranges by weather radars of moderate power (0.5–1 MW) and antenna diameter (5–10 m) [22].

ECHOES FROM BIRDS. Radar echoes from birds are of interest because they are an important source of radar clutter and can be utilized to help minimize the hazard of bird-aircraft collisions and because of their utility for providing information about bird migration. The fact that birds are an important source of radar clutter was not widely recognized for many years.

The U.S. Air Force experienced material damage costing about $25 million in 1973 due to collisions between aircraft and birds [83]. Radar can play a useful role in minimizing the bird hazard in two ways. It can be used to obtain information on bird movements and migration and thus provide a

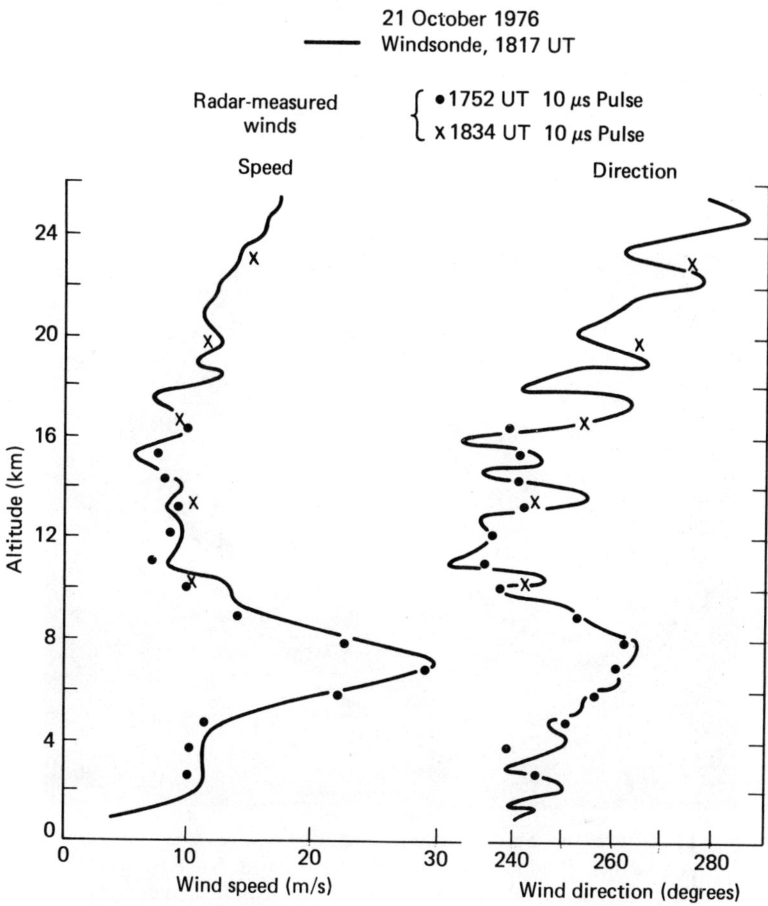

FIGURE 6-15. Profiles of wind speed and direction as determined by Chatanika radar and windsonde. (From Balsley et al., "Winds Derived from Radar Measurements in the Arctic Troposphere and Stratosphere," *Journal of Applied Meteorology*, 1977 [5].)

basis for predicting the bird hazard and planning flight operations in advance. Also radar can provide real-time warning which can be used as a basis for diverting aircraft from bird concentrations and for reducing aircraft speed at low altitudes.

Birds are an important part of natural ecosystems, especially in the case of the Arctic, the oceans, coastal areas, and many lakes and marshes, and by monitoring bird movements and numbers in such areas radar can play a significant environmental-monitoring role. The Arctic, areas near the north and south ends of the Alaska pipeline, and the Gulf of Alaska are especially significant in this respect [29,30]. The decision to proceed with the pipeline and with exploration for oil in continental shelf areas involves a responsibility for adequate environmental protection and monitoring, and radar can assist in meeting this responsibility.

Surveillance radars involve the use of a rather narrow continuously rotating beam which forms a radar map of the surroundings on a PPI display scope. Echoes from migrating birds can often be conveniently distinguished by taking time-exposure photographs, as illustrated in Fig. 6-16, where

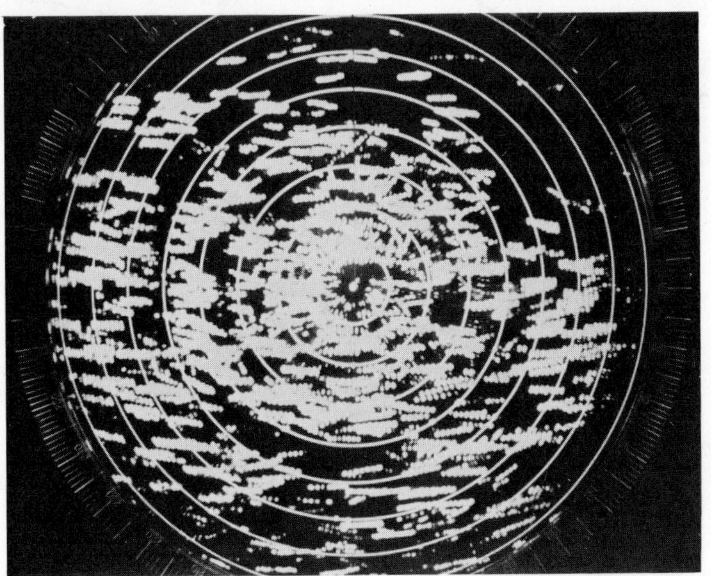

FIGURE 6-16. Spring eastward migration of birds over Oliktok, Alaska (near Prudhoe Bay): 40-nmi (74-km) range, 5-nmi (9.26-km) range marks on PPI display; 5.6-min time exposure, geographic north up, 2005 ADT, May 28, 1972.

flocks of birds appear as streaks having lengths consistent with the length of the time exposure and the velocity of the birds. Radar also has some capability for identifying birds as to size or type by recording amplitude or Doppler "signatures" [32]. For this purpose the radar beam is held in a fixed direction or moved only to follow the bird's movements, rather than rotated continuously. A principal feature of amplitude signatures is modulation at the bird's wingbeat frequency, which varies inversely as the size of the bird. Figure 6-17 shows the amplitude signature of a bird which was identified by its signature, and a knowledge of what birds frequented the area, as a great blue heron. The wingbeat frequency in this case has the low value of 2.7 Hz. Near the other extreme in wingbeat rate in Colorado is the broad-tailed hummingbird with a wingbeat rate near 40 Hz. Radar also has capability for detecting airborne insects and monitoring their movements.

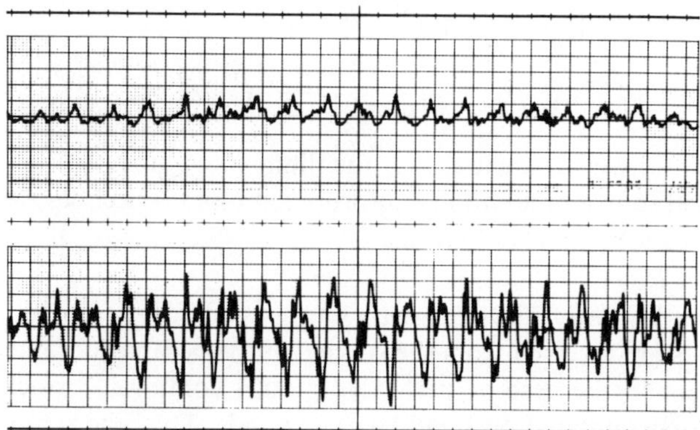

FIGURE 6-17. Amplitude signature of great blue heron, Boulder, Colorado, 2015 MDT, May 28, 1973. Horizontal scale: 5 small div. = 1 s. The upper trace represents the signature after filtering by a 2- to 200-Hz bandpass filter, and the lower trace is the unfiltered signature.

6-4-5 Remote Sensing of Oceans

Awareness of the importance of the oceans and of sea ice has been increasing in recent years. Oceans cover about 71% of the surface of the earth. Exploration for and development of oil and other mineral resources in the Arctic has been a major factor contributing to increased interest in the sea ice of the Arctic. Another reason for the interest is the possibility that if the ice cover of the Arctic decreases significantly, due either to natural or man-made causes, an unstable condition may develop which would result in further major melting, with profound implications for the world's climate and sea level. The remoteness, vastness, and inaccessibility of the oceans and the sea ice of the Arctic make them highly suitable subjects for remote sensing. Finally the subjects of the oceans and sea ice are appropriate for illustrating the application of three remote sensing techniques, utilizing reflected radiation, radiometry, and radar.

While the oceans are vast and complex in nature, the marine phenomena that can be sensed remotely seem comparatively few in number when compared to the situation for the land. The applications of remote sensing to the land are virtually without limit as to number and variety.

In considering radar reflection from a surface, it is convenient to use a normalized radar cross section σ°, having units of m^2/m^2 or, in other words, being nondimensional. Thus it can be expressed as a dB value. A cross section $(\sigma^\circ)_{dB}$ of -10 means that $\sigma^\circ = 0.1$. σ° is referred to actual surface area, but a cross section referred to projected area is sometimes used. In terms of σ°

the radar equation [Eq. (6-20)] takes the form

$$W_r = \frac{1}{(4\pi)^2} \int \frac{W_t G_t A_r \sigma^\circ \, dA}{R^4} \tag{6-70}$$

where the integral is evaluated over the illuminated area.

A good source of additional information on radar echoes from the land and sea has been provided by Long [56].

Ocean Waves. Since the early days of radar, it has been known that radar echoes are received from the surface of the oceans and other bodies of water, except in the case of perfectly smooth conditions which are most likely to be encountered on small lakes or ponds. The echoes are commonly regarded as a nuisance, but Crombie [18] showed that echoes at high frequencies could provide useful information about ocean waves, which occur over a range of wavelengths, frequencies, and amplitudes. The Bragg scattering mechanism plays an important role in the backscatter of radio waves from the ocean, and Crombie correctly interpreted the echoes he observed on this basis. Echoes are received from structure having a periodicity L in the radial direction from a radar of wavelength λ when $L = \lambda/2$. Waves moving radially toward and away from the radar can cause such echoes. When the Doppler frequency of the echoes are recorded and displayed, a result like that of Fig. 6-18 is commonly observed. The echoes from the approaching and receding waves appear with clearly defined Doppler shifts of opposite sense. The Doppler displacements are observed to vary with the square root of the HF carrier frequency, which suggests that ocean gravity waves, for which $v = \sqrt{gL/2\pi}$ [50], are being recorded (g is 9.8 m/s and L is wavelength). The technique has attracted attention as a means of monitoring sea state over the wide expanse of the oceans by use of HF or lower-frequency radio waves. This approach has been studied at San Clemente Island, off the coast of California, a location allowing coverage via ionospheric reflection of the North Pacific, which is of importance in connection with oil tanker shipments from Alaska. A complication is that the Doppler-shifted echoes come from a continuum of waves traveling toward and away from the radar and having a range of wavelengths. The echoes thus do not necessarily arise from the predominant wave system. Echoes from the predominant system would have a larger amplitude than the other echoes, with the large echoes corresponding to the direction the predominant system is moving but with an echo corresponding to waves traveling in the opposite direction as well. In the case of ionospheric reflection, furthermore, the lower frequencies (300 kHz to 5 MHz) which would resonate with the predominant wave system are highly attenuated. Also the practical difficulties of building a system to cover this frequency range, including the ability to scan in azimuth, are great. In the frequency range that is most feasible for ionospheric reflection,

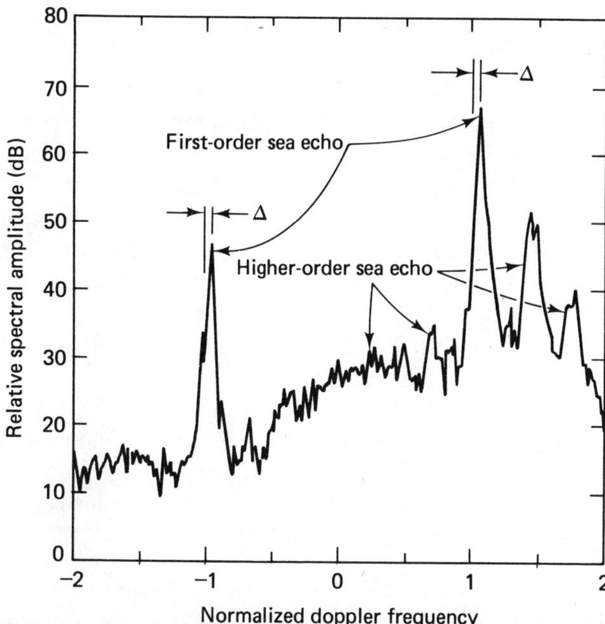

FIGURE 6-18. Averaged radar sea-echo Doppler spectrum at 9.4 MHz. The Doppler-frequency scale is normalized with ± 1 corresponding to Doppler shifts of ± 0.313 Hz from the carrier. **Δ** represents the shift of the record due to an underlying current. (From D. E. Barrick, J. M. Headrick, R. W. Bogle, and D. D. Crombie, "Sea Backscatter at HF: Interpretation and Utilization of the Echo," *Proc. IEEE*, vol. 62, pp. 673–687, June 1974.)

Doppler-shifted echoes are received from ocean waves but from waves having lengths between perhaps 25 and 6.5 m. These waves are generally present in a fully developed condition, give a σ° of -17 dB, and provide no information about sea state. These echoes are useful for calibration purposes, and an asymmetric shift of the two echoes about zero Doppler frequency can supply information on ocean currents. With respect to detecting the predominant wave system and obtaining information about sea state, a possibility of much interest has been that higher-order sea echoes, illustrated in Fig. 6-18, for ionospherically propagated transmissions may be useful [80]. Radars which do not depend on ionospheric reflection can monitor the long ocean waves of interest from land without the complication of needing to use higher-order echoes to a distance of about 200–300 km. One possible approach has been to use LORAN-A transmissions [79]. Another approach is to recognize the the virtue of obtaining data on ocean currents and to exploit the asymmetric shift of Doppler echoes mentioned above. Barrick et al. [7] have developed mobile HF radars operating near 25 MHz for this application. These units have provided maps of surface ocean currents to distances of 70 km from shore.

Before discussing observations and techniques further, it is desirable to mention some elementary considerations about the reflections of waves from a surface. A perfectly smooth surface reflects incident electromagnetic waves in a specular manner and in the forward direction only. The reflection coefficients are different for horizontally and vertically polarized transmissions and are given by Eqs. (2-42) and (2-43). If a surface is rough, however, energy is reflected or scattered in other directions, including backwards to a radar transmitter as well as the forward direction. A commonly accepted criterion for roughness is the Rayleigh criterion, which can be explained with the help of Fig. 6-19, taking it to refer to vertically polarized waves. Consider two rays A and B such that ray A, upon reflection, follows a path longer than that of ray B by π rad. The two rays thus interfere destructively for forward reflection. Therefore it can be argued that some of the incident energy is scattered in other than the forward direction. The amount Δr by which the path of ray A exceeds that of B is given by

$$\Delta r = 2\Delta h \sin \theta$$

and the corresponding phase difference $\Delta\phi$ will be set equal to π so that

$$\Delta\phi = \frac{4\pi}{\lambda} \Delta h \sin \theta = \pi$$

from which

$$\Delta h = \frac{\lambda}{4 \sin \theta} \tag{6-71}$$

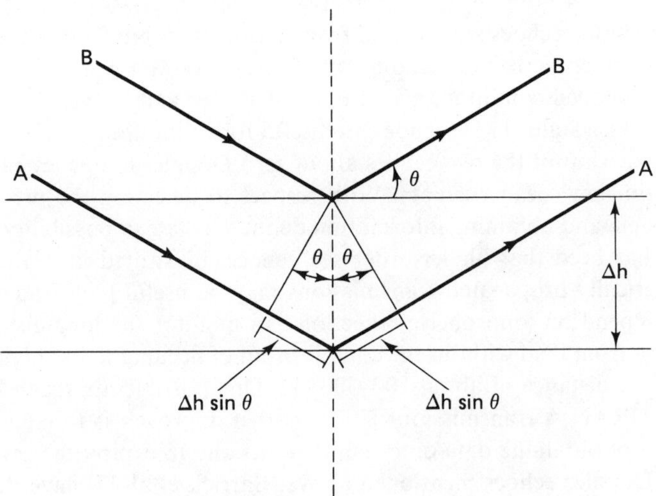

FIGURE 6-19. Basis for Rayleigh criterion.

can be taken as the criterion for roughness. The Rayleigh criterion is usually stated as a criterion for smoothness in which case if

$$\Delta h < \frac{\lambda}{8 \sin \theta} \tag{6-72}$$

the surface is considered to be smooth. The integer in the denominator is arbitrary and might be larger than 8.

Returning to a consideration of the reflection of electromagnetic waves from the ocean, note that the Bragg scattering mechanism appears to be effective at microwave as well as HF frequencies but that at microwaves the Bragg reflection must come from smaller capillary waves which are superimposed on the larger gravity waves. The surface of the sea is thus viewed as a composite surface having both large and fine structures. In the case of illumination from overhead at angles θ' measured from the vertical that are not too large, a tangent plane model is sometimes assumed to apply. This model involves reflection from surfaces, with maximum reflection taking place when $\tan \theta'$ equals the slope of the surface (when the surfaces are perpendicular to the line of sight). For radiation incident at small angles above the horizontal, the tangent plane model does not apply, but reflections from smooth reflecting surfaces or facets embedded in the wave structure are believed to play a role [56]. In both cases, Bragg scattering is probably important as well. The tangent plane model, Bragg scattering, and Rayleigh scattering are all mentioned by Brown et al. [13] as possible explanations for the observed ability of L-band coherent imaging (side-looking) radars on aircraft to image ocean waves and other features. The Rayleigh scattering mechanism would involve scattering from a large number of small surface perturbations. This mechanism had not received serious consideration previously as an explanation for sea scatter but has been proposed as an explanation for radar scatter from within Death Valley, California, where a range of roughness uncomplicated by vegetation is encountered [68].

The intensity of backscatter from the sea surface is less for horizontal polarization than for vertical polarization for rather calm seas and for near grazing incidence, although the reflection coefficient for horizontal polarization [Eq. (2-42)] is larger than that for vertical polarization [Eq. (2-43)]. This seeming contradiction is explained by noting that the reflection coefficients apply to reflection from a smooth surface and that such reflection is in the forward direction. The explanation of why the backscatter coefficient is less for horizontal polarization is believed by Long [56] to involve a local interference process, similar to that of Sec. 2-4 except that the reflection takes place very close to a crest of a capillary wave, as suggested in Fig. 6-20. The difference in path lengths is slight, but ray A is reversed in phase by 180° on reflection and is thus 180° out of phase with ray B. The magnitude of the forward reflection coefficient is greatest for the horizontally polarized wave,

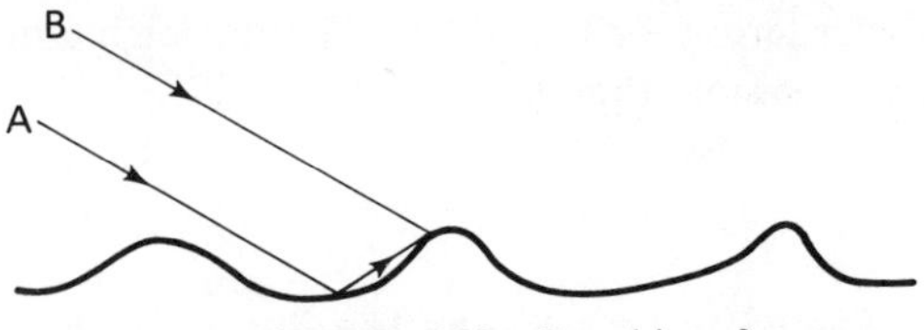

FIGURE 6-20. Local interference.

so the interference is most pronounced for horizontal polarization. For sufficiently rough seas, however, the backscatter for horizontal polarization may be equal to or greater than that for vertical polarization. Figures 6-19 and 6-20 are similar, but they refer to different situations. A reversal of phase upon reflection was not assumed for Fig. 6-19, and the figure would thus tend to apply to vertical polarization and for reflection at an angle θ that is not too small. Also the path-length difference for Fig. 6-20 is smaller than that for Fig. 6-19 (and is neglected) because θ is assumed smaller for Fig. 6-20 and $\Delta r = 2\Delta h \sin \theta$. In neither case is a precise analysis given; the two figures and accompanying discussions merely illustrate the rudiments of the ideas involved.

Radar systems with sufficiently high spatial resolution can display individual wave crests, as in the case of the *L*-band coherent imaging radar mentioned previously [13] that was used on aircraft and which will also be installed on the Seasat-A satellite. The *X*-band harbor surveillance radar operated by the U.S. Coast Guard near the Golden Gate, San Francisco can also display the individual wave crests, as shown in Fig. 6-21.

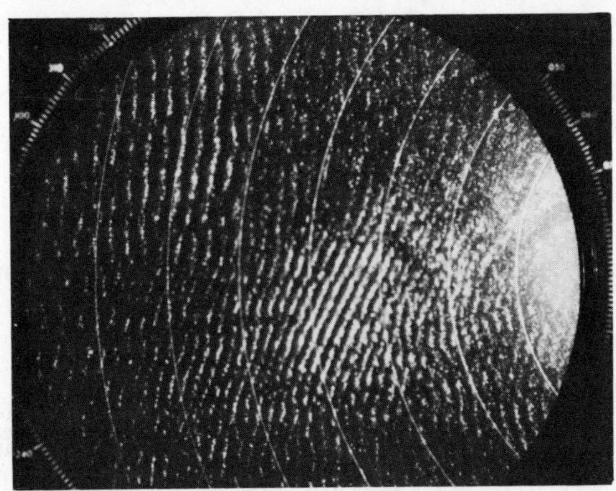

FIGURE 6-21. Ocean waves as recorded by the U.S. Coast Guard *X*-band Pt. Bonita radar (outside the Golden Gate, California) on an off-centered 2-nmi (3.7-km) display. The range marks are spaced $\frac{1}{4}$ nmi (463 m) apart, and the first mark showing on the right is the 1.5-nmi (2.77-km) mark. The pulse length employed was 0.05 μs. 1000 PST, Dec. 9, 1975, one-rotation exposure.

Sea Ice. As in the case for ocean waves, satellite, airborne, and ground-based techniques are all capable of playing a useful role in monitoring and studying sea ice. The use of satellites is essential for monitoring the vast remote areas of the polar seas. By comparison, aircraft can cover only relatively limited areas on an infrequent basis, but they have provided valuable data and can be used to advantage in conjunction with satellites. Ground-based techniques are extremely limited in coverage area, but radar installations at a few locations of particular interest can provide continuous all-weather coverage which satellites and aircraft cannot supply.

The Landsat satellites can supply superb high-resolution images of sea ice distribution. Figure 6-22 shows a Landsat image of the Bering Strait area. Of interest are the areas of open water or polynyas south of the Diomede Islands and of the Seward Peninsula. Figures 6-23 and 6-24 show echoes received by the U.S. Air Force ACW radar located at Tin City, Alaska, near the edge of the Bering Strait, at about the same time as the image of Fig. 6-22. The figures illustrate the characteristics of two types of video signals. MTI video emphasizes echoes from matter exhibiting motion, especially water waves but also small ice floes that are bobbing about under the influence of wind and waves. A logarithmic, short-time-constant video em-

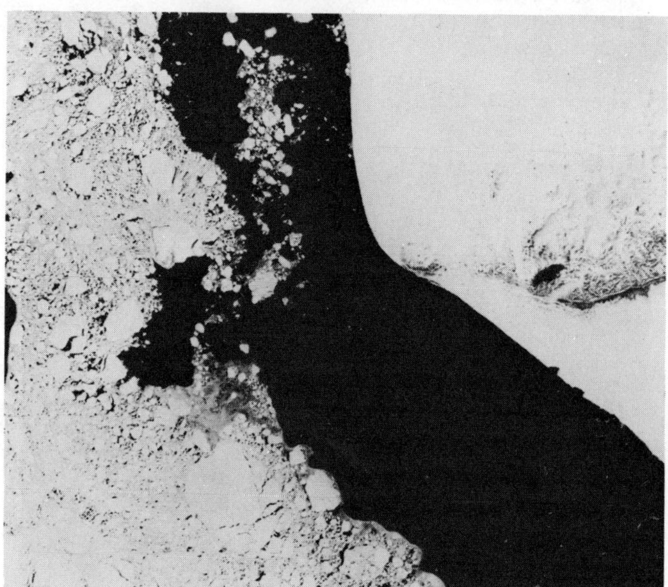

FIGURE 6-22. Landsat satellite (MSS-7) image of Bering Strait. The dark areas are open water, and the white areas are ice or snow. Cape Prince of Wales shows slightly to the right of center, and Little Diomede Island, of the United States, and Big Diomede Island, of the U.S.S.R., show to the south of center, with a polynya or area of open water to the south of the islands. 1040 ADT, May 16, 1975, geographic north up.

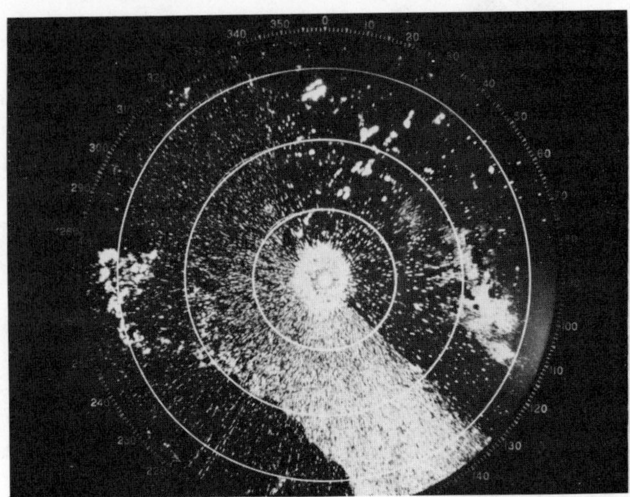

FIGURE 6-23. U.S. Air Force Tin City, Alaska *L*-band radar screen showing MTI video signals, 1305 ADT, May 21, 1975. The photograph is a 5-min time exposure, and the range marks are spaced 10 nmi (18.5 km) apart. Geomagnetic north is up (about 20° to the east of geographic north). The area is the same as that shown in the Landsat image of Fig. 6-22 but was taken several days later. Much the same large area of open water shows on both Figs. 6-22 and 6-23, the echo from the water appearing as the prominent bright area on Fig. 6-23.

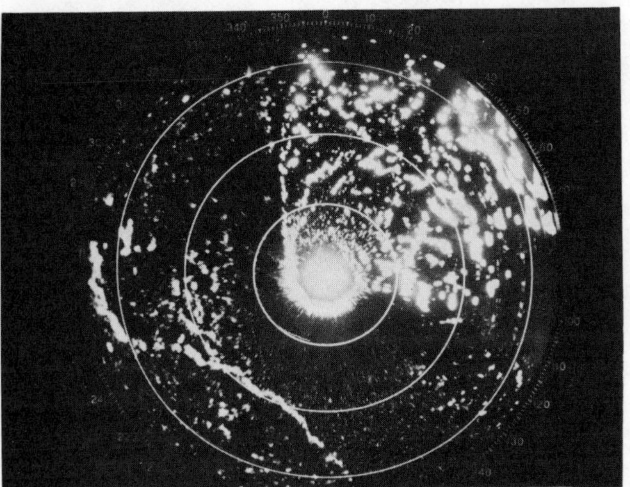

FIGURE 6-24. U.S. Air Force Tin City, Alaska *L*-band radar screen showing logarithmic short-time-constant video signals, 1300 ADT, May 21, 1975. Except for the different type of video signal, Fig. 6-24 was recorded in the same way as Fig. 6-23.

phasizes echoes from stationary objects, especially from boundaries and discontinuities such as those between open water and sea ice.

The NOAA satellites also provide coverage of the area [57] as part of their worldwide coverage with less spatial resolution but with the advantage of having an infrared scanner that can record images in darkness. The very high-resolution radiometer (VHRR) of the NOAA satellites forms images in one visible (0.6- to 0.7-μm) and one infrared (10.5- to 12.5-μm) band. The ground resolution is 1 km, and twice-a-day coverage is provided, once in daylight and once in darkness.

Microwave scanning radiometers have the advantage over infrared systems of being able to operate in the presence of clouds and overcast but tend to have less ground resolution than infrared systems. Studies carried out with NASA Convair 990 aircraft showed that not only could sea ice and open water be distinguished on the basis of their different emissivities but first-year and multiyear ice could be distinguished as well [15,34]. The basis for the distinction is that at the frequencies utilized the atmosphere in the Arctic has little effect and the brightness temperature which the radiometer records is determined primarily by ϵT_s, where ϵ is emissivity and T_s is actual surface temperature [see Eq. (6-58)]. The actual temperatures tend to vary slowly with time and distance, and contrasts in emissivity are useful for distinguishing water and ice, etc.

The Nimbus-5 [35] and Nimbus-6 [36] satellites are equipped with electrically scanning microwave radiometers (ESMRs). The Nimbus-5 radiometer operates at a wavelength of 1.55 cm with a maximum ground resolution at nadir of 25×25 km. The Nimbus-6 system is similar but operates at a wavelength of 0.81 cm. Because of the low resolution, the systems are useful primarily for recording large-scale features such as the boundary of polar ice and the major areas of first-year and multiyear ice. At the wavelength of 1.55 cm, first-year ice is reported to have an emissivity near 0.95, while multiyear ice has an emissivity near 0.89, and the emissivity of calm sea water at 271 K is near 0.45.

The satellite systems described above are passive systems using reflected sunlight or thermal radiation. Active systems transmit electromagnetic waves that are reflected from objects or regions of interest. Included in the category of active systems are laser profilometers, radar altimeters, side-looking and surveillance radars, and scatterometers. The profilometers and altimeters direct a beam straight downwards and obtain information about variations in height or relief. The laser profilometer can obtain detailed information about height variations and roughness of the upper surface of the sea ice. The type of laser used has been a modulated helium-neon laser providing 25 mW at a wavelength of 0.6328 μm and producing a beamwidth of less than 10^{-4} rad. Radar altimeters have lower resolution and are best suited for sea surface measurements rather than for sea ice applications. A system for obtaining remotely the thickness of sea ice is very much desired but has not been perfected as yet.

Side-looking radars and scatterometers can provide information about sea ice from aircraft and satellites. Side-looking radars are pulsed systems operating in the microwave frequency range and can penetrate fog and overcast. The simplest type of side-looking radar utilizes a real aperture capable of providing moderate resolution along the track of an aircraft. Higher along-track resolution can be provided by the greater sophistication and complexity of synthetic aperture systems, which process signals in the Doppler-frequency domain. The resolution of such systems allows recording data concerning characteristics such as the degree of packing, relative cover of floes of different sizes, degree of hummocking, and amount, size, and orientation of free water areas and channels. Some ability to obtain information on the age of ice in rough categories is based largely on experience with correlating ice condition with the intensity and nature of signal returns.

Scatterometers are CW systems which direct beams in particular directions and record echoes as a function of angle of incidence, polarization, and, in some cases, transmitter frequency and Doppler frequency. The resolution provided by scatterometers is less than that for synthetic-aperture side-looking radars. The ability to record data as a function of the parameters mentioned, however, is an important advantage, and scatterometers also have capability for determining the age or type of sea ice in rough categories.

The Seasat-A satellite has most of the different types of instrumentation mentioned here, including a radar altimeter (13.5 GHz), scatterometer (14.5 GHz), synthetic-aperture side-looking radar (1.275 GHz), scanning multifrequency microwave radiometer, and visible and infrared radiometers. It is designed primarily for obtaining data over the oceans, including the polar seas.

Compared to the extensive and far-ranging studies of sea ice carried out by aircraft and satellites, little data on sea ice have been provided by land-based radars and for good reason—the limited area coverage possible with land-based radars. The most extensive program has been that carried out by Tabata [77] off the Okhotsk Sea (north) coast of Hokkaido Island of Japan. A network of three stations, located at heights of 440, 300, and 200 m above sea level, has been used to cover an area about 70 km in width and 250 km in length. The Okhotsk Sea coast is covered with sea ice from about the beginning of January to the end of March. The frequency employed has been 5.54 GHz. Ice movement is monitored by recording the movement of distinctive features of the ice. Signal intensities have been classified into 16 steps by computer to aid in associating signal level with surface topography of ice. At Pt. Barrow, Alaska, Sackinger [67] has recorded sea ice echoes by use of an *X*-band marine radar, and Flock [31] has used the *L*-band ACW radar at the edge of the Bering Strait (Figs. 6-23 and 6-24).

6-4-6 Remote Sensing of Land

BARE GROUND. Radar backscatter data obtained by an L-band airborne imaging radar from Death Valley, California is of interest because the backscatter signal from that location can be correlated with surface roughness without having to cope with the complication of vegetation [68]. The valley surface was classified into seven numbered geologic units in order of decreasing roughness, and radar backscatter and surface characteristics were compared. Unit I, the most strongly and isotropically backscattering surface, is the Devil's Golf Course, and otherwise flat surface covered by extremely rough, irregular rock salt shaped into weird forms, including reentrant cavities. Units II and III also consist of what is called a salt pan but have less surface roughness. The alluvial fans extending into the valley from the surrounding mountains provide rock and gravel of a range of sizes. Unit IV consists of younger deposits of gravel and boulders; Unit VI consists of a rather smooth varnished surface known as desert pavement. The latter is composed of the oldest fan-gravel deposits, which are small fragments "fitted together" to form the smooth surface. Unit VII involves flat, white dry-lake or floodplain deposits and has the lowest backscatter of all units. A radius of curvature a of 0.1 λ was found to be the dividing line between what appeared to be Rayleigh scattering ($a < 0.1\lambda$, with a signal intensity variation as a^6) and scatter varying as a^2.

The possibility of remotely sensing soil water content by radar means is dependent on the variation of the electrical properties of soil with water content. The variation of complex relative dielectric constant K_c of soil with water content is as shown in Fig. 6-25. From knowledge of K_c, the reflection coefficients for horizontally and vertically polarized waves can be calculated using Eqs. (2-42) and (2-43) and assuming a smooth surface. Surface roughness, vegetation, and inhomogeneous soil complicate the determination of soil moisture. Ulaby and Batlivala [82] report that a frequency of 4 GHz and an angle of incidence between 7 and 15° from nadir (the vertical) are optimum radar characteristics for mapping soil moisture of bare soil. Essentially the same frequency and angles turn out to be suitable for measuring soil moisture in the presence of crops [81]. It should be noted that backscatter at other than normal incidence is dependent on the presence of some roughness. The determination of soil moisture was found to be most reliable when based on the incoherent power reflection coefficient. [Equations (2-42) and (2-43) are for field intensity reflection coefficients.] The incoherent power reflection coefficient ρ_p is related to emissivity ϵ' by

$$\rho_p = 1 - \epsilon' \tag{6-73}$$

This relation follows from the fact that absorptivity α' and emissivity ϵ' are equal in magnitude and $\rho_p + \alpha' = 1$.

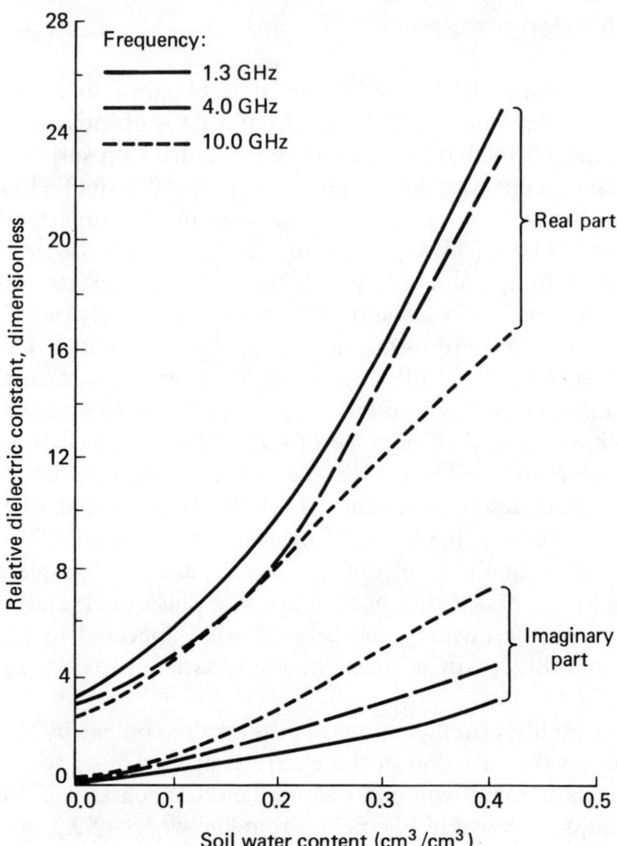

FIGURE 6-25. Relative dielectric constant of loam as a function of moisture content. (From J. Cihlar and F. T. Ulaby, "Dielectric Properties of Soils as a Function of Moisture Content," *CRES Tech. Report 177-47*, University of Kansas Center for Research, Inc., 1974; also reproduced in "Microwave Remote Sensing of Soil Water Content," *Remote Sensing Lab Tech. Report 264-6*, 1975, by the same authors.)

Landsat and Side-Looking Radar Capabilities. The applications of Landsat satellites and side-looking radars (SLARs) to the remote sensing of land are virtually without limit as to number and variety. Each system has advantages which complement those of the other, and the two types of systems are discussed here under a common heading. In recognition of the desirability of comparing and combining synthetic-aperture radar (SAR) data with Landsat data, the Goodyear Aerospace SAR program has used north-south flight paths with the radar antenna illuminating land to the west in order to correspond as closely as possible with Landsat data. Data from

Landsat and SAR systems have been combined in single images [45]. The following categories of Landsat (formerly ERTS) experiments [65] gives some feeling for the range of applications:

1. Agriculture/forestry
2. Environmental quality/ecology
3. Cartography/geography/demography
4. Geology
5. Hydrology/water resources
6. Interpretive technique developments (includes land use)
7. Meteorology
8. Oceanography
9. Sensor technology

Essentially the same categories (with the possible exception of meteorology, which is not the subject of this section) apply to SLAR, including SAR. A bibliography of references emphasizing applications to the land is included at the end of this volume. The list includes a number of references which are not cited specifically within this chapter as well as some that are.

The application of remote sensing techniques in general to agriculture and forestry has been considered by the National Academy of Sciences [60], and the following categories have been listed:

1. Identification and area measurements of the major agricultural crop types
2. Mapping of soil and water temperatures
3. Mapping of surface water, including snowpack
4. Mapping of disease and insect invasion
5. Mapping of gross forest types
6. Mapping of forest-fire boundaries
7. Assessment of crop and timber-stand vigor
8. Determination of soil characteristics and soil moisture condition
9. Delineation of rangeland productivity
10. Mapping of areas of high potential forest-fire hazard
11. Mapping of major soil boundaries

Landsat satellites provide coverage of essentially the entire earth on a regular schedule, with a repeat period of 18 days, and Landsat imagery is readily available to users at moderate cost. Landsat data are not subject to the target shadowing effect of radar, and the four Landsat frequency bands facilitate the recognition of vegetation and soil types, water characteristics, etc. The aircraft carrying SAR fly at only moderately high altitudes, and SAR has resolution 5 to 10 times finer than that of Landsat satellites (Fig. 6-26). The relatively low grazing angle that causes shadows in SAR imagery emphasizes features of the surface terrain and is thus advantageous for recording geological structure, which may be related to the occurrence of minerals. It is not practical for SAR to cover extremely large areas on a regular schedule as for Landsat satellites, but aircraft can be scheduled when and where needed. As SAR responds principally to terrain features and only the general type of vegetation cover, it is not essential that SAR systems operate on a frequent schedule. The sensitivity of Landsat satellites to changes in vegetation and moisture conditions, however, causes its 18-day repeat period to be advantageous. In accordance with the criterion for smoothness [Eq. (6-72)], specular reflection, as contrasted to diffuse reflection, is encountered much more commonly at the microwave frequencies utilized by SAR than in Landsat imagery. SAR tends to record strong echoes from metallic structures, and it provides readily recognizable images of cultural features, including cities and towns. A major advantage of SAR is its ability to pene-

FIGURE 6-26. Synthetic-aperture radar imagery, showing vegetation cover of the Amazon basin. Scale: 1:25,000. (Courtesy Goodyear Aerospace Corporation.)

trate overcast and fog (assuming conditions are not so bad as to preclude flying the aircraft) and to operate in darkness. The Landsat sensors are limited by clouds and fog and require sunlight for operation.

Remote sensing techniques allow the detection of large-scale geological features, including faults, that may not be readily discernible otherwise. Applications to hydrology involve the mapping of snow cover, the progress of snowmelt and runoff in the spring, and the areal extent of flooding.

An early well-known application of SLAR was to the mapping of Darien province of Panama, which is overcast and invisible from the air most of the time. This particular operation was conducted in 1968 by the Westinghouse Electric Corporation, using a real-aperture side-looking radar. It was in 1970 that SAR systems were released from military security classification. Extensive applications of remote sensing techniques have been made to other South American nations, especially Brazil, which operated one of only four permanent Landsat ground stations outside the United States as of 1977 [69]. Landsat programs in Brazil have involved soil maps, crop forecasts, geologic structures related to the occurrence of minerals, the extent of deforestation, and coastal navigational hazards. The extent of the damage to coffee trees due to the frost of July 1975 was demonstrated by a Landsat image. SAR, developed by Goodyear Aerospace Corporation, has been used in Brazil in conjunction with ground investigations in Project Radam (for Radar Amazon). It has discovered a major previously unknown tributary of the Amazon, shown that the forest cover and soils of the Amazon basin are not as homogeneous as supposed, and discovered some areas of fertile soil. The Landsat and Radam programs are highly regarded in Brazil [69].

The interpretation of Landsat and SAR data is facilitated by the use of false-color processing. Data from the four Landsat channels can be combined and displayed in a color presentation which, for example, shows healthy vegetation as red (Fig. 6-27). Even though SAR systems utilize only one radar frequency, different optical frequencies can be used to illuminate the data film and to thus expose a standard three-color image film. Empirical studies show that the use of color presentations allows increased capability for discerning features of the image. (The objective of the color processing is not to reproduce the true colors of the terrain.) The combination of Landsat and SAR data in a false-color presentation has also proved to be highly advantageous (Fig. 6-28) [45].

VEGETATION. The response of vegetation to radiation in the visible and infrared frequency ranges is illustrated in Fig. 6-29 [33]. High absorption and low reflection and transmission of incident radiation are shown in the ultraviolet and visible frequency range, but with the minimum of absorption and maximum of reflection and transmission in the green that gives vegeta-

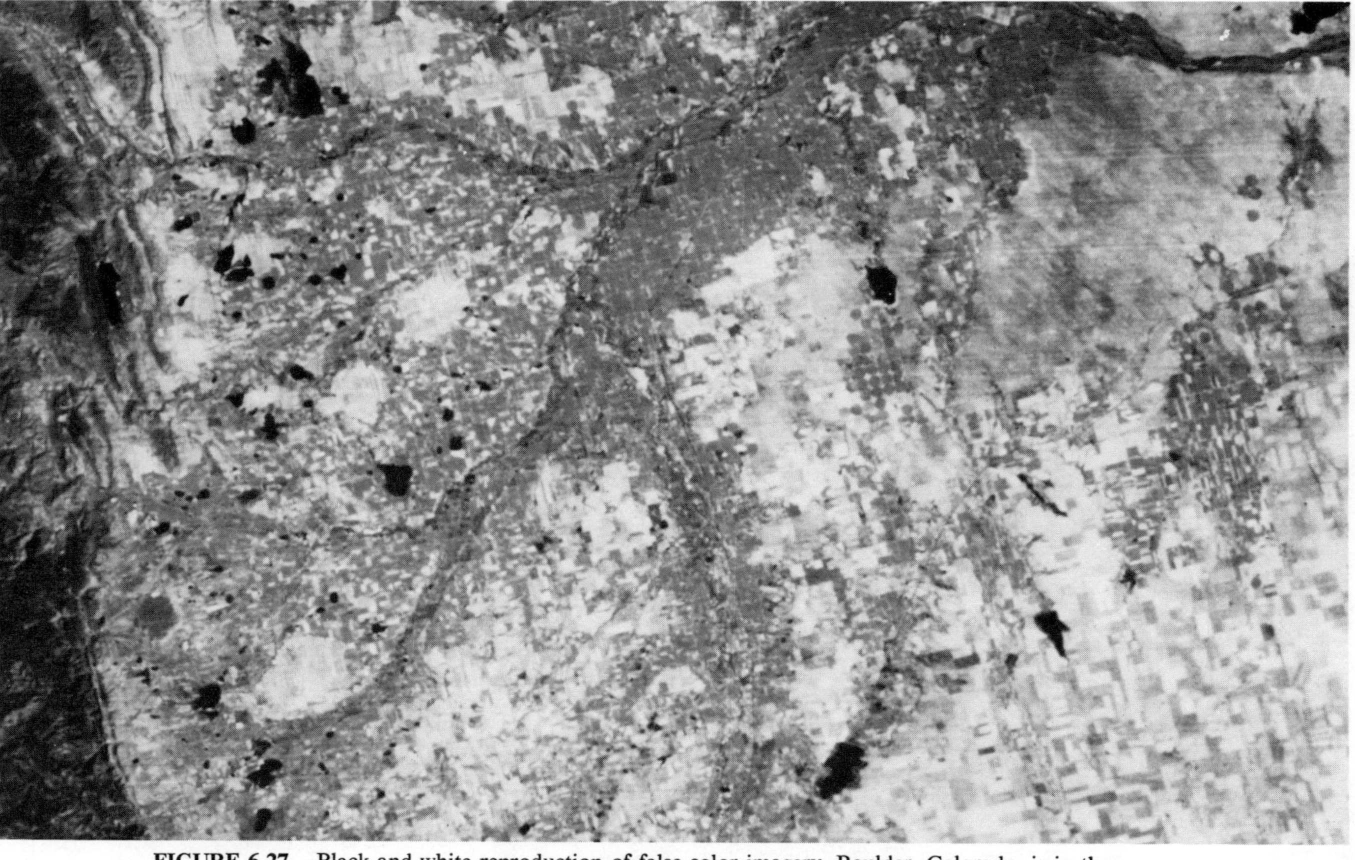

FIGURE 6-27. Black-and-white reproduction of false-color imagery. Boulder, Colorado, is in the lower left corner, the Front Range of the Rockies is along the left (west) border, and Boulder Creek flows to the northeast to join the South Platte River. The dark areas other than the mountains are mostly irrigated croplands that show as red in the color image. The rectangular forms in the lower right that show as bluish and white in the color version represent dry-land farming areas.

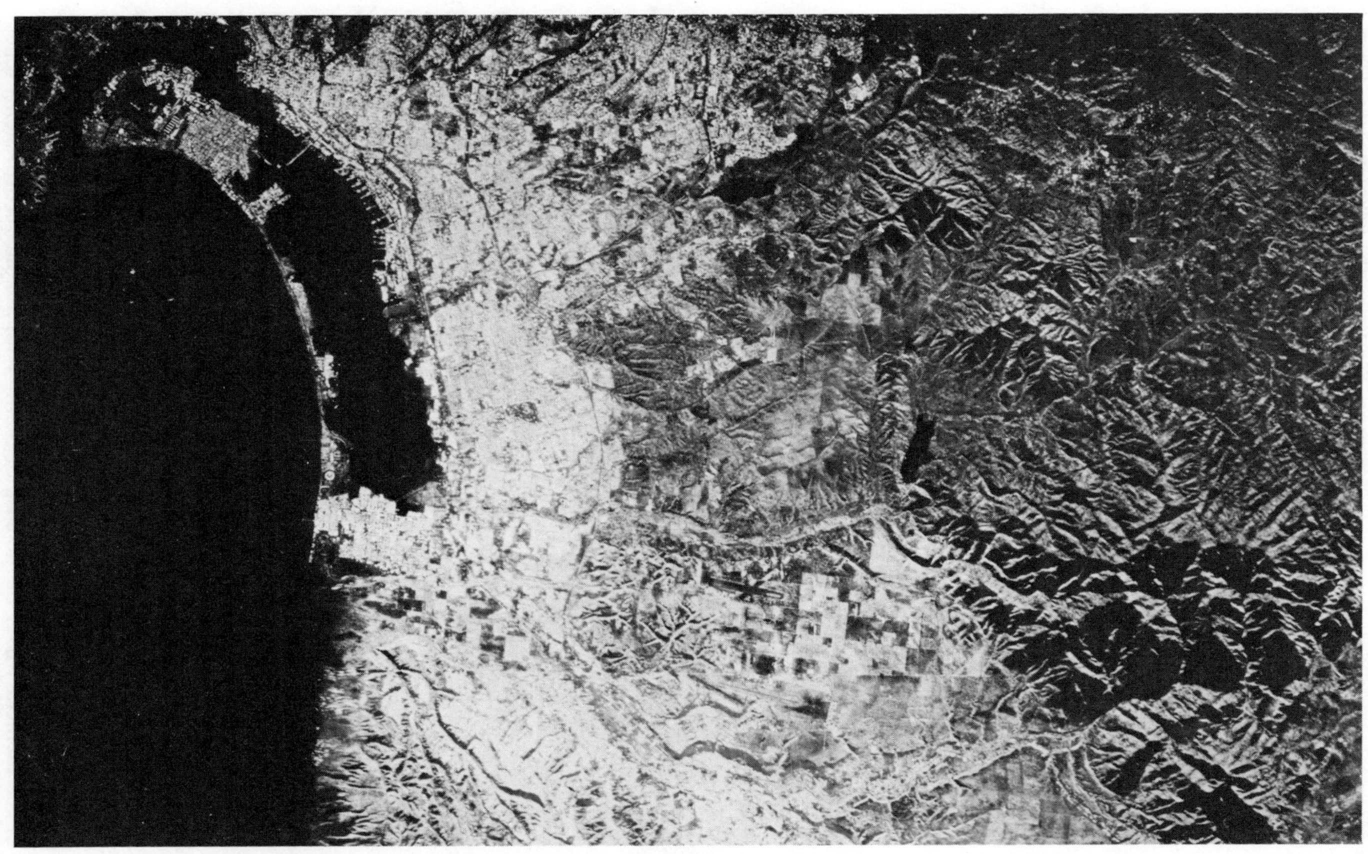

FIGURE 6-28. Imagery of San Diego, California area, derived from both Landsat and synthetic-aperture radar data. Scale: 1 : 250,000 (Courtesy of Goodyear Aerospace Corporation.)

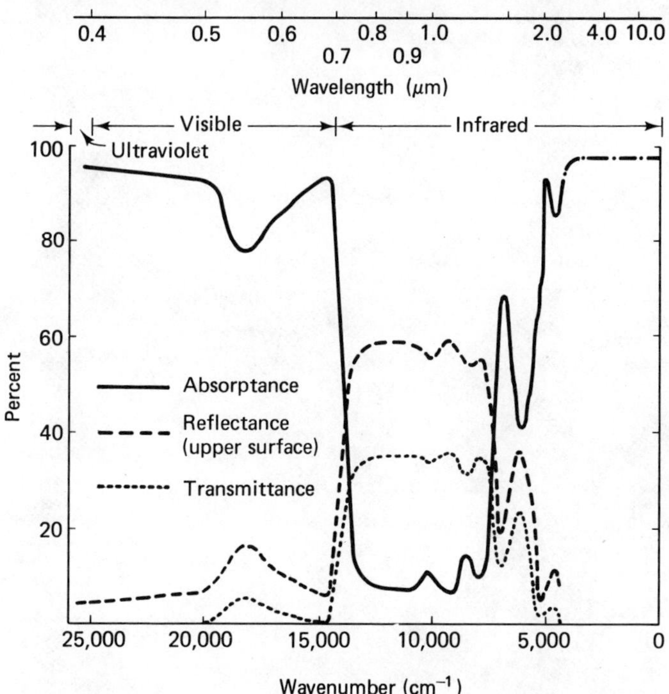

FIGURE 6-29. Absorption, reflection, and transmission of leaf of *Nerium oleander*. (From Gates, "Physical and Physiological Properties of Plants," 1970 [33].)

tion its green appearance. In the near infrared (0.7–1.5 μm) rather high reflection and transmission and low absorption are shown, but absorption increases and reflection decreases again in the far infrared. The response of deciduous leaves of course changes with the development of the leaves in the spring and the end of the growing season in the fall. A feature of major interest from the remote sensing viewpoint is a decrease of the near-infrared reflection coefficient in the case of vegetation subject to stress and loss of vigor, whether due to insects, pathogens, frost, mineral deficiency or toxicity, drought, or flooding.

The response of vegetation to electromagnetic radiation involves the combined effect of chlorophyll content, water content, and structure. In the case of longer wavelengths than considered above, microwave wavelengths, for example, response no longer involves chlorophyll specifically but does involve water content and structure. The reflection of incident radiation is also a function of the angle of incidence, polarization, and the wavelength of the incident wave. In the case of plants with simple vertical stems reflection should be greatest for vertically polarized waves at near-grazing incidence. Plants having nearly horizontal leaves may also give strong return for vertically polarized waves as in that case the leaves may be only a small fraction of a wavelength in extent, in the direction of the electric field of the wave, and may scatter radiation throughout a large angle rather than reflecting specularly. Of course the actual construction and form of vegetation are highly complex and variable. In the case of crops, however, plantings in rows provide some degree of regularity. Ulaby [81] concluded that the use of dual frequencies (one around 4–5 GHz and one at 7 GHz or higher), dual polarization (VV and cross), and angles of incidence between 30° and 70° or 80° from the vertical are to be recommended for the determination of crop types by radar.

6-4-7 The Ionosphere

In the previous sections we have considered remote sensing of the troposphere and the surfaces of the sea and land. Remote sensing techniques can also be applied to extraterrestrial bodies [25], the earth's upper atmosphere, and subsurface features [49]. We forego radar astronomy and geophysical prospecting techniques for subsurface exploration but do include here some brief remarks about remote sensing of the ionosphere, which is part of the upper atmosphere.

The ionosonde was discussed in Chap. 4 because of its relation to ionospheric propagation, and incoherent scatter was presented there as a basic phenomenon of interest. Both ionosondes and radars designed to detect incoherent scatter are properly regarded as remote sensing tools. Indeed the ionosonde appears to be the first practical radar system of significance to be developed, and since the achievements in 1925 by Appleton and Barnett in England and in 1926 by Breit and Tuve in the United States it has been used for remotely sensing the ionosphere. The ionosonde has the limitation that it can provide no information about electron density above the peak of the *F* layer, as once the electromagnetic wave emitted from the radar has penetrated beyond the peak no possibility of reflection exists. Top-side sounders, basically ionosondes placed in satellites and viewing the ionosphere from above, can provide information about densities above the peak but not below. Incoherent scatter radars do not have the limitations of ionosondes or top-side sounders and can obtain returns from both below and above the peak of the *F* layer.

Radar remote sensing techniques have been applied to detailed studies of ionospheric characteristics, including the electron density structure, ionospheric motions (tidal oscillations, vertical drifts, winds, traveling ionospheric disturbances), and the equatorial and auroral electrojets (Sec. 3-2-6). Incoherent scatter facilities have been located in Massachusetts (the Millstone radar facility, operating at 440 and 1295 MHz); at Jicamarca near Lima, Peru on the magnetic equator (49.92 MHz); at Arecibo, Puerto Rico (430

MHz); at Nancy, France (originally a bistatic and later a quadristatic system at 1935 MHz); near Fairbanks, Alaska in the auroral zone (the Chatanika radar, 1290 MHz); and at Malvern, England (400 MHz).

The Jicamarca radar facility, originally a joint project of the Department of Commerce laboratories, Boulder, Colorado and the Instituto Geofisico del Peru and now operated by the institute, utilizes a square array of dipoles 300 m on a side. The $\lambda/2$ dipoles are located 0.3 λ above a ground screen, with 96 dipoles on a side. A second set of orthogonal dipoles allows the use of either or both of the two linear polarizations or, by phasing the two linear polarizations 90° from each other, left or right circular polarization. Altogether the system involves the use of 18,432 $\lambda/2$ dipoles. The peak power utilized is 4 MW, and the system noise temperature T_{sys} is 6000 K. The facility, shown in Fig. 6-30, is located in a valley in the foothills of the Andes. The Arecibo facility is constructed in a limestone sinkhole in Puerto Rico. A peak power of 2.5 MW is used, and T_{sys} is 300 K. As mentioned in Sec. 5-5-3, the reflector is 305 m in diameter (viewed from above) and spherical in shape. The Chatanika radar [53] is a joint project of the Geophysical Institute of the University of Alaska and the Stanford Research Institute. It utilizes a 26.8-m steerable antenna, a peak power of 5 MW, and a T_{sys} of 300 K. Further information about the incoherent scatter radars and data they have supplied are given by Evans [24] and in a special issue of *Radio Science* [6,48].

Two of the methods used for determining electron densities by the incoherent scatter technique will now be described. The power profile method

FIGURE 6-30. Jicamarca radar observatory near Lima, Peru.

involves using the relation

$$W_r = \frac{CN(h)\sigma(h)}{h^2} \tag{6-74}$$

where W_r is the received echo power, C is a constant, h is altitude, and $N(h)$ and $\sigma(h)$ are the electron density and effective cross section of an electron expressed as function of height. $\sigma(h)$ is given in terms of σ_0 for an isolated electron, 1×10^{-28} m^2, (Prob. 6-13) by

$$\sigma(h) = \frac{\sigma_0}{(1 + \alpha^2)(1 + T_e/T_i + \alpha^2)} \tag{6-75}$$

where T_e is electron temperature, T_i is ion temperature, and α is $4\pi D/\lambda$, where D is the Debye length, $\varepsilon_0 kT_e/4\pi Ne^2$, where ε_0 is the permitivity of free space, k is Boltzmann's constant, and e is the charge of the electron. For a sufficiently long wavelength α is small enough to be neglected, but the T_e/T_i ratio must always be taken into account. This ratio may be determined in practice by comparing the spectrum of the echo with curves like those illustrated in Fig. 6-31. A further problem arises in determining the spectrum, however, in that the desired frequency spectrum of the echo is contaminated by the frequency spectrum of the transmitted pulses. This problem is solved by determining an autocorrelation function, using pairs of pulses with one

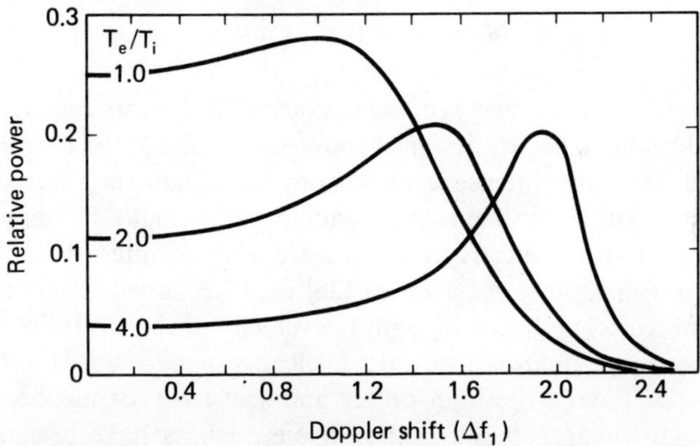

FIGURE 6-31. Ionic component of spectrum of incoherent scatter echoes as a function of T_e/T_i ratio for $\alpha = 0.1$. The Doppler-frequency scale has been normalized by dividing by the Doppler shift that would be encountered by an approaching ion at the mean thermal speed of the ions. An electronic component, consisting of a narrow *plasma line* displaced from the transmitted frequency by the plasma frequency f_p, can also be observed under certain conditions. (From Evans, "Theory and practice of ionosphere study by Thomson scatter radar," *Proc. IEEE*, vol. 57, pp. 496–530, April 1969 [24].)

delayed from the other, and then taking the Fourier transform of the autocorrelation function [52]. A second method involves recording Faraday rotation ϕ of the received echoes as a function of height and taking the derivative of ϕ with respect to h to get electron density N [Eq. (4-108)]. The Faraday rotation is determined by transmitting left and right circularly polarized waves and recording two orthogonal components of the echo pulses for the two polarizations. The expression for Faraday rotation [Eq. (4-105)] is equivalent to

$$\phi = \frac{\theta_r - \theta_l}{2} \text{ rad} \tag{6-76}$$

and this quantity can be determined from measurements of the orthogonal components

$$X_r = V_r \sin \theta_r \tag{6-77a}$$

$$Y_r = V_r \cos \theta_r \tag{6-77b}$$

$$X_l = V_l \sin \theta_l \tag{6-77c}$$

$$Y_l = V_l \cos \theta_l \tag{6-77d}$$

as determined by a phase detector circuit which compares the received echoes with a stable reference source from which the transmitter pulses are derived. In particular it develops that (Prob. 6-12)

$$\theta_r - \theta_l = \tan^{-1} \left(\frac{\overline{X_r Y_l}}{\overline{Y_r Y_l}}\right) \tag{6-78}$$

where the bars over the two products indicate their mean values.

The Jicamarca radar, designed to record incoherent scatter echoes, can also record the more intense echoes from the equatorial electrojet. These echoes may create a problem when attempting to make incoherent scatter observations, but the electrojet echoes are also of interest in themselves. A two-stream instability mechanism [26] explains some of the echoes, but non-two-stream echoes, moving with the velocity of the electrojet, also occur [3]. For observing equatorial or auroral electrojet echoes, it is not necessary to use a radar having the high power and sensitivity of incoherent scatter radars. Radar auroral, or auroral electrojet, echoes have been studied by noncoherent radars, but coherent radars have the advantage that they can record the velocity of the ionization irregularities that cause the echoes [5]. One important application of auroral radar has been the monitoring of the northern auroral oval by an HF step-frequency sounder located at Resolute Bay, Northwest Territories, Canada, near the geomagnetic pole [9]. An auroral radar located at Anchorage, Alaska transmits information to the Space Environment Laboratory, Boulder, Colorado [37].

A well-established characteristic of radio auroral echoes is that they are received from ionization irregularities that are field-aligned, namely structures that have considerable extent along the length of the earth's magnetic field lines and a small extent perpendicular to the lines. Thus the line of sight from the radar transmitter must be close to perpendicular to the earth's field lines in the echoing region for VHF-UHF radars. In Alaska strict perpendicularity cannot be obtained, but radar beams directed toward magnetic north can come close to perpendicularity and can receive echoes from the *E* region of the auroral ionosphere at ranges on the order of 500–900 km. It is not possible in the auroral zone to receive coherent echoes at VHF-UHF frequencies from overhead structures as in the case of the equatorial region where a vertically pointing radar can be located directly under the electrojet and receive echoes.

The situation is different for HF transmissions. HF waves experience considerably more refraction in the ionosphere than higher-frequency waves, and HF waves tend to be refracted so as to become perpendicular to magnetic field lines. Thus the HF step-frequency radar at Resolute Bay has received echoes from the entire auroral oval. The radar steps through frequencies in the 6- to 32-MHz range and utilizes a log-periodic antenna which rotates in 30° increments. The refraction conditions are not sufficiently favorable for all frequencies to receive echoes, but one or more of the frequencies stepped through will normally provide echoes. The use of a radar for monitoring the auroral oval brings up the degree of correspondence between radar auroral echoes and visual aurora. A very high correlation exists between the two phenomena as a function of time. The spatial correlation is not quite so high; the locations of visual aurora and radar echoing regions are not precisely identical. They are sufficiently close, however, to allow radar to monitor the auroral oval very well. Auroral radar echoes are sometimes classified into discrete (localized) and diffuse (widespread in range) categories. It appears that the region of diffuse radio aurora in the afternoon and evening is the location of the eastward electrojet current at that time and that the echo intensity is related to electrojet current intensity [37]. Thus auroral radar echoes may correlate quite closely with two other major manifestations of auroral activity, namely visual aurora and geomagnetic field variations, the latter due to the electrojet currents.

There was some concern initially that the Chatanika radar would be hampered in making observations of the auroral ionosphere by the incoherent scatter techniques by strong, discrete radio-auroral echoes. It turned out, however, that the radar was not so hampered, and it has provided valuable data about the auroral ionosphere [6,48,53]. The Ballistic Missile Early Warning System (BMEWS) 400-MHz radar, though not designed for the purpose, has sufficient sensitivity to receive incoherent scatter echoes [28].

The high power and sensitivity of the incoherent scatter radars has proved to be useful for studying the winds and structure of the troposphere and stratosphere as well as the characteristics of the ionosphere, thus providing a connecting link between ionospheric and tropospheric studies [1,4,88]. It was the observations of low-altitude winds at the Jicamarca radar observatory that led to the radar observations of winds reported in Sec. 6-4-4.

6-5 CONCLUSION

The discussion of remote sensing has given emphasis to radar and other techniques employing electromagnetic waves. A major aspect of remote sensing, but one that is largely beyond the scope of this treatment, is that of data processing, including pattern recognition. Huge amounts of environmental data can be collected by the sensors mentioned in the chapter, but exploiting the data to the utmost requires careful attention to the techniques used for recording and analyzing the data.

The means for remotely sensing the environment have been greatly augmented in recent years. The developments in this field have been referred to as a quiet revolution [66] that affects not only the active workers in the field but large numbers of people in all walks of life as well.

PROBLEMS

6-1 A radiometer utilizing the Dicke type of receiver has an antenna noise temperature of 100 K, a receiver that has a noise figure of 2 dB, a transmission line with a loss of 1 dB, a 4-MHz bandwidth, and an integration time of 5 s. Find the minimum temperature change that can be detected.

6-2 In applying a radiometer method for determining attenuation due to rainfall, a brightness temperature T_b of 136 K is recorded at a frequency of 15 GHz. Find the estimated attenuation for a 15-GHz signal passing through the region in question if the intrinsic temperature T_i is taken to be 273 K.

6-3 a. The sun, subtending an angle of 0.224 deg^2 at the earth is viewed with a 15-GHz antenna having beamwidths in both orthogonal directions of 2°. Find the brightness temperature recorded if the sun has a temperature of 1.2×10^4 K at this frequency and the sky is clear.

b. Calculate the brightness temperature if conditions remain the same otherwise but a cloud having an optical depth of unity and an intrinsic temperature of 200 K is between the antenna and the sun.

6-4 a. A pulse-Doppler radar operates at a wavelength of 10 cm with a pulse width of 1 μs and a repetition rate of 1000 pps. Determine the range and velocity resolution and the maximum unambiguous range and velocity, assuming an observation or dwell time for measuring Doppler frequency of 1 s.

b. For what closest range R, where $R > R_m$, would a target give an echo that would appear on the radar display as if it were at a range of 20 nmi? For what closest radial velocity v_R, where $v_R > v_{Rm}$, might the velocity appear to be only 10 knots?

c. Which of the results of part a will change if the pulse length is reduced to 0.25 μs, other parameters remaining the same as in part a? Calculate the new value or values.

d. Which of the results of part a will change if the pulse repetition rate is increased to 1500 pps, other parameters remaining the same as in part a? Calculate the new value or values.

6-5 a. Determine the maximum distance at which a target of $\sigma = 0.01\ \text{m}^2$ can be detected by a 3.2-cm radar having a peak power of 20 kW, a pulse length of 1 μs, a receiver noise figure of 8 dB, and a paraboloidal antenna having an aperture 1.22 m in diameter. Assume that the receiver bandwidth matches the pulse width, that the effective area for calculating gain is 0.6 of the aperture area, and that $(S/N)_{\min} = 2$.

b. Repeat for a peak power of 1 kW, all other conditions remaining as in part a.

c. Repeat for a pulse width of 0.05 μs (with a corresponding receiver bandwidth), all other conditions remaining as in part a.

6-6 It is desired to employ an FM-CW radar for monitoring the lower atmosphere to a height of 3000 m with a range resolution of 5 m and a minimum range of 20 m. Assuming that available equipment allows meeting these goals, what should the values of B and T be if Δf_B, corresponding to $\Delta R = 5$ m, is taken to be 20 Hz? (See Fig. 6-4.) What will be the values of f_B for the maximum and minimum ranges?

6-7 a. Problem 6-6 referred to the use of an FM-CW radar for measuring range only. If it is desired to measure velocity as well as range, it may be necessary to accept lower range resolution in order to measure velocities of interest. If a range resolution ΔR of 60 m is accepted and if a B/T ratio of 6×10^8 is utilized, find the maximum unambiguous velocity for a radar wavelength of 10 cm in m/s and mi/h. *Hint:* Use Eq. (6-33) with $n = 1$, and consider that the maximum unambiguous value of $\Delta\phi$ is π rad.

b. For a range R of 2000 m, calculate f_B for a stationary target. Then determine by what amount the recorded frequency will differ for a target at the same range but moving with a velocity of 6 m/s, assuming the same system parameters as in part a.

6-8 MTI systems have blind speeds, for which the signal output theoretically goes to zero as for a stationary target. Calculate the lowest blind speed for a pulse repetition rate f_r of 1000 pps and a radar wavelength of 10 cm. *Hint:* Take the difference between the voltage given by Eq. (6-36) and the corresponding voltage for $t + T$, where $T = 1/f_r$.

6-9 Find the signal-to-noise ratio for rainfall of 10 mm/h at a range of 75 km for an S-band ($\lambda = 10$ cm) radar having a peak power of 20 kW, a 1-μs pulse width, a receiver noise figure of 6 dB, and a paraboloidal antenna 1.83 m in diameter. Use the equation including the Probert-Jones correction, use 0.8

of the aperture area for directivity and 0.6 of the aperture area for gain, and take $K_c = 78.45 - j11.19$. [Refer to Sec. 2-4 if necessary. Note that Eq. (6-68) for Z gives Z in mm^6/m^3 for r in mm/h but that Z in Eq. (6-67) must be in m^6/m^3. Consider a circular-aperture antenna for which $\theta_{\text{HP}} = \phi_{\text{HP}}$ and assume that $\Omega_A = \frac{4}{3}\theta_{\text{HP}}\phi_{\text{HP}}$.]

6-10 Assuming the complex relative dielectric constant K_c for sea water at 20°C at $\lambda = 10$ cm to be $69 - j39$, calculate the reflection coefficients for horizontal and vertical polarization for reflection at an angle θ, measured from the surface, of 5°, assuming a perfectly smooth surface. [Use Eqs. (2-42) and (2-43), taking $K_c = K - jK_{\text{im}}$ and noting that $K_{\text{im}} = \sigma/\omega\epsilon_0$.]. For these same conditions, calculate a value of the height difference Δh between two points on the surface which would cause the surface to be considered rough. [The Rayleigh criterion for smoothness is that $\Delta h \sin\theta < \lambda/2$. For roughness, let us take $\Delta h \sin\theta = \lambda/4$.]

6-11 The velocity v of ocean waves in deep water is given by $v = \sqrt{gL/2\pi}$. For such waves having a wavelength of 152.4 m, calculate the wave velocity and period of the waves and determine the frequency of a radar that could be used for monitoring the waves, assuming line-of-sight operation, that the ocean waves are traveling directly toward the radar, and that Bragg scattering from the predominant waves is the mechanism responsible for the echoes. By what amount would the received echoes be shifted in frequency with respect to the transmitted signal? [g is the acceleration of gravity, 9.8 m/s, and L is wavelength.]

6-12 Starting with the quantities X_r, Y_r, X_l, and Y_l, as given by Eqs. (6-77), prove that $\theta_r - \theta_l$ can be determined by the use of Eq. (6-78). (The nature of the incoherent scatter process is such that the phase angles of the echoes, θ_r and θ_l, are random variables, but $\theta_r - \theta_l$ retains the expected value for *rc* and *lc* waves.)

6-13 Show that σ_0, the radar cross section for a single, isolated electron, is 1×10^{-28} m^2.

REFERENCES

[1] Aso, T., K. Sato, and R. M. Harper, "Arecibo middle atmosphere experiment," *Geophysical Research Letters*, vol. 4, pp. 10–12, 1977.

[2] Atlas, D., "Advances in radar meteorology," in *Advances in Geophysics*, Vol. 10. New York: Academic Press, 1964.

[3] Balsley, B. B., "Some characteristics of non-two-stream irregularities in the equatorial electrojet," *Journal of Geophysical Research*, vol. 74, pp. 2333–2347, May 1, 1969.

[4] Balsley, B. B., and W. L. Ecklund, "VHF power spectra of the radar aurora," *Journal of Geophysical Research*, vol. 77, pp. 4746–4760, Sept. 1, 1972.

[5] BALSLEY, B. B., N. CIANOS, D. T. FARLEY, and M. J. BARON, "Winds derived from radar measurements in the arctic troposphere and stratosphere," *Journal of Applied Meteorology*, vol. 16, pp. 1235–1239, Nov. 1977.

[6] BARON, M. J., "Electron densities within aurorae and other auroral *E*-region characteristics," *Radio Science*, vol. 9, pp. 341–348, Feb. 1974.

[7] BARRICK, D. E., M. W. EVANS, and B. L. WEBER, "Ocean surface currents mapped by radar," *Science*, vol. 198, pp. 138–144, Oct. 14, 1977.

[8] BARTON, D. K., *Radar System Analysis*. Englewood Cliffs, NJ: Prentice-Hall, 1964.

[9] BATES, H. F., S. I. AKASOFU, D. S. KIMBALL, and J. C. HODGES, "First results from the north polar auroral radar," *Journal of Geophysical Research*, vol. 78, pp. 3857–3864, July 1, 1973.

[10] BATTAN, L. J., *Radar Observation of the Atmosphere*. Chicago: University of Chicago Press, 1973.

[11] BEAN, B. R., E. J. DUTTON, and B. D. WARNER, "Weather effects on radar," in *Radar Handbook* (M. Skolnik, ed.). New York: McGraw-Hill, 1970.

[12] BONHAM, L. L., and L. V. BLAKE, "Radar echoes from birds and insects," *Scientific Monthly*, vol. 82, pp. 204–209, April 1956.

[13] BROWN, W. E., Jr., C. ELACHI, and T. W. THOMPSON, "Radar imaging of ocean wave patterns," *Journal of Geophysical Research*, vol. 81, pp. 2657–2667, May 20, 1976.

[14] BROWNING, K. A., and C. D. WATKINS, "Observations of clear air turbulence by high power radar," *Nature*, vol. 227, pp. 260–263, 1970.

[15] CAMPBELL, W. J., P. GLOERSEN, W. J. WEBSTER, T. T. WILHEIT, and R. D. RAMSEIER, "Beaufort Sea ice zones as delineated by microwave imagery," *Journal of Geophysical Research*, vol. 81, pp. 1103–1110, Feb. 20, 1976.

[16] CHADWICK, R. B., K. P. MORAN, R. G. STRAUCH, G. E. MORRISON, and W. C. CAMPBELL, "Microwave radar wind measurements in the clear air," *Radio Science*, vol. 11, pp. 795–802, Oct. 1976.

[17] CRANE, R. K., "Prediction of the effects of rain on satellite communication systems," *Proc. IEEE*, vol. 65, pp. 456–474, March 1977.

[18] CROMBIE, D. D., "Doppler spectrum of sea echo at 13.56 Mc/s," *Nature*, vol. 175, pp. 681–682, 1955.

[19] DERR, V. E., and C. G. LITTLE, "A comparison of remote sensing of the clear atmosphere by optical, radio, and acoustic radar techniques," *Applied Optics*, vol. 9, pp. 1976–1992, Sept. 1970.

[20] DICKE, R. H., "The measurement of thermal radiation at microwave frequencies," *Review of Scientific Instruments*, vol. 17, pp. 268–275, July 1946.

[21] DIFRANCO, J. V., and W. L. RUBIN, *Radar Detection*. Englewood Cliffs, NJ: Prentice-Hall, 1968.

[22] DOVIAK, R. J., and C. T. JOBSON, "Mapping clear air winds with dual-Doppler radars," in *Proc. URSI Commission F Open Symposium, LaBaule, France*, April 28–May 6, 1977, pp. 505–510. Issy-Les-Moulineaux, France: Centre National d'Etudes des Telecommunications, 1977.

[23] EASTWOOD, E., *Radar Ornithology*. London: Methuen, 1967.

[24] EVANS, J. V., "Theory and practice of ionosphere study by Thomson scatter radar," *Proc. IEEE*, vol. 57, pp. 496–530, April 1969.

[25] EVANS, J. V., and T. HAGFORS, *Radar Astronomy*. New York: McGraw-Hill, 1968.

[26] FARLEY, D. T., "A plasma instability resulting in field-aligned irregularities in the ionosphere," *Journal of Geophysical Research*, vol. 68, pp. 6083–6097, Nov. 15, 1963.

[27] FISHLOCK, D. (ed.), *A Guide to Earth Satellites*. New York: American Elsevier, 1971.

[28] FLOCK, W. L., "Environmental studies for radar operations in the auroral zone, Part 1," *Final Report*, Contract AF04 (647) 179. Fairbanks: Geophysical Institute, University of Alaska, 1962.

[29] FLOCK, W. L., "Radar observations of bird migration at Cape Prince of Wales," *Arctic*, vol. 25, pp. 83–98, June 1972.

[30] FLOCK, W. L., "Radar observations of bird movements along the Arctic coast of Alaska," *The Wilson Bulletin*, vol. 85, pp. 259–275, Sept. 1973.

[31] FLOCK, W. L., "Monitoring open water and sea ice in the Bering Strait by radar," *IEEE Trans. on Geoscience Electronics*, vol. GE-15, pp. 196–202, Oct. 1977.

[32] FLOCK, W. L., and J. L. GREEN, "The detection and identification of birds in flight, using coherent and noncoherent radars," *Proc. IEEE*, vol. 62, pp. 745–753, June 1974.

[33] GATES, D. M., "Physical and physiological properties of plants," in *Remote Sensing with Special Reference to Agriculture and Forestry*, pp. 224–252. Washington, D.C.: National Academy of Sciences, 1970.

[34] GLOERSEN, P., T. C. CHANG, T. T. WILHEIT, and W. J. CAMPBELL, "Polar sea ice observations by means of microwave radiometry," in *Advanced Concepts and Techniques in the Study of Snow and Ice Resources*, pp. 541–550. Washington, D.C.: National Academy of Sciences, 1974.

[35] Goddard Space Flight Center, NASA, *The Nimbus 5 User's Guide*. Greenbelt, MA: NASA, Nov. 1972.

[36] Goddard Space Flight Center, NASA, *The Nimbus 6 User's Guide*. Greenbelt, MA: NASA, Feb. 1975.

[37] GRAY, A. M., and W. L. ECKLUND, "The Anchorage, Alaska real-time auroral radar monitor: system description and some preliminary analyses," *NOAA Technical Report ERL 306-AL9*. Boulder, CO: NOAA Environmental Research Laboratories, Sept. 1974.

[38] GREEN, J. L., J. M. WARNOCK, R. H. WINKLER, and T. E. VANZANDT, "Studies of winds in the upper troposphere with a sensitive VHF radar," Geophysical Research Letters, vol. 2, pp. 19–21, Jan. 1975.

[39] GRODY, N. C., "Remote sensing of atmospheric water content from satellites using microwave radiometry," *IEEE Trans. on Antennas and Propagation*, vol. AP-24, pp. 155–162, March 1976.

[40] HALL, F. F., "Temperature and wind structure studies by acoustic echo-sounding," in *Remote Sensing of the Troposphere* (V. E. Derr, ed.), pp. 18-1 to 18-25. Washington, D.C.: Supt. of Documents, U.S. Government Printing Office, 1972.

[41] HARDY, K. R., "Studies of the clear atmosphere using high power radar," in *Remote Sensing of the Troposphere* (V. E. Derr, ed.), pp. 14-1 to 14-34. Washington, D.C.: Supt. of Documents, U.S. Government Printing Office, 1972.

[42] HARDY, K. R., and I. KATZ, "Probing the clear atmosphere with high power, high resolution radars," *Proc. IEEE*, vol. 57, pp. 468–480, April 1969.

[43] HARDY, K. R., and H. OTTERSTEN, "Radar investigation of convective patterns in the clear atmosphere," *Journal of Atmospheric Sciences*, vol. 26, pp. 666–672, 1969.

[44] HARDY, K. R., D. ATLAS, and K. M. GLOVER, "Multiwavelength backscatter from the clear atmosphere," *Journal of Geophysical Research*, vol. 71, pp. 1537–1552, March 15, 1966.

[45] HARRIS, G., and L. C. GRAHAM, "Landsat-radar synergism," XIII Congress of the International Society for Photogrammetry, Helsinki, 1976; also reproduced by Goodyear Aerospace Corp., Litchfield Park, AZ.

[46] HICKS, J. J., and J. L. ANGELL, "Radar observations of breaking gravitational waves in the visually clear atmosphere," *Journal of Applied Meteorology*, vol. 7, pp. 114–121, 1968.

[47] HUDSON, R. D., Jr., *Infrared System Engineering*. New York: Wiley, 1969.

[48] HUNSUCKER, R. D., "Simultaneous riometer and incoherent scatter radar observations of the auroral *D* region," *Radio Science*, vol. 9, pp. 335–340, Feb. 1974.

[49] KELLER, G. V., and F. C. FRISCHKNECHT, *Electrical Methods in Geophysical Prospecting*. Elmsford, NY: Pergamon, 1966.

[50] KINSMAN, B., *Wind Waves*. Englewood Cliffs, NJ: Prentice-Hall, 1965.

[51] LACK, D., and G. C. VARLEY, "Detection of birds by radar," *Nature*, vol. 156, p. 446, Oct. 13, 1945.

[52] LATHI, B. P., *An Introduction to Random Signals and Communication Theory*. Scranton, PA: International Textbook Company, 1968.

[53] LEADABRAND, R. L., M. J. BARON, J. PETRICEKS, and H. F. BATES, "Chatanika, Alaska, auroral-zone incoherent scatter facility," *Radio Science*. vol. 7, pp. 747–756, July 1972.

[54] LIGDA, M. G. H., "Radar storm observation," in *Compendium of Meteorology* (T. F. Malone, ed.), pp. 1265–1282. Boston: American Meteorology Society, 1951.

[55] LITTLE, C. G., "Acoustic methods for remote probing of the lower atmosphere," *Proc. IEEE*, vol. 57, pp. 571–578, April 1969.

[56] LONG, M. W., *Radar Reflectivity of Land and Sea*. Lexington, MA: Heath, 1975.

[57] MCLAIN, E. P., "Environmental research and applications using the very

high resolution radiometer (VHRR) on the NOAA-2 satellite—a pilot project in Alaska," in *Climate of the Arctic* (G. Weller and S. A. Bowling, eds.), pp. 415–429. Fairbanks: Geophysical Institute, University of Alaska, 1975.

[58] MOORE, R. K., "Microwave remote sensors," in *Manual of Remote Sensing* (R. G. Reeves, ed.), pp. 399–531. Falls Church, VA: American Society of Photogrammetry, 1975.

[59] NATHANSON, F. E., *Radar Design Principles*. New York: McGraw-Hill, 1969.

[60] National Academy of Sciences, *Remote Sensing with Special Reference to Agriculture and Forestry*. Washington, D.C.: National Academy of Sciences, 1970.

[61] PROBERT-JONES, J. R., "The radar equation in meteorology," *Quarterly Journal of the Royal Meteorological Society*, vol. 88, pp. 485–495, 1962.

[62] Radiation Laboratory, Massachusetts Institute of Technology, Radiation Laboratory Series, 28 vols. New York: McGraw-Hill, 1947.

[63] REEVES, R. G. (ed.), *Manual of Remote Sensing*, Vols. I and II. Falls Church, VA: American Society of Photogrammetry, 1975.

[64] RICHTER, J. H., "High resolution tropospheric radar sounding," *Radio Science*, vol. 4, pp. 1261–1268, Dec. 1969.

[65] ROUSE, J. W., Jr., and S. RITER, "ERTS experiments," *IEEE Trans. on Geoscience Electronics*, vol. GE-11, pp. 3–76, Jan. 1973.

[66] RUDD, R. D., *Remote Sensing: A Better View*. Belmont, CA: Wadsworth, 1974.

[67] SACKINGER, W. M., "Sea ice and permafrost problems," *Annual Report, 1973–1974*, pp. 41–48. Fairbanks: Geophysical Institute, University of Alaska, 1974.

[68] SCHABER, G. C., C. BERLIN, and W. E. BROWN, "Variation in surface roughness within Death Valley, CA: geologic evaluation of 25-cm wavelength images," *Geological Society of America Bulletin*, vol. 87, pp. 29–41, Jan 1976.

[69] *Science*, "Remote sensing, I: Landsat takes hold in South America; II: Brazil explores its Amazon wilderness; III: The tools continue to improve," vol. 196, pp. 511–516, April 29, 1977.

[70] SKOLNIK, M. I., *Introduction to Radar Systems*. New York: McGraw-Hill, 1962.

[71] SKOLNIK, M. I. (ed.), *Radar Handbook*. New York: McGraw-Hill, 1970.

[72] SNIDER, J. B., and E. R. WESTWATER, "Radiometry," in *Remote Sensing of the Troposphere* (V. E. Derr, ed.), pp. 15-1 to 15-33. Washington, D.C.: Supt. of Documents, U.S. Government Printing Office, 1972.

[73] STAELIN, D. H., et al., "Microwave spectroscopic imagery of the earth," *Science*, vol. 197, pp. 991–993, Sept. 2, 1977.

[74] STARR, J. R., and K. A. BROWNING, "Observations of lee waves by highpower radar," *Quarterly Journal of the Royal Meteorological Society*, vol. 98, pp. 73–85, 1972.

[75] STRAUCH, R. G., "Theory and Application of FM-CW Doppler Radar," Ph.D. dissertation. Boulder, CO: University of Colorado, 1976.

[76] STRAUCH, R. G., and A. COHEN, "Atmospheric remote sensing with laser radar," in *Remote Sensing of the Troposphere* (V. E. Derr, ed.), pp. 23-1 to 23-35. Washington, D.C.: Supt. of Documents, U.S. Government Printing Office, 1972.

[77] TABATA, T., "Sea-ice reconnaissance by radar," *Journal of Glaciology*, vol. 15, no. 73, pp. 215–223, 1975.

[78] TAYLOR, W. L., "Atmospherics and severe storms," in *Remote Sensing of the Troposphere* (V. E. Derr, ed.), pp. 17-1 to 17-17. Washington, D.C.: Supt. of Documents, U.S. Government Printing Office, 1972.

[79] TEAGUE, C. C., "An experimental determination of the value of σ° for 7-second ocean waves," presented at the 1974 USNC-URSI Fall Meeting, Boulder, CO, Oct. 14–17, 1974.

[80] TYLER, G. L., W. E. FAULKERSON, A. M. PETERSON, and C. C. TEAGUE, "Second-order scattering from the sea; 10 meter radar observation of the Doppler continuum," *Science*, vol. 177, pp. 349–351, 1972.

[81] ULABY, F. T., "Radar response to vegetation," *IEEE Trans. on Antennas and Propagation*, vol. AP-23, pp. 36–45, Jan. 1975.

[82] ULABY, F. T., and P. P. BATLIVALA, "Optimum radar parameters for mapping soil moisture," *IEEE Trans. on Geoscience Electronics*, vol. GE-14, pp. 81–93, April 1976.

[83] USAF BIRD STRIKE SUMMARIES, issued annually. Norton AFB, CA: Directorate of Aerospace Safety, Air Force Inspection and Safety Center.

[84] VANZANDT, T. E., J. L. GREEN, K. S. GAGE, and W. L. CLARK, "Vertical profiles of refractivity turbulence structure constant: comparison of observations by the Sunset radar with a new theoretical model," submitted to *Radio Science*, 1977.

[85] WESTWATER, E. R., "An analysis of the correction of range errors due to atmospheric refraction by microwave radiometric techniques, *ESSA Tech. Report IER 37-ITSA37*. Washington, D.C.: U.S. Department of Commerce, 1967.

[86] WESTWATER, E. R., and O. N. STRAND, "Inversion techniques," in *Remote Sensing of the Troposphere* (V. E. Derr, ed.), pp. 16-1 to 16-13. Washington, D.C.: Supt. of Documents, U.S. Government Printing Office, 1972.

[87] WILSON, D. A., and L. J. MILLER, "Atmospheric motion by Doppler radar," in *Remote Sensing of the Troposphere* (V. E. Derr, ed.), pp. 13-1 to 13-34. Washington, D.C.: Supt. of Documents, U.S. Government Printing Office, 1972.

[88] WOODMAN, R. F., and A. GUILLEN, "Radar observations of winds and turbulence in the stratosphere and mesosphere," *Journal of the Atmospheric Sciences*, vol. 31, pp. 493–505, March 1974.

[89] WULFSBERG, K. N., and E. E. ALTSHULER, "Rain attenuation at 15 and 35 GHz," *IEEE Trans. on Antennas and Propagation*, vol. AP-20, pp. 181–187, March 1972.

APPLICATIONS OF REMOTE SENSING TO LAND

1. Badgley, P. C. and W. L. Vest, "Orbital remote sensing and natural resources," *Photogrammet. Eng.*, vol. XXXII, pp. 780–790, Sept. 1966.

2. Baker, L. R. and R. M. Scott (coauthors-editors), "Electro-optical remote sensors with related optical sensors," in *Manual of Remote Sensing*, R. C. Reeves (ed.), pp. 325–366. Falls Church, VA: American Society of Photogrammetry, 1975.

3. Barrett, E. C. and L. F. Curtis (eds.), *Environmental Remote Sensing: Applications and Achievements.* London: Edward Arnold (Publishers), Ltd.; New York: Crane, Russak & Company, Inc., 1974.

4. Bauer, M. E., P. H. Swain, R. P. Mroczynski, P. E. Anuta, and R. B. MacDonald, "Detection of southern corn leaf blight by remote sensing techniques," Proc. Seventh Symposium on Remote Sensing of Environment, pp. 693–704, 1971.

5. Beckmann, P. and A. Spizzichino, *The Scattering of Electromagnetic Waves from Rough Surfaces.* New York: The Macmillan Company, 1963.

6. Crandall, C. J., "Radar mapping in Panama," *Photogrammet. Eng.*, vol. XXXV, pp. 641–646, July 1969.

7. Dellwig, L. F., H. C. MacDonald, and J. N. Kirk, "The potential of radar in geological exploration," *Proc. Fifth Symposium on Remote Sensing of Environment*, pp. 747–763, 1968.

8. Driscoll, R. S., "Multispectral scanner imagery for plant community classification," *Proc. Eighth Symposium on Remote Sensing of Environment*, pp. 1259–1278, 1972.

BIBLIOGRAPHY

9. Estes, J. E. and L. W. Senger (eds.), *Remote Sensing Techniques for Environmental Analysis.* Santa Barbara, CA: Hamilton Publishing Company, 1974.

10. Hay, C. U., "Agricultural inventory techniques with orbital and high-altitude imagery," *Photogrammet. Eng.*, vol. XL, pp. 1283–1293, Nov. 1974.

11. Highway Research Board, National Academy of Sciences, Remote Sensing and its Application to Highway Engineering, Special Report, 102. Washington, D.C.: National Academy of Sciences, 1969.

12. Holz, R. K. (ed.), *The Surveillant Science, Remote Sensing of the Environment.* Boston: Houghton Mifflin Company, 1973.

13. Johnson, P. L. (ed.), *Remote Sensing in Ecology.* Athens, GA: University of Georgia Press, 1969.

14. Levine, D. (ed.), "Combinations of photogrammetric and radargrammetric techniques," in *Manual of Photogrammetry*, Third Edition. Falls Church, VA: American Society of Photogrammetry, pp. 1003–1048, 1966.

15. Lintz, J. and D. S. Simonett (eds.), *Remote Sensing of Environment.* Reading, MA: Addison-Wesley Publishing Company, 1976.

16. Lowe, D. S. (author-editor), "Imaging and nonimaging sensors," Chapter 8 in *Manual of Remote Sensing*, R. G. Reeves (ed.), pp. 367–397. Falls Church, VA: American Society of Photogrammetry, 1975.

17. Meyer, W. and R. I. Welch (coauthors-editors), "Water resources assessment," Chap. 19 in *Manual of Remote Sensing*, R. G. Reeves (ed.), pp. 1479–1551. Falls Church, VA: American Society of Photogrammetry, 1975.

18. Moore, R. K., "Radar scatterometry—an active remote sensing tool," *Proc. Fourth Symposium on Remote Sensing of Environment*, pp. 339–373, 1966.

19. Moore, R. K., "Imaging radars for geoscience use," *IEEE Trans. on Geoscience Electronics,* vol. GE-9, pp. 155–164, July, 1971.

20. Moore, R. K. and F. T. Ulaby, "The radar radiometer," *Proc. IEEE*, vol. 57, pp. 587–590, April 1969.

21. National Academy of Sciences, *Remote Sensing with Special Reference to Agriculture and Forestry*. Washington, D.C.: National Academy of Sciences, 1970.

22. *Photogrammetric Engineering and Remote Sensing* (periodical), American Society of Photogrammetry, 105 N. Virginia Avenue, Falls Church, VA 22046. (Formerly *Photogrammetric Engineering.*)

23. *Proceedings of The IEEE, Special Issue on Remote Environmental Sensing*, vol. 57, No. 4, April 1969.

24. Reeves, R. G., A. N. Kover, R. J. P. Lyon, and H. T. V. Smith (coauthors-editors), "Terrain and minerals: assessment and evaluation," Chap. 16 in *Manual of Remote Sensing*, R. G. Reeves (ed.), pp. 1102–1351. Falls Church, VA: American Society of Photogrammetry, 1975.

25. Reeves, R. G. (ed.), *Manual of Remote Sensing*, Vols. I and II. Falls Church, VA: American Society of Photogrammetry, 1975.

26. *Remote Sensing of Environment* (periodical), Elsevier North-Holland, Inc., 52 Vanderbilt Ave., New York, NY 10017.

27. ROWAN, L. C., "Application of satellites to geologic exploration," *Am. Scientist*, vol. 63, pp. 393–403, July–Aug. 1975.

28. RUDD, R. D., *Remote Sensing: A Better View*. Belmont, CA: Watsworth Publishing Co., 1974.

29. SABINS, F. F., "Geologic applications of remote sensing," in *Remote Sensing of Environment*, J. Lintz and D. S. Simonett (eds.), pp. 508–571. Reading, MA: Addison-Wesley Publishing Company, 1976.

30. *Science*, "Remote sensing, I: Landsat takes hold in South America; II: Brazil explores its Amazon wilderness; III: The tools continue to improve," vol. 196, pp. 511–516, April 29, 1977.

31. SHAHROKHI, F. (ed.), *Remote Sensing of Earth Resources*, Vol. 1. Tullahoma, Tennessee Space Institute, 1972.

32. SHORT, N. M., P. D. LOWMAN, Jr., S. C. FREDEN, and W. A. FINCH, Jr., *Mission to Earth: Landsat Views The World*. Washington, D.C.: National Aeronautics and Space Administration, 1976. (For sale by Supt. of Documents, U.S. Government Printing Office, Washington, D.C. 20402.)

33. SIMONETT, D. S., "Remote sensing of cultivated and natural vegetation: cropland and forest land," in *Remote Sensing of Environment*, J. Lintz and D. S. Simonett (eds.), pp. 442–481. Reading, MA: Addison-Wesley Publishing Company, 1976.

34. SIMONETT, D. S. and S. A. MORAIN, "Remote sensing from spacecraft as a tool for investigating arctic environments," in *Arctic and Alpine Environments*, W. H. Osburn and H. F. Wright (eds.). Bloomington, IN: Indiana University Press, 1968.

35. SKIBITZKE, H. E., "Remote sensing for water resources," in *Remote Sensing of Environment*, J. Lintz and D. S. Simonett (eds.), pp. 572–592. Reading, MA: Addison-Wesley Publishing Company, 1976.

36. THORLEY, G. A. (author-editor), "Forest lands: inventory and assessment," in *Manual of Remote Sensing*, R. G. Reeves (ed.), pp. 1353–1426. Falls Church, VA: American Society of Photogrammetry, 1975.

37. VEZIROGLU, T. N., *Remote Sensing, Energy-related Studies*. Washington, D.C.: Hemisphere Publishing Co., 1975.

38. WAIT, J. R., *Electromagnetic Waves in Stratified Media*. New York: The Macmillan Company, 1962.

39. WAIT, J. R., (ed.), *Electromagnetic Probing in Geophysics*. Boulder, CO: The Golem Press, 1971.

A

B

F

G

H

I

J

K

L

M

N

O

P

R

S

T